Statistics for Industry and Technology

Quality Improvement Through Statistical Methods

Bovas Abraham
Editor

In association with
N. Unnikrishnan Nair

1998
Birkhäuser
Boston • Basel • Berlin

Bovas Abraham
Institute for Improvement in Quality
and Productivity
University of Waterloo
Waterloo, Ontario N2L 3G1
Canada

Library of Congress Cataloging-in-Publication Data

Quality improvement through statistical methods / Bovas Abraham,
    editor ; in association with N. Unnikrishnan Nair.
       p.   cm. -- (Statistics for industry and technology)
    Includes bibliographical references and index.
    ISBN 0-8176-4052-5 (Boston : alk. paper). -- ISBN 3-7643-4052-5
    (Basel : alk. paper)
    1. Quality control--Statistical methods. I. Abraham, Bovas,
    1942- . II. Unnikrishnan Nair, N.   III. Series.
    TS156.Q3Q36   1998
    658.5'62--dc21   97-51995
                                                   CIP

Printed on acid-free paper
© 1998 Birkhäuser Boston

Birkhäuser

ISBN 0-8176-4052-5
ISBN 3-7643-4052-5
Typeset by the Editor in L\^AT_EX.
Cover design by Vernon Press, Boston, MA.
Printed and bound by Hamilton Printing, Rensselaer, NY.
Printed in the U.S.A.

9 8 7 6 5 4 3 2 1

Contents

Preface   xvii

List of Contributors   xix

List of Tables   xxv

List of Figures   xxix

PART I: STATISTICS AND QUALITY

1. Scientific Learning
   G. E. P. Box   **3**

   1.1 Mathematical Statistics or Scientific Statistics?, 3
      1.1.1 Inadequacy of the mathematical paradigm, 4
      1.1.2 Scientific learning, 5
   1.2 Statistics for Discovery, 7
   1.3 Can Statistics Departments be Reformed?, 8
   1.4 The Role of Computation, 9
      1.4.1 Deductive aspects, 9
      1.4.2 Inductive aspects, 10
      1.4.3 Learning, 10
      References, 11

2. Industrial Statistics and Innovation
   R. L. Sandland   **13**

   2.1 Introduction, 13
   2.2 Innovation, 14
   2.3 Industrial Statistics and Quality, 15
   2.4 The Current Industrial Statistics Paradigm, 17
   2.5 Product Innovation, 17
   2.6 Process Innovation, 19
   2.7 Link to Technology and Business Strategies, 20
      References, 21

**3. Understanding QS-9000 With Its Preventive
and Statistical Focus**
Bovas Abraham and G. Dennis Beecroft   **23**

    3.1 Introduction, 23
    3.2 Quality Evolution, 24
    3.3 ISO 9000 Control Focus, 25
    3.4 QS-9000 Prevention Focus, 26
    3.5 Statistical Requirements in QS-9000, 30
    3.6 Summary and Concluding Remarks, 32
        References, 33

4. A Conceptual Framework for Variation Reduction
Shams-ur Rahman   **35**

    4.1 Statistical Thinking, 35
    4.2 Variation Reduction, 37
        4.2.1 Quality by inspection, 38
        4.2.2 Quality by process control, 39
        4.2.3 Quality by design, 39
    4.3 Conceptual Framework, 42
    4.4 Conclusion, 43
        References, 43

**5. The Role of Academic Statisticians in Quality
Improvement: Strategies of Small and Medium Sized
Companies**
*John F. Brewster, Smiley W. Cheng, Brian D. Macpherson
and Fred A. Spiring*   **45**

    5.1 Background, 45
    5.2 Workshop Training for Industry, 46
    5.3 Workshop Activities, 47
    5.4 Workshop Preparation and Delivery, 50
    5.5 The Pros and Cons of Our Workshop Involvement, 52
    5.6 Statistical Consultation Services, 54
    5.7 Quality Resource Centre, 56
    5.8 Recent Initiatives, 56
    5.9 Conclusion, 57

PART II: STATISTICAL PROCESS CONTROL

**6. Developing Optimal Regulation and Monitoring
Strategies to Control a Continuous
Petrochemical Process**
A. J. Ferrer, C. Capila, R. Romero and J. Martin   **61**

6.1 Introduction, 61
6.2 Polymerization Process, 62
    6.2.1 Process description, 62
    6.2.2 Model formulation and fitting, 63
6.3 EPC Component: Developing the Control Rule, 63
    6.3.1 The Minimum Mean Square Error Controller
          (MMSEC), 64
    6.3.2 The Two-Step Ahead Forecasting Controller
          (TSAFC), 64
    6.3.3 Clarke's Constrained Controller (CCC(r)), 65
6.4 Comparison of Control Strategies, 66
6.5 Robustness Properties of the Controllers, 67
6.6 EPC Component: Performance Under Assignable
    Causes, 69
6.7 SPC Component: Developing Monitoring Procedures, 70
6.8 Concluding Remarks, 72
    References, 73
    Appendix - Tables and Figures, 74

7. Capability Indices when Tolerances are Asymmetric
Kerstin Vännman **79**

7.1 Introduction, 79
7.2 Capability Indices for Asymmetric Tolerances, 81
7.3 The Estimator of $C_{pa}(u,v)$ and Its Distribution, 87
7.4 The Estimator of $C_{pv}(u,v)$ and Its Distribution, 88
7.5 Some Implications, 89
7.6 Concluding Remarks, 93
    References, 94

8. Robustness of On-line Control Procedures
M. S. Srivastava **97**

8.1 Introduction, 97
8.2 Robustness Study When Inspection Cost is Zero, 99
    8.2.1 Effect of non-normality, 102
8.3 Robustness Study When Inspection Cost is Not Zero, 103
    8.3.1 Effect of non-normality, 104
8.4 Conclusion, 106
    References, 107

9. An Application of Filtering to Statistical Process Control
A. Thanvaneswaran, B. D. MacPherson and B. Abraham **109**

9.1 Introduction, 109

9.2 Control Charts for Autocorrelated Data, 111
   9.2.1 Modeling the autocorrelations using ARIMA
      (p,d,q) models, 111
   9.2.2 An application of the EWMA statistic to
      autocorrelated data, 111
9.3 Optimal Filter and Smoother, 112
   9.3.1 Optimal filter, 112
   9.3.2 Optimal smoother, 114
9.4 Applications, 115
   9.4.1 ARMA(0,0,1) process or MA(1) process, 115
   9.4.2 ARMA(1,1) process, 117
   9.4.3 A numerical example, 117
9.5 Summary and Concluding Remarks, 118
References, 119

**10. Statistical Process Control and Its Applications in
Korean Industries**
Sung H. Park and Jae J. Kim   **121**

10.1 Introduction: Definition of SPC, 121
10.2 Flow of an SPC System, 123
10.3 Team Effort for Quality Improvement in SPC, 124
10.4 Computerization of SPC and Its Difficulties, 125
10.5 Case Study, 128
10.6 Summary and Concluding Remarks, 133
References, 133

**11. Multivariate Statistical Process Control of a Mineral
Processing Industry**
Nihal Yatawara and Jeff Harrison   **135**

11.1 Introduction, 135
11.2 PCA and PLS Methods, 136
   11.2.1 Principal component analysis, 136
   11.2.2 Method of partial least squares (PLS), 137
11.3 PLS Monitoring Methods and Diagnostics, 138
11.4 Application of PLS for HPRC Data, 139
   11.4.1 The processing plant and data, 139
   11.4.2 Application of PLS for crusher performance
      monitoring, 140
11.5 Conclusion, 141
References, 142
Appendix - Figures, 143

**12. Developing Objective Strategies for Monitoring
Multi-Input/Single-Output Chemical Processes**
B. K. Pal and V. Roshan Joseph   **151**

12.1 Introduction, 151
12.2 General Formulation, 153
12.3 A Case Study, 154
12.4 Conclusions, 162
    Reference, 162

13. Properties of the Cpm Index for Seemingly Unstable Production Processes
Olle Carlsson   **163**

13.1 Introduction, 163
13.2 Different Models, 165
    13.2.1 Random effects model, 165
    13.2.2 Fixed effects model, 167
13.3 Confidence Intervals, 168
13.4 Probability of Conforming Items, 170
13.5 Summary and Discussion, 171
    References, 171

14. Process Capability Indices for Contaminated Bivariate Normal Population
Subrahmaniam Kocherlakota and Kathleen Kocherlakota   **173**

14.1 Introduction, 173
14.2 Contaminated Bivariate Normal Distribution, 175
    14.2.1 The distribution of $(S_1^2/\sigma_1^2, S_2^2/\sigma_2^2)$, 175
    14.2.2 PDF of $(S_1^2/\sigma_1^2, S_2^2/\sigma_2^2)$, 177
14.3 Behavior of $\widehat{CP}_1$, $\widehat{CP}_2$ Under Contamination, 178
    14.3.1 Distribution function of $\widehat{CP}_1$, $\widehat{CP}_2$, 178
    14.3.2 Moments of $\widehat{CP}_1$, $\widehat{CP}_2$, 179
14.4 Tests of Hypotheses, 184
    References, 186

15. Quality Improvements by Likelihood Ratio Methods for Surveillance
Marianne Frisen and Peter Wessman   **187**

15.1 Introduction, 187
    15.1.1 Notations and specifications, 188
15.2 Methods, 188
15.3 Results and Conclusions, 189
    15.3.1 Robustness with respect to intensity, 189
    15.3.2 Delay of an alarm, 190
    15.3.3 Predicted value, 191
15.4 Summary and Concluding Remarks, 191
    References, 192

16. Properties of the Taguchi Capability Index for Markov Dependent Quality Characteristics
Erik Wallgren   **195**

16.1 Introduction, 195
16.2 Earlier Results, 196
16.3 The Taguchi Index for Autocorrelated Variables, 197
16.4 The Model, 197
16.5 Estimation of C_{pmr}, 198
16.6 A Simulation Study, 199
16.7 Summary and Concluding Remarks, 202
     References, 202

PART III: DESIGN AND ANALYSIS OF EXPERIMENTS

17. Fast Model Search for Designed Experiments with Complex Aliasing
Hugh A. Chipman   **207**

17.1 Introduction, 207
17.2 Priors, 209
17.3 Efficient Stochastic Search, 210
17.4 Automatic Selection of Prior Parameters, 211
17.5 A More Difficult Problem, 215
17.6 Discussion, 219
     References, 220

18. Optimal 12 Run Designs
K. Vijayan and K. R. Shah   **221**

18.1 Introduction, 221
18.2 Notation and Preliminaries, 222
18.3 Optimal Design, 225
18.4 E^*-optimality $(p \geq 9)$, 226
18.5 D-optimality $(p \geq 9)$, 229
18.6 E^* and D-optimality for $p = 6$, 7, 8, 232
18.7 Summary and Concluding Remarks, 235
     References, 235

19. Robustness of D-Optimal Experimental Designs for Mixture Studies
David W. Bacon and Rodney Lott   **237**

19.1 Introduction, 237
19.2 Linear Empirical Model Forms, 238
19.3 Experimental Designs for First Order Linear
     Mixture Models, 238

19.4 Experimental Designs for Second Order Linear
Mixture Models, 239

19.5 Experimental Design Criteria, 240

19.6 Robustness of *D*-Optimal Mixture Designs, 240

19.6.1 Case 1, 241

19.6.2 Case 2, 242

19.6.3 Case 3, 243

19.6.4 Case 4, 245

19.7 Some General Observations, 246

References, 247

20. A Bivariate Plot Useful in Selecting a Robust Design
Albert Prat and Pere Grima   **249**

20.1 Introduction, 249

20.2 Background, 250

20.3 Description of the Problem, 250

20.4 Proposed Methodology, 251

20.4.1 Hypothesis about the model for the response, 252

20.4.2 Estimation of the model parameters, 254

20.4.3 Model analysis: Response variance and expected
value, 255

20.4.4 Choosing the optimum values for design factors:
The distance-variance plot, 257

20.5 Conclusions, 258

References, 261

**21. Process Optimization Through Designed Experiments:
Two Case Studies**
A. R. Chowdhury, J. Rajesh and G. K. Prasad   **263**

21.1 Introduction, 263

21.2 Case Study 1—Frame Soldering Process Improvement, 264

21.2.1 Experimentation, 264

21.2.2 Design and analysis, 265

21.3 Case Study 2—Electro-plating Process, 267

21.3.1 Experimentation, 269

21.3.2 Analysis, 270

21.4 Summary and Conclusions, 274

References, 274

22. Technological Aspects of TQM
C. Hirotsu   **275**

22.1 Introduction, 275

22.2 Use of Nonnumerical Information, 276

22.2.1 Natural orderings in parameters, 276

22.3 Classification of Factors, 277

22.4 A General Theory for Testing Ordered Alternatives, 278

22.5 Testing Ordered Alternatives in Interaction Effects, 280

22.6 Analysis of Generalized Interactions, 281

22.7 Applications, 281

    22.7.1 Gum hardness data, 281

    22.7.2 Taste testing data, 282

    22.7.3 Three-way ordinal categorical data, 283

22.8 Concluding Remarks, 285

References, 286

23. On Robust Design for Multiple Quality Characteristics

Jai Hyun Byun and Kwang-Jae Kim   **289**

23.1 Introduction, 289

23.2 Desirability Function Approach, 290

23.3 Robust Design with Multiple Characteristics, 291

23.4 An Illustrative Example, 293

23.5 Summary and Concluding Remarks, 294

References, 296

24. Simultaneous Optimization of Multiple Responses Using a Weighted Desirability Function

Sung H. Park and Jun O. Park   **299**

24.1 Introduction, 299

24.2 Desirability Function Approach, 301

    24.2.1 Desirability function, 301

    24.2.2 One-sided transformation (maximization of $\hat{y}_i$), 301

    24.2.3 Two-sided transformation, 302

24.3 Weighted Desirability Function Approach, 303

24.4 Example, 304

24.5 Conclusion, 310

References, 310

25. Evaluating Statistical Methods Practiced in Two Important Areas of Quality Improvement

Subir Ghosh and Luis A. Lopez   **313**

25.1 Introduction, 313

25.2 Misleading Probability Plots in Factor Screening, 314

25.3 Analysis with Missing Data, 316

25.4 Conclusions and Remarks, 321

References, 323

PART IV: STATISTICAL METHODS FOR RELIABILITY

26. Applications of Generalized Linear Models in Reliability Studies for Composite Materials
M. V. Ratnaparkhi and W. J. Park **327**

26.1 Introduction, 327
26.2 Models for Fatigue Life of a Composite Material, 328
   26.2.1 Fatigue failure experiments, 328
   26.2.2 Notation and terminology, 329
   26.2.3 Deterministic and probability models for the fatigue data, 329
26.3 Application of GLIM: Estimation of Parameters (k, b, c) of Graphite/Epoxy Laminates, 331
   26.3.1 Fatigue data for graphite/epoxy laminates and related data analysis problems, 331
   26.3.2 Generalized linear models for the analysis of fatigue data, [McCullagh and Nelder (1989)], 333
   26.3.3 Results, 334
   26.3.4 Discussion, 335
26.4 Summary and Concluding Remarks, 336
   References, 336

27. Bivariate Failure Rates in Discrete Time
G. Asha and N. Unnikrishnan Nair **339**

27.1 Introduction, 339
27.2 Scalar Failure Rate, 340
27.3 Vector Failure Rate, 343
27.4 Conditional Failure Rate, 344
27.5 Inter-Relationships and Their Implications, 348
27.6 Summary and Concluding Remarks, 349
   References, 349

28. A General Approach of Studying Random Environmental Models
Pushpa L. Gupta and Ramesh C. Gupta **351**

28.1 Introduction, 351
28.2 The General Approach – Univariate Case, 352
28.3 The General Approach – Bivariate Models, 354
28.4 Bivariate Models Under Series System, 356
28.5 Crossings of Survival Curves and Mean Residual Life Functions, 359
28.6 Summary and Conclusions, 360
   References, 360

29. Testing for Change Points Expressed in Terms of the Mean Residual Life Function
Emad-Eldin A. A. Aly   **363**

29.1 Introduction, 363
29.2 Testing for Change Points, 365
29.3 Proof of Lemma 29.2.1, 367
29.4 Summary and Conclusions, 368
    References, 369

30. Empirical Bayes Procedures for Testing the Quality and Reliability With Respect to Mean Life
Radhey S. Singh   **371**

30.1 Introduction and Development, 371
    30.1.1 Preliminaries, 371
    30.1.2 Bayesian approach, 372
    30.1.3 Empirical Bayes approach when the prior
        distribution is unknown, 373
30.2 Development of Empirical Bayes Procedures, 374
30.3 Asymptotic Optimality of the EB Procedures and
        Rates of Convergence, 375
30.4 Extension of the Results and Concluding Remarks, 375
    References, 376
    Appendix, 377

31. On a Test of Independence in a Multivariate Exponential Distribution
M. Samanta and A. Thavaneswaran   **381**

31.1 Introduction, 381
31.2 Likelihood and Sufficiency, 383
31.3 A Test of Independence, 385
31.4 Power Results, 387
31.5 Summary, 388
    References, 388

PART V: STATISTICAL METHODS FOR QUALITY IMPROVEMENT

32. Random Walk Approximation of Confidence Intervals
D. J. Murdoch   **393**

32.1 Introduction, 393
32.2 Covering the Region by Gibbs Sampling, 395
32.3 Calculation of Interval Endpoints, 397

32.4 Examples, 399

    32.4.1 BOD example, 399

    32.4.2 Osborne's Gaussian mixture, 400

32.5 Conclusions, 403

References, 404

33. A Study of Quality Costs in a Multicomponent and Low Volume Products Automated Manufacturing System
Young-Hyun Park **405**

33.1 Introduction, 405

33.2 Screening Procedures: A Review, 406

    33.2.1 Single screening procedure, 406

    33.2.2 Screening procedures in a multi-stage system, 407

33.3 Model Development, 408

    33.3.1 Multi-stage manufacturing system, 408

    33.3.2 Quality costs, 409

33.4 Concluding Remarks, 411

33.5 References, 412

34. Estimating Dose Response Curves
Sat N. Gupta and Jacqueline Iannuzzi **415**

34.1 Introduction, 415

34.2 Methods, 417

34.3 Discussion, 419

34.3 Conclusion, 421

References, 422

35. On the Quality of Preterm Infants Formula and the Longitudinal Change in Mineral Contents in Human Milk
Brajendra C. Sutradhar, Barbara Dawson and James Friel **423**

35.1 Introduction, 424

35.2 Analysis of Macrominerals in Human Milk, 425

    35.2.1 Longitudinal effects, 425

    35.2.2 Gestation effects, 430

35.3 Analysis of Trace Elements in Human Milk, 431

    35.3.1 Longitudinal effects, 431

    35.3.2 Gestation effects, 433

35.4 Analysis of Ultratrace Minerals in Human Milk, 434

    35.4.1 Longitudinal and gestational effects, 434

35.5 Summary and Conclusion, 436

References, 436

Subject Index
439

Preface

This book is based on the papers presented at the International Conference 'Quality Improvement through Statistical Methods' in Cochin, India during December 28-31, 1996. The Conference was hosted by the Cochin University of Science and Technology, Cochin, India; and sponsored by the Institute for Improvement in Quality and Productivity (IIQP) at the University of Waterloo, Canada, the Statistics in Industry Committee of the International Statistical Institute (ISI) and by the Indian Statistical Institute.

There has been an increased interest in Quality Improvement (QI) activities in many organizations during the last several years since the airing of the NBC television program, "If Japan can...why can't we?" Implementation of QI methods requires statistical thinking and the utilization of statistical tools, thus there has been a renewed interest in statistical methods applicable to industry and technology. This revitalized enthusiasm has created worldwide discussions on Industrial Statistics Research and QI ideas at several international conferences in recent years.

The purpose of this conference was to provide a forum for presenting and exchanging ideas in Statistical Methods and for enhancing the transference of such technologies to quality improvement efforts in various sectors. It also provided an opportunity for interaction between industrial practitioners and academia. It was intended that the exchange of experiences and ideas would foster new international collaborations in research and other technology transfers.

Several industrial statisticians from around the world were involved in the organization and the program of the conference. For example, the International Committee for the Conference included B. Abraham (Canada, Chair), M. T. Chao (Taiwan), C. Hirotsu (Japan), J. F. Lawless (Canada), N. U. Nair (India), V. N. Nair (USA), B. K. Pal (India), S. H. Park (Korea), A. Prat (Spain), R. L. Sandland (Australia), K. Vijayan (Australia), and C. F. J. Wu (USA).

This book is organized into five parts and contains thirty-five chapters. Part I, 'Statistics and Quality', contains non-technical papers dealing with statistical education, innovation, training in industry, and quality systems such as QS-9000 and ISO 9000. Part II deals with 'Statistical Process Control', covering several aspects of process control such as monitoring, assessment, control, and multivariate quality characteristics. Several papers contain real examples and case studies indicating how statistics can be applied in industry to solve real problems. This section also contains some methodological papers on process capability indices. Part III covers topics on 'Design and Analysis of Experiments' including Robust Designs. Some of these papers discuss actual case studies; others include methodology and examples to illustrate the methodology. Part IV is 'Statistical Methods for Reliability' and Part V contains papers on 'Statistical Methods for Quality Improvement' that includes areas such as quality costs, confidence intervals, and regression methods.

This collection is intended for Quality Improvement practitioners and industrial statisticians. Engineers and applied statisticians involved in QI activities will find the case studies and examples beneficial to carrying out similar projects in their own environments. Academic statisticians who are interested in utilizing real examples and recognizing potential application contexts can use this as an excellent resource. More mathematically oriented readers will find the methodological papers appealing.

I am grateful to all the authors for sending their articles in on time and to all the referees for their valuable efforts. Because the papers came from various parts of the world with different computer systems, substantial editing and even retyping of some papers were necessary. Our aim was to keep the content accurate and we feel that the finished product is of high quality. We hope that the respective authors feel the same way.

I appreciate the support received from the Institute for Improvement in Quality and Productivity (IIQP) at the University of Waterloo, Canada; Cochin University of Science and Technology, India; the Indian Statistical Institute through the Bangalore Centre; and the United States Army Research Office through George Washington University.

I would like to thank Giovanni Merola, Hugh Chipman, N. Unnikrishnan Nair, Astra Goodhue, and Beverly Rodgers for helping with different aspects of the conference and the preparation of this volume. I would also like to thank Nandanee Basdeo, Anita Mallett, Linda Lingard, Lucy Simpson and Tracy Taves for their patient typesetting. I wish to acknowledge the assistance from the series editor Prof. N. Balakrishnan, and Birkhäuser personnel Mr. Wayne Yuhasz (Executive Editor) and Ms. Lauren Lavery (Assistant Editor).

BOVAS ABRAHAM
Waterloo, Ontario, Canada

DECEMBER 1997

Contributors

Abraham, Bovas Institute for Improvement in Quality and Productivity, University of Waterloo, 200 University Ave. W., Waterloo, Ontario N2L 3G1, Canada
e-mail: *babraham@setosa.uwaterloo.ca*

Aly, Emad-Eldin Department of Statistics and Operations Research, Kuwait University, P.O. Box 5969, Safat 13060, Kuwait
e-mail: *emad@kuc01.kuniv.edu.kw*

Asha, G. Department of Statistics, Cochin University of Science and Technology, Cochin 682022, India

Bacon, David W. Department of Chemical Engineering, Queen's University, Kingston, Ontario K7L 3N6, Canada
e-mail: *bacond@post.queensu.ca*

Beecroft, Dennis Institute for Improvement in Quality and Productivity, University of Waterloo, 200 University Ave. W., Waterloo, Ontario N2L 3G1, Canada
e-mail: *dbeecroft@watdragon.uwaterloo.ca*

Box, George E. P. Centre for Quality and Productivity Improvement, University of Wisconsin-Madison, Madison, Wisconsin 53706, USA

Brewster, John Department of Statistics, University of Manitoba, Winnipeg, Manitoba R3T 2N2, Canada

Byun, Jai-Hyun Department of Industrial Engineering, Gyeongsang National University, Chinju, Gyeongnam 660-701, South Korea
e-mail: *jbyun@nongae.gsnu.ac.kr*

Capila, C. Department of Statistics, Polytechnic University of Valencia, Apartado 22012, Valencia 46071, Spain

Carlsson, Olle ESA Department of Statistics, University of Orebro, P.O. Box 923, Fakultetsgatan 1, Orebro S-70182, Sweden
e-mail: *olle.carlsson@netserver.hoe.se*

Cheng, S. Department of Statistics, University of Manitoba, Winnipeg, Manitoba R3T 2N2, Canada
e-mail: *smiley_cheng@umanitoba.ca*

Chipman, Hugh A. Department of Statistics and Actuarial Science, University of Waterloo, 200 University Ave. W., Waterloo, Ontario N2L 3G1, Canada
e-mail: *hugh.chipman@uwaterloo.ca*

Chowdhury, Ashim R. SQC and OR Unit, Indian Statistical Institute - Bangalore Centre, R.V. College Post, 8th Mile - Mysore Rd, Bangalore - Karnataka 560059, India

Dawson, B. Department of Mathematics and Statistics, Memorial University, St. John's, Newfoundland A1C 5S7, Canada

Ferrer, Alberto J. Department of Statistics, Polytechnic University of Valencia, Apar-tado 22012, Valencia 46071, Spain
e-mail: *aferrer@eio.upv.es*

Friel, James Department of Biochemistry, Memorial University, St.John's, Newfoundland A1C 5S7, Canada

Frisen, Marianne Department of Statistics, University of Goteborg, Goteborg, S-41180, Sweden
e-mail: *Marianne.Frisen@statistics.au.se*

Ghosh, Subir Department of Statistics, University of California - Riverside, Riverside, California 92521-0138, USA
e-mail: *ghosh@ucrac1.ucr.edu*

Grima, P. Departament d'Estadistica i Investigacio Operativa, Universitat Politecnica de Catalunya, Av. Diagonal - 647 - ETSEIB, Barcelona 08028, Spain

Gupta, Ramesh C. Department of Mathematics and Statistics, University of Maine, Orono, Maine 04469-5752, USA
e-mail: *rcgupta@maine.edu*

Gupta, Pushpa L. Department of Mathematics and Statistics, University of Maine, Orono, Maine 04469-5752, USA

Gupta, Sat N. Department of Mathematics and Statistics, University of Southern Maine, Portland Maine 04103, USA
e-mail: *rxm381@usm.maine.edu*

Harrison, J. Department of Mathematics and Statistics, Curtin University, GPO Box U1987, Perth - Western Australia 6001, Australia

Hirotsu, Chihiro Department of Mathematical Engineering, University of Tokyo, Bunkyo-ku, Tokyo 113, Japan
e-mail: *hirotsu@stat.t.u-tokyo.ac.jp*

Iannuzzi, J. Department of Mathematics and Statistics, University of Southern Maine, Portland Maine 04103, USA

Joseph, V. R. SQC and OR Unit, Indian Statistical Institute - Bangalore Centre, R.V. College Post, 8th Mile - Mysore Rd, Bangalore - Karnataka 560059, India

Kim, K. J. Pohang Institute of Technology, Pohang, South Korea

Kim, Jae J. Department of Statistics, Seoul National University, Seoul 151742, Korea

Kocherlakota, Subrahmaniam Department of Statistics, University of Manitoba, Winnipeg, Manitoba R3T 2N2, Canada
e-mail: *kocherl@ccu.umanitoba.ca*

Kocherlakota, Kathleen Department of Statistics, University of Manitoba, Winnipeg, Manitoba R3T 2N2, Canada

Lopez, L. A. National University of Colombia, Colombia, South America

Lott, Rodney B. Department of Chemical Engineering, Queen's University, Kingston, Ontario K7L 3N6, Canada

Macpherson, Brian D. Department of Statistics, University of Manitoba, Winnipeg, Manitoba R3T 2N2, Canada
e-mail: *brian_macpherson@umanitoba.ca*

Martin, J. Department of Statistics, Polytechnic University of Valencia, Apartado 22012, Valencia 46071, Spain

Murdoch, Duncan J. Department of Mathematics and Statistics, Queen's University, Kingston, Ontario K7L 3N6, Canada
e-mail: *dmurdoch@mast.queensu.ca*

Nair, N. Unnikrishnan Department of Statistics, Cochin University of Science and Technology, Cochin 682022, India
e-mail: *stats@md2.vsnl.net.in*

Pal, B. K. SQC and OR Unit, Indian Statistical Institute - Bangalore Centre, R.V. College Post, 8th Mile - Mysore Rd, Bangalore - Karnataka 560059, India
e-mail: *bkp@isibang.ernet.in*

Park, Jun O. Department of Statistics, Seoul National University, San 56-1 Shinrim-Dong, Kiwanak-Ku, Seoul 151-742, Republic of Korea

Park, Sung H. Department of Computer Science and Statistics, Seoul National University, San 56-1 Shinrim-Dong, Kiwanak-Ku, Seoul 151-742, Republic of Korea
e-mail: *parksh@alliant.snu.ac.kr*

Park, Young Hyun Department of Industrial Engineering, Kang Nam University, San 6 2 Gugalri, Kiheung Kun, Yongin - Kyunaggi-Do 449-702, Korea
e-mail: *yhpark@kns.kangnam.ac.kr*

Park, W. J. Department of Mathematics and Statistics, Wright State University, Dayton, Ohio 45435, USA

Prasad, G. Krishna SQC and OR Unit, Indian Statistical Institute - Bangalore Centre, R.V. College Post, 8th Mile - Mysore Rd, Bangalore - Karnataka 560059, India

Prat, Albert Departament d'Estadistica i Investigacio Operativa, Universitat Politecnica de Catalunya, Av. Diagonal - 647 - ETSEIB, Barcelona 08028, Spain
e-mail: *prat@eio.upc.es*

Rahman, Shams-ur Graduate School of Management, University of Western Australia, Nedlands, Perth-Western Australia 6907, Australia
e-mail: *srahman@eiel.uwa.edu.au*

Rajesh, Juglum SQC and OR Unit, Indian Statistics Institute - Bangalore Centre, R.V. College Post, 8th Mile - Mysore Rd, Bangalore - Karnataka 560059, India

Ratnaparkhi, Makarand Department of Mathematics and Statistics, Wright State University, Dayton, Ohio 45435, USA
e-mail: *mratnap@desire.wright.edu*

Romero, R. Department of Statistics, Polytechnic University of Valencia, Apartado 22012, Valencia 46071, Spain

Samanta, M. Department of Statistics, University of Manitoba, Winnipeg, Manitoba R3T 2N2, Canada
e-mail: *samanta@ccm.umanitoba.ca*

Sandland, Ron Division of Mathematical and Information Sciences, CSIRO, Locked Bag 17, North Ryde, NSW 2113, Australia
e-mail: *ron.sandland@dmssyd.dms.csiro.au*

Shah, Kirti R. Department of Statistics and Actuarial Science, University of Waterloo, 200 University Ave. West, Waterloo, Ontario N2L 3G1, Canada
e-mail: *kshah@math.uwaterloo.ca*

Spiring, F. Department of Statistics, University of Manitoba, Winnipeg, Manitoba R3T 2N2, Canada

Singh, Radhey S. Department of Mathematics and Statistics, University of Guelph, Guelph, Ontario N1G 2W1, Canada
e-mail: *rsingh@msnet.mathstat.uoguelph.ca*

Srivastava, Muni S. Department of Statistics, University of Toronto, Toronto, Ontario M5S 1A1, Canada
e-mail: *srivasta@ustat.utoronto.ca*

Sutradhar, Brajendra C. Department of Mathematics and Statistics, Memorial University, St. John's, Newfoundland A1C 5S7, Canada
e-mail: *bsutradh@kean.ucs.mun.ca*

Thavaneswaran, A. Department of Statistics, University of Manitoba, Winnipeg, Manitoba R3T 2N2, Canada
e-mail: *thavane@ccm.umanitoba.ca*

Vannman, Kerstin Division of Quality Technology, Lulea University, Lulea S-97187, Sweden
e-mail: *kerstin.vannman@ies.luth.se*

Vijayan, Kaipillil Department of Mathematics, University of Western Australia, UWA 6907, Nedlands, Western Australia 6907, Australia
e-mail: *vijayan@maths.uwa.edu.au*

Wallgren, Erik Department of Statistics, University of Orebro, ISA–Itogskolan, Orebo S-70182, Sweden
e-mail: *Erik.Wallgren@netserver.hoe.se*

Wessman, P. Department of Statistics, University of Goteborg, Goteborg S-41180, Sweden

Yatawara, Nihal Department of Mathematics and Statistics, Curtin University, GPO Box U1987, Perth - Western Australia 6001, Australia
e-mail: *nihal@cs.curtin.edu.au*

Tables

Table 3.1   Four major quality eras   **25**

Table 6.1   Average of the closed-loop Mean Square Error
             (MSE × 1E-3) of the MI during one day of
             production following a moderate process/model
             fractional mismatch error using the TSAF
             controller and the CC controller with $r = .02$ **74**

Table 6.2   Average of the Mean Square Error
             (MSE) of MI (x 1E-3) and average of the
             maximum T adjustment for EPC rules based
             on 50 simulations, during one day of
             production after the assignable cause occurs   **74**

Table 8.1   Comparison of AAI and $1 + g$, the
             entries BL are from Box and Luceño's Table 2   **102**

Table 8.2   Comparison of average adjustment interval.
             Normal vs Mixture Normal   **105**

Table 8.3   Comparison of $1 + g = (\text{MSD}/\lambda^2\sigma_a^2) + (1 - 1/\lambda^2)$.
             Normal vs Mixture Normal   **106**

Table 9.1   MSE for the EWMA   **116**

Table 10.1   SW-kerosene color data   **129**

Table 10.2   Correlation matrix, mean and standard
             deviation   **130**

Table 10.3   Results of stepwise regression   **130**

Table 10.4   Data of $L_9(3^4)$ experiment for color quality   **131**

Table 11.1   Annotations for data sets   **140**

Table 12.1   Data on SO_2 gpl   **155**

Table 12.2   ANOVA for SO_2 gpl   **155**

Table 12.3 β-correction   **156**

Table 12.4   Flow data   **158**

Table 12.5   Ready reckoner chart for the control of

SO$_2$ in 4 ± 1 gpl **161**

Table 14.1 Expectation and variance of
$\widehat{CP}_i$ $(i = 1, 2)$ for Case I **181**

Table 14.2 Correlation coefficient of
$\widehat{CP}_1$, $\widehat{CP}_2$ for Case I **181**

Table 14.3 Expectation and variance of
$\widehat{CP}_i$ $(i = 1, 2)$ for Case II **183**

Table 14.4 Correlation coefficient of
$\widehat{CP}_1$, $\widehat{CP}_2$ for Case II **184**

Table 14.5 Values of c, determined from the bivariate
χ^2 distribution such that
$P\{\widehat{CP}_1 > c, \ \widehat{CP}_2 > c \mid c_0 = 1.3\} = .05$ **185**

Table 14.6 Power function for Case II using cut-off
points based on the bivariate χ^2 distribution **185**

Table 15.1 Methods **189**

Table 17.1 Screening experiment with Plackett-Burman
12-run design and response data **208**

Table 17.2 Quantiles of an inverse Gamma distribution
with $\lambda = 1$ **212**

Table 17.3 Models identified under strong and weak
heredity priors. The most probable models of
each size are given **216**

Table 17.4 Models identified under independence prior.
The most probable models of each size are given **216**

Table 17.5 Posterior probabilities on models, weak
heredity case. Second "unweighted" probability
column corresponds to a weak heredity prior with
all nonzero probabilities equal to 0.5 **218**

Table 18.1 Estimation efficiencies for the overall design
and for individual effects **226**

Table 20.1 Spreadsheet calculation to obtain the values
for variance and the distance to the target for
Examples 2 and 4 **259**

Table 20.2 The first row of the table of values for the
model in Example 3 **260**

Table 21.1 Analysis variance (based on S/N ratio) **266**
Table 21.2 Nested ANOVA **268**
Table 21.3 Factors and levels **270**
Table 21.4 Physical layout of the experiment **270**

Table 21.5	Mean and s.d (in microns)	**271**
Table 21.6	ANOVA using $\log_{10}(\text{s.d})^2_{ij}$ as response	**271**
Table 21.7	ANOVA using $20\log_{10}[Xij/(\text{s.d}_{ij})]$ as response	**272**
Table 21.8	ANOVA using $-10\log_{10}(1/10)\Sigma|X_{ijk} - 30|$ as response	**273**
Table 22.1	Gum hardness data	**276**
Table 22.2	Taste testing data	**277**
Table 22.3	Estimating block interaction pattern	**282**
Table 22.4	Number of cancer patients	**283**
Table 22.5	Pooled data throughout age	**285**
Table 23.1	Injection molded housing experiment: S/N ratios, individual desirabilities and overall desirability	**295**
Table 23.2	Factor effects on overall desirability	**295**
Table 23.3	Mean responses	**296**
Table 23.4	Important factors for mean responses and overall desirability	**296**
Table 24.1	Experimental design	**306**
Table 24.2	Each weight proportional to CV	**307**
Table 24.3	Different optimal conditions under G and WG	**307**
Table 25.1	Estimated factorial effects	**314**
Table 25.2	Comparison of models III, IV, and V	**316**
Table 25.3	Comparison of models I and II	**317**
Table 25.4	Comparison of MSE's, R^2's and R^2_{adj}'s for the three methods in Examples 1–5	**322**
Table 26.1	Ultimate static strength $y(0)$ in ksi of G/E[0/90/ $\pm$ 45]s (29 specimens)	**332**
Table 26.2	Fatigue data, number of cycles (n) at levels of S for T300/5208 G/E[0/90/ $\pm$ 45]s and summary statistics	**332**
Table 26.3	Samples means, s.d.'s and c.v.'s for logarithms of (n) of Table 26.2	**333**
Table 26.4(a)	Model description, estimates of k and b, and the deviance	**334**
Table 26.4(b)	95% confidence intervals for k and b	**334**
Table 28.1	Crossing point of $\lambda^*(t)$ and $\lambda(t)$ for various distributions under Gamma and Inverse Gaussian Environmental Effects	**358**

Table 31.1   Estimated powers for tests of independence using
              simulation   **387**
Table 31.2   Approximate powers of tests of independence
              using noncentral chi-square approximations   **387**

Table 32.1   Approximate sample sizes needed from a
              Beta$[P+1)/2, (P+1)/2, L, U]$ distribution
              so that the mode of the distribution of the
              sample minimum is less than $L + 0.01(U - L)$ **398**

Table 34.1   Test results for milk samples spiked with various
              levels of amoxicillin   **416**
Table 34.2   Description of linear relationships for each
              model type   **417**
Table 34.3   Values for the logit, probit and Gompertz
              transformations at each concentration   **417**
Table 34.3   Fitted probabilities for each model with and
              without 0 PPB data   **418**
Table 34.5   95% One-sided lower confidence limit on
              sensitivity at 10 PPB   **419**
Table 34.6   Expected frequencies of models and GOF
              statistics   **420**

Table 35.1   Regressions estimates and their significance
              level (in parenthesis) for the gestation and
              time as a specific factor for weeks 1 through 8   **430**
Table 35.2   The values of the adhoc normal z^* test
              statistic for testing the $H_0 : F_{Y_1} = F_{Y_2}$
              for weeks 1 through 8   **434**

Figures

Figure 1.1   Flow diagram   **4**
Figure 1.2(a)   Traditional method of teaching   **10**
Figure 1.2(b)   A model for modern teaching   **11**

Figure 3.1   Quality hierarchy of customer needs   **26**
Figure 3.2   ISO 9001 test/inspection plan   **28**
Figure 3.3   QS-9000 control plan   **28**

Figure 4.1   Statistical thinking and quality improvement   **37**
Figure 4.2   The stages of product development   **38**
Figure 4.3   Development, application, and emphasis of
                  quality management approaches over time   **38**
Figure 4.4   Conformance to specification and quality loss   **40**
Figure 4.5   The role of Taguchi methods of experimental
                  design in QFD   **41**
Figure 4.6   Framework for variability reduction   **42**

Figure 6.1   Illustration of the resulting MSE of the
                  output MI ($\times$ 1E-4) and the maximum adjustment
                  ∇T under different constraint parameter values
                  r of the constrained controller   **75**
Figure 6.2   MSE of the output MI ($\times$ 1E-3) under
                  different control strategies (normal conditions)   **75**
Figure 6.3   Closed-loop stability regions for fractional
                  mismatch errors in transfer function model
                  parameters ($w_i' - w_i/w_i$ for $i = 0, 1$) using the
                  CC(.02) and TSAF controllers   **76**
Figure 6.4   MI measurement error: (a) Shewhart chart for
                  deviations from predicted performance $\{a_t\}$;
                  (b) EWMA chart for $\{a_t\}$ **76**
Figure 6.4   (continued) MI measurement error:
                  (c) Shewhart chart for deviations from target $\{e_t\}$;
                  (d) predicted deviations from target when no
                  temperature adjustments are applied;

(e) feedback-control actions ∇T;

(f) manipulated variable series T **77**

Figure 7.1   Each of the indices S_{pn} in (7.4) and
              $C_p(v)$ in (7.7) as a function of δ for
              $T > M$, $\beta = 0.5$, $\gamma = 0.10$, 0.25 and
              $v = 1.4$. $S_{pn} \geq C_p(1) \geq C_p(4)$ for
              $-1.5 \leq \delta \leq 0.5$ **83**

Figure 7.2   Each of the indices S_{pn} in (7.4),
              C_{pa} in (7.11), and $C_{pv}(1, v)$ in (7.9) as a
              function of δ for $T > M$, $\beta = 0.10$,
              0.25 and $v = 1.4$ **85**

Figure 7.3   The probability that the process is considered
              capable as a function of $a = (\mu - T)/\sigma$,
              i.e., $P(C_{pa}^*(u, v) > c_{0.05})$, for $-1 \leq a \leq 1$,
              given that $T > M$, $\beta = 0.5$, $\gamma = 0.1$,
              $n = 20$, and $\alpha = 0.05$, where $c_{0.05}$
              is determined by $P(C_{pa}^*(u, v) > c_{0.05}) = 0.05$,
              given that $a = 0$ and $\gamma = 1/6$. In each case
              for a fixed value of u the curves correspond to
              $v = 1, 2, 3, 4$ counted from the top   **91**

Figure 7.4   To the left the probability that the process
              is considered capable as a function of
              $a = (\mu - T)/\sigma$, i.e., $P(C_{pv}^*(u, v) > C_{0.05}$,
              for $-1 \leq a \leq 1$, given that $T > M$,
              $\beta = 0.5$, $\gamma = 0.1$, $n = 20$, and
              $\alpha = 0.05$, where $C_{0.05}$ is determined by
              $P(C_{pv}^*(u, v) > C_{0.05}) = 0.05$, given that
              $a = 0$ and $\gamma = 1/6$. For $u = 1$ the curves
              correspond to $v = 1, 2, 3, 4$ counted from the top.
              To the right the probability that the process is
              considered capable as a function of a using
              $C_{pa}^*(0, 4)$ and $C_{pv}^*(1, 3)$ **92**

Figure 9.1   Chemical process observations with the optimal
              smoother   **118**

Figure 10.1   The goal and key factors of SPC   **122**
Figure 10.2   Flow of an SPC system   **123**
Figure 10.3   Organization of small groups for quality
              improvement   **124**
Figure 10.4   Flow-chart of quality problem solving in SPC   **126**
Figure 10.5   SPC integrated system   **127**
Figure 10.6   Causes-and-effects diagram for color quality   **128**

Figure 10.7   Response contours of $\hat{y}_1$ and $\hat{y}_2$
                  (a) Response contour of $\hat{y}_1$
                  (b) Response contour of $\hat{y}_2$ **132**

Figure 11.1(a)   Schematic of the HPRC plant showing process
                  measurement instruments   **143**
Figure 11.1(b)   Schematic of the ore crushing process   **143**
Figure 11.2   Histogram of DEGRIT for crusher 1B each
                  data set   **144**
Figure 11.3   F-value by latent variables extracted for dual
                  crushers reference data with 1 response variable and
                  BYPASS as an X variable   **145**
Figure 11.4   F-value by latent variables extracted for dual
                  crushers reference data with 2 response variables
                  including a BYPASS factor   **145**
Figure 11.5   Monitoring and SPE charts for reference set
                  – crusher 1B (1 response variable)   **146**
Figure 11.6   Plots of latent variable weights for single
                  crusher 1B (1 response variable)   **147**
Figure 11.7   Monitoring and SPE charts for 3/8/95 data
                  – crusher 1B (1 response variable)   **148**
Figure 11.8   Monitoring and SPE charts for 12/9/95 data
                  – crusher 1B (1 response variable)   **149**

Figure 12.1   A typical graph of a characteristic in a
                  chemical process   **152**
Figure 12.2   Block diagram of a multi input/single output
                  process   **153**
Figure 12.3   Behaviour of the process   **157**
Figure 12.4   Time to get-correction with $X_1 = 14.8$
                  kg/hr and $X_2 = 3000$ kg/hr   **159**
Figure 12.5   Behaviour of process during adjustments
                  ($Y_0 < 4$ gpl and $X_2 = 3000$ kg/hr)   **160**
Figure 12.6 SO_2 gpl values before and after
                  implementation of the ready reckoner   **161**

Figure 16.1   EPCR for autocorrelation $(-.9, .9)$.
                  Process A. Sample sizes 20, 50 and 100. NCL = 90%.
                  Left panel: autocorrelation ignored.
                  Right panel: autocorrelation included   **200**
Figure 16.2   EPCR for autocorrelation $(-.9, .9)$.
                  Process A. Sample sizes 20, 50 and 100. NCL = 99%.
                  Left panel: autocorrelation ignored.
                  Right panel: autocorrelation included   **201**

Figure 16.3   EPCR for autocorrelation $(-.9, .9)$.
               Process B. Sample sizes 20, 50 and 100. NCL = 90%.
               Left panel: autocorrelation ignored.
               Right panel: autocorrelation included   **201**
Figure 16.4   EPCR for autocorrelation $(-.9, .9)$.
               Process B. Sample sizes 20, 50 and 100. NCL = 99%.
               Left panel: autocorrelation ignored.
               Right panel: autocorrelation included   **202**

Figure 17.1   Predicted response under original design
               for various multiples of parameter r **214**
Figure 17.2   Marginal probability that an effect is
               large, weak heredity case   **217**
Figure 17.3   The most probable models ordered by size.
               The vertical bars give total probability for all
               models of this size and the dots give probabilities
               for specific models conditional on that model size   **217**
Figure 17.4   Time series plot of relative probabilities
               of models on the first time each is visited.
               The solid line represents the cumulative probability
               of all models visited so far   **219**

Figure 19.1   Normalized variance of the predicted response
               from a fitted ratio model for two mixture ingredients   **242**
Figure 19.2   Normalized variance of the predicted response
               from a fitted ratio model for three mixture ingredients **243**
Figure 19.3   Normalized variance of the predicted response
               from a fitted ratio model for three mixture ingredients **245**
Figure 19.4   Normalized variance of the predicted response
               from a fitted extended first degree Scheffe polynomial
               model for three mixture ingredients   **246**

Figure 20.1   The interaction of x with z allows us to
               choose the values of x (in this case it will be one
               coded with '+') which reduces the effect of the
               variability of z **253**
Figure 20.2   The quadratic relation between z and y
               allows us to reduce the transmission of variability to
               y by choosing the right values of x (in this case,
               the greatest possible value)   **253**
Figure 20.3   Distance-variance plot for Examples 2 and 4   **260**
Figure 20.4   Distance-variance plot for Example 3   **261**

Figure 21.1   Average response plot   **266**

Figure 21.2   Hierarchical design   **268**
Figure 21.3   Average response plots with
               $\log_{10}(\text{s.d})_{ij}^2$ as response   **272**
Figure 21.4   Average response plots with
               $20\log_{10}[Xij/(\text{s.d}_{ij})]$ as response   **272**
Figure 21.5   Average response plots with
               $-10\log_{1/10}\Sigma|X_{ijk}-30|$ as response   **273**

Figure 24.1   Transformation (24.1) for various values of r **302**
Figure 24.2   Transformation (24.2) for various values of
               s and t **303**
Figure 24.3   The performance of d_1 **308**
Figure 24.4   The performance of d_2 **308**
Figure 24.5   The performance of d_3 **309**
Figure 24.6   The performance of d_4 **309**

Figure 32.1   The Beta densities for $g(\mathcal{S})$ when
               the model and $g(\cdot)$ are both linear, for
               $P = 1, 5, 10$ or 20 parameters   **397**
Figure 32.2   Random walk of 2000 points for BOD model   **399**
Figure 32.3   Predictions from BOD model   **400**
Figure 32.4   Random walk of 2000 points for Osborne model   **401**
Figure 32.5   Predictions from Osborne model   **402**

Figure 35.1   Concentrations of Zn, Ca, Mo and Sn in the milk   **427**
Figure 35.2   Lag 1 correlations   **428**

PART I

STATISTICS AND QUALITY

1

Scientific Learning*

George E. P. Box

University of Wisconsin-Madison, Madison, WI

Abstract: It is argued that the domination of Statistics by Mathematics rather than by Science has greatly reduced the value and the status of the subject. The mathematical "theorem–proof paradigm" has supplanted the "iterative learning paradigm" of scientific method. This misunderstanding has affected university teaching, research, the granting of tenure to faculty and the distributions of grants by funding agencies. Possible ways in which some of these problems might be overcome and the role that computers can play in this reformation are discussed.

Keywords and phrases: Iterative learning, mathematical statistics, orthogonal arrays, scientific statistics

1.1 Mathematical Statistics or Scientific Statistics?

An important issue in the 1930's was whether statistics was to be treated as a branch of Science or of Mathematics. To my mind unfortunately, the latter view has been adopted in the United States and in many other countries. Statistics has for some time been categorized as one of the Mathematical Sciences and this view has dominated university teaching, research, the awarding of advanced degrees, promotion, tenure of faculty, the distribution of grants by funding agencies and the characteristics of statistical journals.

---

*An earlier version of this paper appeared as 'Scientific Statistics, Teaching, Learning and the Computer' in the proceedings volume 'Proceedings in Computational Statistics 1996' published by Physica-Verlag. It is included here with permission from Physica-Verlag.

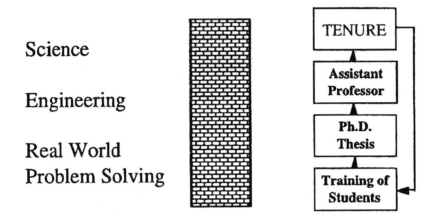

Figure 1.1: Flow diagram

All this has, I believe, greatly limited the value and distorted the development of our subject. A "worst case" scenario of its consequences for teaching, awarding of degrees, and promotion of faculty is illustrated in the flow diagrarn in Figure 1.1.

I think it is fair to say that statisticians are not presently held in high regard by technologists and scientists, who feel, with some justification, that whatever it is that many mathematical statisticians have been doing, it has little to do with their problems. Enormous areas of investigation, discovery and development in industry and the universities are thus presently deprived of the refreshing effect of a major catalyst.

It is this view of the irrelevancy of our subject that has perhaps led to the consequences discussed recently by the then-president of the American Statistical Association in an article entitled "Statistics Departments under Siege" [Iman (1994)] in which he speaks of the closings of some university departments and the decline of others. It may still not be too late to reverse this tide and the sea change in computing power that has been transforming our world can help to do this.

1.1.1   Inadequacy of the mathematical paradigm

The present training of many statisticians renders them competent to **test** a tentative solution to a problem once it has been reached but not to take part in the long, painstaking and often exciting job of discovering that solution. In other words they are not equipped to take part in the process of investigation itself. How can this be?

A purely mathematical education is centered on the one shot paradigm - "Provide me with all the assumptions and conditions associated with some proposition and if it's true I will provide a proof." Not surprisingly this mind-set has also become the paradigm for mathematical statistics - "Provide me with the hypothesis to be tested, the alternative hypothesis and all the other assumptions you wish to make about the model and I will provide an 'optimal' decision procedure." For experimental design this becomes - "Identify the response(s) to be considered, the variables on which they depend, the experimental region to be explored, the model to be fitted, and I will provide you with an alphabetically 'optimal' design."

Perhaps strangest of all is the insistence by some mathematical statisticians that the subject be based on the application of the principle of "coherence". Since this principle can only be maintained within a fixed model (or models) known a priori it excludes the possibility of new departures not mathematically inherent in the premises. Its strict application to the process of investigation would thus bring scientific discovery to a halt.

These mathematical straight jackets are of little use for scientific learning because they require the investigator to provide a priori of all the things he doesn't know.

1.1.2 Scientific learning

The paradigm for scientific learning has been known at least since the times of Aristotle (384-322 B.C.) and was further discussed for example by the great philosopher Robert Grosseteste (1175-1253 A.D.). It is also inherent in the so called Shewhart-Deming cycle: Plan - Do - Check - Act. This iterative inductive-deductive process is not esoteric but is part of our every day experience. For example suppose I park my car every morning in my own particular parking place. As I leave my place of work I might go through a series of inductive-deductive problem solving cycles something like this

Model:	Today is like every day
Deduction:	My car will be in my parking place
Data:	It isn't!
Induction:	Someone must have taken it

Model:	My car has been stolen
Deduction:	My car will not be in the parking lot
Data:	No. It is over there!
Induction:	Someone took it and brought it back

Model:	A thief took it and brought it back
Deduction:	My car will be broken into
Data:	No. It's unharmed and it's locked!
Induction:	Someone who had a key took it

Model:	My wife used my car
Deduction:	She has probably left a note
Data:	Yes. Here it is!

This iterative process is inherent in all discovery and is, I am sure, part of the long and arduous struggle that creative mathematicians must go through to *arrive* at their propositions - propositions which are eventually published and proved deductively with the elegance and precision of a magician pulling a rabbit out of a hat.

Studies of the human brain over the last few decades have confirmed that, for example, the great mathematician Henri Poincaré and the eminent psychologist William James had long ago suspected: that the brain is divided into two parts constructed to perform jointly this inductive-deductive iteration. For the majority of people the left brain is particularly concerned with *deduction*, analysis and rational thinking, while the right brain is much more concerned with *induction*, pattern recognition and creative insight. A continuous conversation between the two takes place via the interconnecting corpus callosum. Thus the generation of *new* knowledge through investigation must take place as an iterative inductive-deductive learning process. It has the necessary property that different starting points and different investigation routes can lead to success and sometimes to different but equally satisfactory solutions. Obviously this dynamic scientific paradigm cannot be squeezed into any static mathematical formulation; for no one can foretell the route that an investigation will take and the ways in which the assumptions, the responses, the variables of interest, and the experimental regions of interest, will all change as the investigation proceeds. Although the statistician cannot himself supply the necessary subject matter knowledge which the scientist- technologist-engineer will be led to inject at each stage of the investigation, nevertheless he can greatly catalyze the iteration by judicious choice of experiments to explore current possibilities and resolve uncertainties. Also the illumination provided by appropriate analysis and in particular graphic analysis can greatly help the inductive ability of the investigator. While it is an essential part of scientific method to rigorously explore the consequences of assumed knowledge its paramount purpose is of course the discovery of new knowledge. There is no logical reason why the former should impede the latter - but it does.

1.2 Statistics for Discovery

Past attempts to break out of the limitations imposed by the one shot mind-set have tended to be absorbed and stifled.

Thus some years ago John Tukey and his followers developed and clearly distinguished between exploratory data analysis on the one hand, and confirmatory data analysis on the other. Many statistics departments at first unsympathetic to these ideas, now assure us that exploratory data analysis is part of their curriculum. But examination often shows that their exploratory data analysis has been reduced to the disinterment and repeated post mortem examination of long dead "data sets" and past experimental designs over which they can no longer have any influence. In these courses it seems unlikely to be mentioned, for instance, that in a live investigation the finding of a suspected bad value should lead the statistician to walk over to the plant or lab where the investigation is going forward to find out what happened. Perhaps the bad value turns out to be just a clerical error, but sometimes what is found can be highly informative. It may lead to certain runs being redone, to new variables associated with abnormal conditions being introduced, and in some cases to new discovery. Whatever happens, is better than a continuance of stationary agonizing.

A second example concerns so-called "response surface methodology". In the late 1940's earlier attempts to introduce statistical design at the Imperial Chemical Industries in England using large preplanned all-encompassing experimental designs had failed. The "one-shot" approach, with the experimental design planned at the start of the investigation when least was known about the system, was clearly inappropriate. The industrial environment where results from an experiment were often available within days, hours, or sometimes even minutes, called for methods matching those of the skilled investigator which allowed for modification of ideas as experimentation progressed. Response surface methods [Box and Wilson, (1951)] were developed in consort with industrial experimenters as one means of filling this need. Fractional factorial experiments were used and where necessary augmented to help in the screening and selection of important factors, steepest ascent was used to allow appropriate relocation of the experimental region, sequential assembly of designs was introduced so that designs could be built up to match the simplicity or complexity of the activity in the region under current exploration and so forth. By these means experimentation and analysis were given movement and provided with some of the adaptive properties necessary to learn about the many different aspects of the investigation.

It was this *dynamic* property that was different from what had gone before. My colleagues and I thought of our efforts to develop such techniques as only a beginning and hoped that our work might inspire others to further develop such methods for experimental learning. However we were doomed to disappointment. It is certainly true that sessions on "Response Surface Methods" are a frequent feature of statistical meetings. But I find these sessions most disheartening. Mathematics has once more succeeded in killing the dynamic features of this kind of experimentation and analysis. In particular one listens to many discussions of the generation of fixed designs in fixed regions in known variables with dubiously "optimal" properties.

So one wonders whether and how this state of affairs can be changed.

1.3   Can Statistics Departments be Reformed?

If we look at the history of our subject we find, I think, that the genuinely new ideas in statistics have usually come from statisticians who were also scientists, or from teamwork with such investigators. Examples are Gauss, Laplace, Gosset, Fisher, Deming, Youden, Tukey, Wilcoxon, Cox, Daniel, Rubin, Friedman, and Effron. A reasonable inference is that, while it is true that the more mathematics we know the better, we must have scientific leadership. Some of the ways that teaching departments might change are:

a) Previous work in an experimental science should be a pre-condition for acceptance as a statistics student.

b) When this condition is not met, suitable remedial courses in experimental science should be required, just as remedial courses in appropriate mathematics might be needed for students from, say, biology.

c) Evidence of effective cooperative work with investigators resulting in new statistical ideas should be a requirement for faculty recruitment and promotion.

d) Ph.D. theses of mainly mathematical interest should be judged by the mathematics department.

e) If statistics departments find that it is not possible for them to teach scientific statistics, then they should encourage engineering, physical. and biological departments and business schools to do so, instead.

1.4   The Role of Computation

The revolution in computer power will, I believe, further catalyze scientific statistics, both in its deductive and inductive phases and will perhaps help to solve our problems. It can also greatly help students to learn about what they need to know.

1.4.1   Deductive aspects

There are many ways in which intensive computation can help the deductive phases of learning. In particular it can allow the investigator to look at the data from many different viewpoints and associated tentative assumptions. One recent application where intensive computation is essential is in the analysis of screening designs. It has recently been discovered that certain two-level orthogonal arrays have remarkable projective properties. For example it turns out [Box et al. (1987), Lin and Draper (1992), Box and Bisgaard (1993)] that the twelve run orthogonal array of Plackett and Burman can be used to form a "saturated" design to screen up to 11 factors with the knowledge that every one of the 165 choices of 3 columns out of 11 produces a full 2^3 design plus a half replicate which is itself a main-effect plan. Thus if, as is often the case in practice, activity is associated with only three or fewer factors, we have the assurance that all main effects and interactions for these factors can be estimated free of aliases. The 12 run design is therefore said to be of projectivity $P = 3$. It can be formally proved [Box and Tyssedal (1994), (1996)] that many other orthogonal arrays but not all of them can be used to produce designs of projectivity 3. However the number of possible choices of 3 factors quickly increases with larger arrays. For instance while the 20×20 orthogonal array of Plackett and Burrnan can be used to screen up to 19 factors at projectivity 3, there are 969 possible 3 dimensional projections producing at least one 2^3 factorial design in the chosen factors. With so many possibilitles it is not to be expected that the important factors can usually be tied down in a single iteration. Furthermore the possibility must be taken account of that more than 3 factors may need to be considered. In recent work [Box and Meyer (1993)] it has been shown how a Bayesian approach may be adopted to compute the posterior probabilities of various numbers of factors being active. The likely combinations are now reconsidered and [Meyer et al (1994)] have shown how intensive computation can select a further subset of experiments to run which maximize the expected change in entropy. After the additional experiments have been run the posterior probabilities can be recalculated and the process repeated if necessary.

1.4.2 Inductive aspects

The creative ability of the human brain is most stimulated by graphic representation of results. It is by devising creative and interactive graphics to display statistical features that the creative mind of the investigator can be stimulated. If the statistician has little experience of iterative learning he may be involved in a dialogue like the following:

Investigator: "You know, looking at the effects on y_1 of variables x_2 and x_3 together with how those variables seem to affect y_2 and y_3 suggests to me that what is going on *physically* is thus and so. I think, therefore, that in the next design we had better introduce the new factors x_4 and x_5 and drop factor x_1."

Statistician: "But at the beginning of this investigation I asked you to list *all* the important variables and you didn't mention x_4 and x_5."

Investigator: "Oh yes, but I had not seen these results then."

1.4.3 Learning

Figure 1.2(a) represents the past pattern for teaching. The students' mind is being used here as a storage and retrieval system. This is a task for which it is not particularly well adapted.

Figure 1.2(b) shows what I believe will be the teaching of the future. Here the teacher acts as a mentor in training the student in unstructured problem solving, somewhat after the pattern adopted in the medical profession in the training of residents. The computer here plays the part of storing and retrieving information while the human mind is set free to do what it does best and what the computer cannot do. which is to be inductively creative.

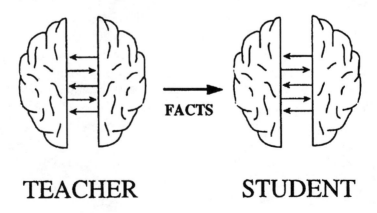

TEACHER　　　　　　**STUDENT**

Figure 1.2(a): Traditional method of teaching

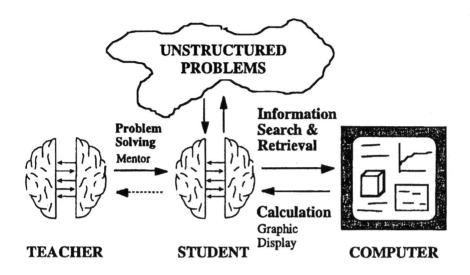

Figure 1.2(b): A model for modern teaching

Acknowledgement. This work was sponsored by a grant from the National Science Foundation #DMI-9414765 and the Universidad de Cantabria, Santander, Spain.

References

1. Iman. R. L. (1994). Statistics Departments Under Siege, *Amstat News*, **212**(6).

2. Box, G. E. P. and Wilson, K. B. (1951). On the experimental attainment of optimum conditions, *Journal of the Royal Statistical Society, Series B*, **13**, 1–38, discussion 39–45.

3. Box, G. E. P., Bisgaard, S. and Fung, C. (1987). *Designing Industrial Experiments: The Engineer's Key to Quality.* Center for Quality and Productivity Improvement; University of Wisconsin; Madison, WI.

4. Lin, D. K. J. and Draper, N. R. (1992). Projection properties of Plackett and Burman designs, *Technometrics*, **4**, 423–428.

5. Box, G. E. P. and Bisgaard, S. (1993). What can you find out from 12 experimental runs?, *Quality Engineering*, **5** (4), 663–668.

6. Box, G. E. P. and Tyssedal, J. (1994). Projective properties of certain orthogonal arrays, Report #116, Center for Quality and Productivity Improvement. University of Wisconsin-Madison; Madison, WI.

7. Box, G. E. P. and Tyssedal, J. (1996). The sixteen run two-level orthogonal arrays, Report #135, Center for Quality and Productivity Improvement. University of Wisconsin-Madison; Madison, WI.

8. Box, G. E. P. and Meyer, R. D. (1993). Finding the active factors in fractionated screening experiments, *Journal of Quality Technology*, **25** (2), 94–105.

9. Meyer, R. D., Steinberg, D. M. and Box, G. E. P. (1996). Follow-up designs to resolve confounding in fractional factorials, *Technometrics*, **38** (4), 303–332.

2

Industrial Statistics and Innovation

R. L. Sandland

CSIRO Mathematical and Information Sciences, Australia

Abstract: Industrial Statistics has been intimately linked with Quality and the Quality Revolution. There are very good historical reasons for this, but caution is indicated; there has recently been a strong backlash against the Quality movement, often because Quality concepts have been misunderstood and misapplied. For example, the measurement component of major Quality awards is usually the worst handled. Deep mastery of variation has failed to take root.

The criteria for the Australian Quality Awards are analysed to provide a template for the creative application of Statistics to learning and to the creation, codification, sharing and deployment of knowledge. This process is intimately linked to innovation, which is increasingly seen as a crucial factor in industrial wealth creation.

The current Industrial Statistics paradigm is critically examined and a program for the re-invention of Industrial Statistics is suggested. The program is based on linkages with product and process innovation, and with technology strategies.

Keywords and phrases: Industrial statistics, innovation, knowledge, quality, Shewhart paradigm

2.1 Introduction

Industrial Statistics has developed as a branch of Statistics largely because of its intimate connection with Quality. The continued health of Industrial Statistics depends on its relevance to the industries to which it is applied. The purpose of this paper is to examine some trends in industry and investigate their relevance to the growth and prosperity of Industrial Statistics. Of particular interest is the emerging link between industrial innovation and commercial success.

2.2   Innovation

Innovation means doing something which is both new and valuable.

It doesn't necessarily entail the creation of a new technology or product. It may simply relate to putting things together in a different way, creating a new product by thinking differently about the market, exploiting knowledge in new ways or altering a process to increase its effectiveness.

Some specific examples of this include:

- electronic recording of customer measurements, cutting and delivering "tailored" jeans for a modest cost premium

- using financial spreadsheets to create new value-added customer services such as scenario generation for mortgage repayment schedules

- doing business in new ways through the Internet

- increasing throughput in a polymer unit using evolutionary operations technology (EVOP)

- developing the hardware and customer interfaces for automatic teller machines, (carried out in Citibank's Research Laboratory)

- using expert knowledge of bread-making to create an automatic bread-making machine that actually works

Innovation is crucial to business success because:

- competition is increasingly global

- product cycles are being dramatically reduced, [Smith & Reinertsen (1991)]

- customer expectations are increasingly geared to a rapidly changing market

2.3   Industrial Statistics and Quality

The focus of Industrial Statistics has been largely on Quality, and there are good reasons for this.

- Quality has measurement as one of its cornerstones.

- Variation and Quality have been shown to have an intimate connection, and variation is grist to the statistician's mill.

- The successful deployment of industrial statistics in Japanese industry gave a strong impetus for its use in western countries.

- Japan's post-war economic miracle was seen to have been based on its Quality focus and achievements. Deming's role in his transformation, and his championing of the role of statistical thinking as an integral part, added further to this view.

- The competitive success of Japanese manufacturing invited a response from their Western competitors. Quality became flavour of the decade during the eighties.

But the pendulum is now swinging away from Quality as the key determinant of business success. Some of the reasons for this can be listed:

- There have been prominent failures among Quality companies.

- There is a backlash against the strong requirements for documentation imposed by the ISO 9000 standards and confusion between certification as an end in itself and thorough going Quality programs in which Quality (and business performance) is the primary focus.

- There is a feeling, at least in some quarters, that the Quality lessons have all been learnt.

It is a commonplace observation that the "measurement" component of the Baldrige Awards and the Australian Quality Awards is by far the worst handled of any of the assessment categories. This suggests that, with significant exceptions, the deep mastery of variation/statistical thinking that Deming so eloquently advocated has not really taken hold. Certainly this is true in Australian industry.

The Australian Quality awards criteria provide a good starting point to gain an insight into what might be considered best practice. [Australian Quality Awards Assessment Criteria (1997)].

The relevant category is "Information and Analysis" Evaluators in the category look for:

- systems for deciding what to measure, what to do with (the data), and how to use it

- Companies driven by "statistical thinking" ie. that use data effectively in decision-making at all levels

The Australian Quality Awards documentation goes on to state that business success is increasingly dependent on the ability of companies to:

- collect data for key stakeholders and core activities

- convert the data to information

- increase knowledge of the business

- deploy the knowledge to continuously improve processes, outputs and results

In fact in this context it is worth examining the meaning of continuous improvement. Continuous improvement is an organisation-wide set of principles and processes focussing on the creation and dissemination of knowledge that facilitates improvements.

It requires:

- learning from data

- recognising the dynamic nature of knowledge and continually challenging the status quo

- understanding the value of experimentation and evolutionary operations

- recognition that continuous improvement is itself a (set of) process(es) capable of improvement

- recognising, valuing and sharing knowledge

The process of transformation from data to information to knowledge is intrinsic to continuous improvement and, at this level, it is also strongly linked to innovation.

In order for the potential of Industrial Statistics to be realised in the industrial context its role needs to be re-examined and reconstructed focussing much more on innovation.

2.4 The Current Industrial Statistics Paradigm

The Shewhart paradigm has completely dominated industry's view of the potential role of Industrial Statistics. The Shewhart paradigm brought with it some profound benefits, in particular, a simple action rule for responding to process data and a platform for the application of the scientific method in industry, Shewhart (1939).

In today's industrial climate the cracks in the Shewhart fabric are very evident. Its continuing usefulness is limited by:

- decades of misapplication (for example, the inappropriate substitution of tolerance limits for control limits)

- its predominantly univariate focus

- the inappropriateness of the underlying assumptions (for example, independence, stationarity) in many industrial contexts

- advances in data capture technology which permit the real time automatic capture of data in complex forms such as images [Sandland (1993)].

- advances in information technology

- new manufacturing and service delivery environments (for example, those producing individually tailored products and services, flexible manufacturing, shorter runs etc)

Box (1993) captures the essence of what is required for the re-invention of Industrial Statistics:

"Every operating system - manufacturing, billing, dispatching and so forth - produces information that can be used to improve it".

The re-invention requires defining roles in both product and process innovation and a good understanding of what makes companies tick in both their technology and business strategies. Of course much of this knowledge relates to the current industrial environment and we need to recognise that it is changing very rapidly.

2.5 Product Innovation

Nonaka and Takeuchi (1995) discuss the product innovation process in some detail. Basically the process is one of knowledge creation and management.

Nonaka and Takeuchi follow other authors in separating tacit and explicit knowledge. Knowledge that can be codified by expressing it in words, numbers and diagrams is explicit knowledge. Knowledge that is of the form of subjective insights, intuition, the result of deep experience, unable to be easily codified is tacit knowledge. An example is the specific "how to" knowledge that a fine craftsperson brings to his or her craft. Nonaka and Takeuchi rightly insist on the importance of tacit knowledge in product innovation. One of their examples is the development of the first truly automatic bread-making machine by Matsushita "Home Bakery" in 1984. Crucial to the success of this product was the capture of tacit knowledge of bread making of the head baker at the Osaka International Hotel by Ikuko Tanka, a software developer on the project team. This was achieved by Ms Tanaka actually becoming apprenticed to the head baker to learn his craft.

Nonaka and Takeuchi (1995) capture the essence of the process involved in knowledge creation and dissemination required in product development in the table given below:-

	tacit knowledge **TO**	explicit knowledge
tacit knowledge	Socialisation	externalisation
FROM explicit knowledge	internalisation	combination

The entries in the table are the process by which knowledge is converted from the row classification to the column classification. For example, socialisation is the process of sharing tacit knowledge among workmates. The interested reader is referred to Nonaka and Takeuchi (1995) for a description of the other processes. However the names are broadly suggestive of the process involved.

All modes of knowledge creation and dissemination are important. Industrial Statistics has a role to play in the externalisation and combination processes, but virtually no role to play in the (at least) equally critical processes of socialisation and internalisation.

Within these processes, more explicit analysis of the role of Industrial Statistics would show roles in:

- measurement of product quality

- experimental strategies for prototyping

- statistical definition of, and procedures for testing for, appropriate quality characteristics

- defining and measuring customer needs

- defining and measuring customer perceptions

These potential inputs link significantly to other technological inputs, for example:

computational modelling of product characteristics via finite element analysis; optimisation; information technology systems, including knowledge-based systems and artificial intelligence.

2.6 Process Innovation

The role of Industrial Statistics in process innovation has never been described more fully and effectively than in Box, Hunter & Hunter (1978).

Evolutionary operation (EVOP) as a tool for process innovation has many positive features. Its principal disadvantage is that it has never achieved paradigmatic status. One possible explanation for this is that the tacit knowledge of EVOP possessed by "statistical masters" has not been effectively shared in working environments, nor has it been made explicit (and thereby more widely useable) in the same sense as Taguchi methods.

Industrial Statistics has considerable work to do here in both of the above processes.

Generally experimentation, monitoring and control will remain the key technologies for Industrial Statistics in process innovation.

There are also important links to other technologies that need to be better understood in order to help us build a stronger role for Industrial Statistics in Process Innovation. These include:

deterministic process modelling/monitoring (generally based on understanding the physics of a particular process and converting this understanding into partial differential equations); scheduling and optimisation; modelling order and dispatch processes; IT systems; and, measurement technology.

In the last-named area the ability to capture data on line, often in very large volumes and in complex forms (such as image and video) is increasing rapidly. Such data creates real challenges for Industrial Statistics in terms of reduction, extraction of information, interpretation, and action. Some of the statistical issues and challenges arising from this new technology are described in Sandland (1993). This area epitomises the need for Industrial Statistics to be aware of new challenges and to be prepared to actively meet them.

2.7   Link to Technology and Business Strategies

These challenges will tend to arise from companies' technology or business strategies. Industrial Statistics is delicately poised with respect to the former because it is an enabling technology: it does not create new products; it does not define new process. It is thus highly vulnerable to being overlooked in technology or business strategies. It is therefore of great importance to Industrial Statistics that the messages of the rapidly developing field of technology management are understood and incorporated in our ideas about the future of Industrial Statistics.

Recent thinking about the management of technical inputs to industrial research and development is well captured in Roussel, Saad and Erickson (1991). They describe "3rd Generation R&D" in which:

- funding varies with technology maturity and impact

- resource allocation to R&D projects are based on a balance of priorities and risks/rewards

- targeting is based on strategic business objectives and takes into account benefits versus costs

- results are measured not just against technological expectations but also against business outcomes

- progress against research objectives is evaluated regularly

This model for R&D management contrasts strongly with earlier models in which R&D set its own research agendas and successful outcomes for the company were achieved more by good luck than good management (first generation R&D), and even the those in which business considerations had a strong input to decisions on a project level but there was no overall approach to determining the research portfolio (second generation R&D).

Third generation R&D is based on the development of a total R&D portfolio for a company which is intimately linked with the company's business strategy. The implications for Industrial Statistics are potentially quite profound.

They include:

- the need to be actively involved in planning processes to establish R&D portfolios

- the need to be responsive to changing business/technology requirements - if the R&D portfolio changes, the role of Industrial Statistics may need to change with it (for example, more work on image analysis and accelerated life testing may be required, and less on SPC)

The overall role of Industrial Statistics would, under this model, be decided on the basis of its contributions to the research projects which constitute the R&D portfolio.

Industrial statisticians will be called upon to display vision (in terms of potential technical inputs to the portfolio projects), global understanding of the business and technology strategies of the company, and negotiation skills to ensure the potential role of Industrial Statistics is realised.

The implications for the academic community of industrial statisticians are less immediate but will follow inexorably from the decisions taken by technical and business managers in corporations. It is incumbent on us as industrial statisticians to understand these trends and provide the necessary intellectual and strategic support to those practising statistics at the industrial coalface.

References

1. *Australian Quality Awards Assessment Criteria*, Australian Quality Awards Foundation 1997.

2. Box, G. E. P. (1993). Quality Improvement - the New Industrial Revolution, *International Statistical Review* **61** 3–19.

3. Box, G. E. P., Hunter, W. G. and Hunter, J. S. (1978). *Statistics for Experimenters*, John Wiley & Sons, New York.

4. Nonaka, I. and Takeuchi, H. (1995). *The Knowledge Creating Company*, Oxford University Press, Oxford.

5. Roussel, P. A., Saad, K. N. and Erickson, T. J. (1991). *Third Generation R&D*, Harvard Business School Press, Boston.

6. Sandland, R. L. (1993). Discussion of: George Box 'Quality Improvement - the New Industrial Revolution', *International Statistical Review* **61**, 21–26.

7. Shewhart, W. A. (1939). *Statistical Method from the Viewpoint of Quality Control*, The Department of Agriculture, Washington.

8. Smith, P. G. and Reinertsen, D. G. (1991). *Developing Products in half the Time*, Van Nosstrand Reinhold Inc., New York.

3

Understanding QS-9000 With Its Preventive and Statistical Focus

Bovas Abraham and G. Dennis Beecroft

University of Waterloo, Waterloo, Ontario, Canada

Abstract: The automotive Quality System Requirements standard QS-9000 is often presented as "ISO 9000 plus specific automotive interpretations". However, QS-9000 and ISO 9000 are fundamentally different in their approach and philosophy. While ISO 9000 defines quality system requirements to control product quality through inspection and control, QS-9000's approach is to prevent poor quality first by focusing on the design and then by controlling the manufacturing processes that produce the product.

Statistics play a very important role in the effective implementation of QS-9000. Data collection and analysis are vital in maximizing the benefits from the quality system. This includes the use of company level data which are to be used to manage and prioritize the business activities. It also includes the collection and analysis of data on customer satisfaction and benchmarks on competition. Manufacturing processes are to be managed and controlled through the extensive use of Statistical Process Control (SPC) data. Suppliers are required to demonstrate minimum process capability and stability requirements in addition to showing continuous improvement in those same characteristics. Measurement systems and process machinery are to be monitored and maintained by continuous monitoring and analysis of measurement data.

Keywords and phrases: ISO-9000, prevention of defects, QS-9000, quality system requirements, statistical requirements

3.1 Introduction

The automotive Quality System Requirements, QS-9000 introduced in August 1994 is a harmonized Quality System Standard that was developed by the

Chrysler, Ford and General Motors Supplier Quality Requirements Task Force. This standard must be implemented by all suppliers of Chrysler, Ford and General Motors. It is described in the manual "Quality System Requirements" with additional information given in the reference manuals: "Statistical Process Control", "Advanced Product Quality Planning and Control", "Measurement Systems Analysis", and "Production Part Approval process". These manuals are published by the Automotive Industry Action Group (AIAG) (1995). To conform with QS-9000 requirements, an automotive supplier must have designed and implemented a quality system that makes effective use of a wide variety of concepts and tools. Sometimes it is presented as "ISO 9000 plus specific automotive interpretation and requirements". However, QS-9000 and ISO 9000 are fundamentally different in their philosophy and approach. While ISO 9000 defines quality assurance system requirements to *control product* quality, QS-9000's approach is very much *prevention* driven by defining requirements to *control and improve the processes* which produce products using a wide variety of statistical concepts and tools. The QS-9000 provides many prescriptions designed to promote Quality Improvement and Consistency in manufacturing.

3.2   Quality Evolution

Garvin's (1988) "Four Major Quality Eras" (see Table 3.1) is very useful in the understanding of the approaches imbedded in ISO 9000 and QS-9000. In the initial era, - "Inspection" quality was viewed as a problem to be solved primary through detection. Gauging and measurement methods were used by quality professionals as they inspected, sorted, counted and graded products. The quality role fell solely on the inspection department. This method was very expensive and time consuming which led eventually to the "Statistical Quality Control" era. The primary concern was control and quality was still viewed as a problem to be solved. The emphasis was product uniformity with reduced inspection. Quality professionals were involved in trouble shooting and the application of statistical methods. While the quality responsibility was being transferred to the manufacturing and engineering departments, the orientation and approach were trying to "control in" quality. Many organizations were unhappy with this approach as it was after the fact and did not have any real impact on quality. If poor quality was produced, some would be shipped to customers in spite of "best efforts".

This then led to the third era - "Quality Assurance" which was revolutionary in concept. The revolutionary thinking was that if one produces poor quality products, regardless of whether one uses 100% inspection or statistical quality control, poor quality products still exist and customers will likely receive them, therefore the only solution is not to *produce* poor quality products. This led to the understanding that the only way to produce quality products is to control the processes that produce the products. The quality Assurance era emphasized the coordination of all functions in the entire supply chain from design to customer, the contribution of all functional groups, especially designers, to prevent quality failures. The method used to address quality was to implement programs and systems. The role of the quality professional was quality measurement, quality planning and program design. The quality responsibility was now shared by all functions with the view to building in product quality.

Table 3.1: Four major quality eras			
Stage 1	**Stage 2**	**Stage 3**	**Stage 4**
Inspection	Statistical Quality Control	Quality Assurance - Process Focus - Prevention	Strategic Quality Management
Inspect in - Quality	Control in - Quality	Build in - Quality	Manage in - Quality
ISO 9000		QS-9000	

In the final era - "Strategic Quality Management" the primary concern is strategic impact with the emphasis on market and customer needs. The methods used in this era are strategic planning, goal setting, and mobilization of the entire organization. The role of the quality professional in this era is goal setting, education and training, consultative work with other departments and program design. In this era everyone in the organization is responsible for quality with top management exercising strong leadership. The orientation and approach is to manage in quality.

3.3   ISO 9000 Control Focus

ISO 9000 Quality System Requirements are at the initial two eras of the quality evolution - "Inspection" and "Statistical Quality Control". This is further reinforced by a recently published document ISO/TC 176/SC 2/N336. Figure 3.1 shows ISO 9001 as the baseline standard requirement which is based on the following:

  i) Say what you do (Quality Policy)

 ii) Document what you say (Document the Policy)

iii) Do what you say that you do (Implement the Procedure)

iv) Demonstrate that you do what you say (Auditing)

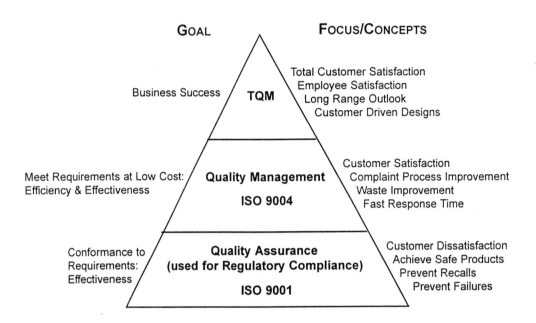

Figure 3.1: Quality hierarchy of customer needs
(Adapted from ISO/TC 176/SC 2/N336)

ISO 9001 defines the quality assurance requirements to be used for Regulatory Compliance focusing on conformance to requirements and customer dissatisfaction. The primary focus is to prevent customers receiving poor quality products thereby preventing recalls and customer failures. ISO 9004 is a Quality Management guideline which addresses customer satisfaction, process improvement, waste reduction and response time. As a guideline, ISO 9004 is not part of the compliance requirements and therefore "optional". Evidence indicates that most companies do not use ISO 9004 to assist in the design and implementation of their quality systems because their "registration requirements" are ISO 9001.

3.4   QS-9000 Prevention Focus

QS-9000 Quality System Requirements focus on both the third and fourth quality evolution eras - Quality Assurance and Strategic Quality Management. This is particularly evident through its focus on prevention, control of processes -

throughout the supply chain, and strategic focus on customers, business planning and training. The QS-9000 task force chose to address all three levels in the Quality Hierarchy of Customer Needs in Figure 3.1 incorporating much of what is in ISO 9004 and the criterion for Total Quality Management (TQM) awards, such as the Malcolm Baldrige National Quality Award [NIST Publications (1996)], as part of the contractual standard.

The differences in approaches of ISO 9000 and QS-9000 are immediately obvious in the QS-9000's stated goal - "is the development of fundamental quality systems that provide for continuous improvement, emphasizing defect prevention and the reduction of waste in the supply chain". This contrasts with the ISO 9001's implied goal to "control product quality".

Management Responsibility for QS-9000 and ISO 9001 are the same requirements for the overall quality responsibility, organization and resources for quality relative to identification of nonconformances, resolution of problems and corrective action. However QS-9000 expanded this management responsibility to include quality and quality improvement as part of the business plan. The quality improvement must include specific initiatives addressing improvements in product quality, service (including timing and delivery) and price for all customers. These improvement initiatives must use the appropriate current quality improvement tools and the supplier must demonstrate knowledge in all of these current tools. It further outlines management's role in ensuring the interface of the organization's departments and functions. The strategic focus is demonstrated by the business planning requirement addressing not only the short term (1 to 2 years) but also the long term plan of 3 years or more. The analysis and use of company level data requirement ensures that suppliers keep focussed not only on their own business performance but also to benchmark both inside and outside their organizations as a means to supporting their ongoing success. Customer satisfaction has also been added as an additional requirement. Suppliers are required to develop processes to determine current levels and trends in customer satisfaction and also use appropriate competitor trends and benchmarks.

Element 4.2, Quality System is one of the most important elements in terms of understanding the differences in approach of ISO 9000 and QS-9000. The requirement of a quality plan is usually addressed in ISO-9001 with an Inspection/Test Plan, see Figure 3.2. This plan defines the inspections or tests to be conducted with the acceptance criteria being specified. These tests and inspections are on the product at different points throughout the manufacturing process - incoming, in process, and final inspection for very specific product parameters to evaluate the acceptability of the product.

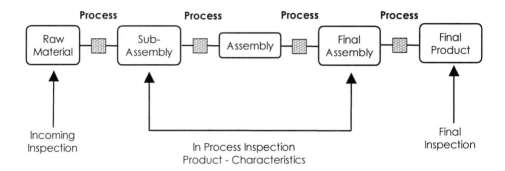

Figure 3.2: ISO 9001 test/inspection plan

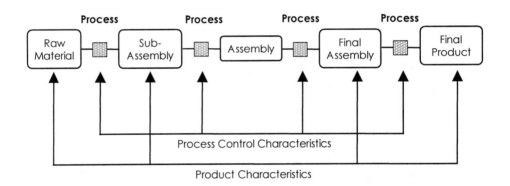

Figure 3.3: QS-9000 control plan

The quality plan for QS-9000 however is a "Control Plan", see Figure 3.3, which is very different from an inspection/test plan. While a control plan may include key product characteristics, it also includes the key process control characteristics to be monitored to ensure that the manufacturing processes are producing only good products. The control plans are developed as an output of the quality planning activity. Quality planning uses multi-function teams of engineering and manufacturing personnel, and sometimes even suppliers and/or customers. Disciplined processes such as potential failure mode and effects analysis (PFMEA), both design and manufacturing, and capability studies are used to assist in the quality planning process. The control plan defines the statistical tools which are to be used to monitor and control the processes. The concept of mistake proofing is introduced as a tool to eliminate the requirement of process monitoring. During the quality planning process, particularly the PFMEA process, potential modes of failure during the manufacturing of the product are identified. An ideal solution is to use error proofing to prevent a particular failure mode.

Process capability studies are required as part of the Production Part Approval Process (PPAP) to be done at the initial planning phase of the product introduction. This is to ensure that the manufacturing system is capable of meeting the product characteristics. Measurement systems analysis is a key component of the control planning activity. Studies, such as Gauge Repeatability and Reproducibility (R&R), are required to ensure the capability and reproducibility of the process and product monitoring systems. The control plan is a "living document" under continuous review as process and product parameters change as a result of improving systems. Review is triggered by both internal and external inputs. External inputs could be from customer feedback or returns and/or quality improvement requirements. Internal inputs include nonconformances found during the manufacturing process and the initiatives identified in the continuous improvement plans.

Element 4.9 Process Control represents a major difference between ISO 9000 and QS-9000. The requirement is to control those processes that directly impact quality. Many organizations that have implemented ISO 9000 quality systems have interpreted this to mean only those inspection and testing functions as it is during these operations that the decisions of product acceptability are being made. The rationale here is that anything that anyone else does to impact product quality would be detected and verified by these test/inspection functions. These test/inspection operations would then require operator instructions or procedures to ensure that these tasks were executed properly. Special training of some operators is sometimes used to ensure product quality depending on the tasks being performed. For example ISO 9000 defines Special Processes as processes "where the results of the processes cannot be fully verified by subsequent inspection and testing of the product..." for example welding or plating. Here the emphasis moves from verification of product to focusing on the capabilities of the people performing the functions. QS-9000, on the other hand, interprets Process Control 4.9 requirements to apply to *all* operations in manufacturing, design, materials and contract review functions. The manufacturing equipment and facilities, in addition to inspection, measuring and test equipment, are given special attention in QS-9000.

It is a requirement that any key process, defined as those controlling key characteristics, must be monitored continually to demonstrate process control, capability and stability. It is further required that the supplier demonstrate continuous improvement in these processes. Preventive maintenance, including predictive maintenance methodologies, is required on these equipments. The tooling and fixturing repair/replacement is to be managed effectively to assure process performance.

The differences in approach and philosophy of ISO 9000 and QS-9000 should now be much clearer. While ISO 9000 is an excellent base to build upon, in order to achieve maximum benefit from a Quality System it must be designed to operate in Phases 3 and 4 of the Quality Evolution.

3.5   Statistical Requirements in QS-9000

Continuous improvement is a key element in QS-9000 and this implies that QS-9000 is knowledge driven (current knowledge to control processes and new knowledge to improve processes). Statistics deals with the acquisition of knowledge by observation. Planning for data collection, the actual data collection, processing of data into information and the communication of acquired knowledge are part of Statistics. Thus it forms a nervous system, which senses the data and communicates the knowledge for the implementation of QS-9000.

Statistical thinking and tools are required in many areas in QS-9000. Some of these are listed below [as seen in the manual "Quality System Requirements", AIAG (1995)]:

(i) **4.1.5: Analysis and use of company level data**
"Shall document trends in quality, operational performance (productivity, efficiency, effectiveness) and current quality levels for key product and service features." Trends does not mean comparing last months numbers with this month's. It usually requires the monitoring of company level information on a run chart or a "chart for individuals". Section 4.1.6 discusses trends in customer satisfaction which points to the need for statistical tools.

(ii) **4.4.5: Design Output - Supplemental**
"Utilization of techniques such as Quality Function Deployment (QFD), Design of Experiments (DOE), Tolerance Studies, Response Surface Methodology (RSM), Geometric dimensioning and tolerancing are required."

(iii) **4.9: Process Control**
"Suppliers shall comply with all customer requirements for designation, documentation and control of special characteristics." The subsections process monitoring and operator instructions (4.9.1), preliminary process capability requirement (4.9.2), and ongoing process performance (4.9.3) involve Statistical Thinking and ttools such as Control Charts, C_{pk}, and P_{pk}. In this section specific requirements for process performance are given in terms of C_{pk} and P_{pk}. It should also be noted that the APQP and PPAP manuals, AIAG (1995) also refer to specific requirements for process performance in terms of C_{pk} and P_{pk}.

(iv) **4.11.4: Measurement System Analysis**
"Evidence is required that appropriate statistical studies have been conducted to analyse the variation present in the results of each type of measuring and test equipment system."

(v) **4.20: Statistical Techniques**
"The supplier shall identify the need for statistical techniques required for establishing, controlling and verifying process capability and product characteristics."(4.20.1) "The supplier shall establish and maintain documented procedures to implement and control the application of the statistical techniques identified in 4.20.1." (4.20.2) Understanding of variation, stability, capability and over adjustment through out the supplier's organization is expected.

(vi) **Section II, Part 2: Continuous Improvement**
"Continuous improvement philosophy shall be fully deployed throughout the supplier's organization." This calls for various statistical tools (C_p, C_{pk}, Control Charts, Design of Experiments, etc.). This part (Continuous Improvement) is unique to QS-9000 and it implies that for the implementation of QS-9000 Statistical Thinking and methods play a major role.

Advanced Product Quality Planning and Control Plan reference manual AIAG (1995) describes a product Quality Planning Timing Chart with five phases: Planning stage; Product Design and Development; Process Design and Development; Product and Process Validation stage; and Feedback, Assesment and Corrective Action stage. In each phase Statistical Thinking and tools are required.

In the planning stage, voice of internal and external customers is utilized (recommendations, complaints, etc.). Such information can be obtained by proper data collection via appropriate questionnaires or customer interviews. Information can also be obtained by analysing historical warranty and quality information. At the product design and development phase issues of design for manufacturability, design sensitivity to manufacturing variation etc. need to be considered and statistical techniques such as Design of Experiments, and Robust Design are very useful. In building specifications, understanding of variation is crucial. In process design and development, Statistical evaluations, Measurement System Analysis and preliminary Process Capability studies are required. Product and process validation stage also calls for Measurement System Analysis plan and evaluation, and Process Capability studies. For appropriate feedback, assessment and corrective action, understanding of variation is essential. Effective plans for dealing with variation are part of this phase. Also assessing customer satisfaction forms an important part.

Because of the overwhelming intermingling of statistical tools in the implementation of QS-9000, the requirements refer to two AIAG reference manuals indicated before, SPC and MSA.

The following statistical concepts and procedures are essential for the proper implementation of QS-9000:

i) Control Charts:
Run charts, charts for individuals, $\overline{X}$ and R charts, attribute charts, etc., need to be implemented depending on the context. Concepts of stability and tampering need to be understood.

(ii) Capability Ratios:
Process capability measures such as C_{pk}, P_{pk} are to be employed for normal and non-normal data.

(iii) Measurement System Analysis:
The concepts of precision, bias, linearity, and stability need to be understood. Repeatability and Reproducibility studies are to be performed to control variation from different sources. Strategies for improving (reducing bias and variability) measurement systems need to be part of the system.

(iv) Designed Experiments:
Factorial and fractional experiments are very useful for identifying important design and process factors. For efficient process improvement (adjusting to target and reducing variation) designed experiments are essential.

(v) Continuous Improvement (CI):
There needs to be a structured approach for CI. This helps to identify and understand families of variation.

3.6   Summary and Concluding Remarks

This paper gave a brief comparison of ISO 9000 and QS-9000, and the prevention and process focus of QS-9000 was emphasized. This quality system cannot function without the factual information and the continual growth of knowledge about all processes used in the design and manufacturing of the product. QS-9000 is knowledge driven and Statistics deals with the acquisition of knowledge by observation. We think of statistics as the nervous system of QS-9000 and it has a major role in the implementation of QS-9000.

Acknowledgements.   The authors would like to acknowledge the support from MMO (Materials Manufacturing Ontario) and NSERC (Natural Sciences and Engineering Research Council) for this project.

References

1. Automotive Industry Action Group (1995). *Advanced Product Quality Planning and Control Plan*: Reference Manual.

2. Automotive Industry Action Group (1995). *Measurement Systems Analysis*.

3. Automotive Industry Action Group (1995). *Potential Failure Mode and Effects Analysis*.

4. Automotive Industry Action Group (1995). *Production Part Approval Process*.

5. Automotive Industry Action Group (1995). *Quality System Requirements QS-9000*.

6. Automotive Industry Action Group (1995). *Statistical Process Control*.

7. Beecroft, G. D. (1995). "QS-9000 - Not "ISO 9000 Plus", *Institute for Improvement in Quality and Productivity (IIQP) Spring Newsletter*, University of Waterloo.

8. Garvin, D. A. (1988). *Managing Quality: The Strategic and Competitive Edge*, The Free Press, New York.

4

A Conceptual Framework for Variation Reduction

Shams-ur Rahman

The University of Western Australia, Perth, Australia

Abstract: The application of statistical tools has a long history of successful use at the operational level of both research and manufacturing organisations. The development of the Total Quality Management (TQM) concept brought with it a renewed emphasis on statistical thinking and its role in the overall management of organisations. Statistical thinking encompasses three important components: knowledge of systems; understanding variation; and use of data to guide decisions. Managers must strive to understand, acquire, and internalise an organisational system based on statistical thinking in order to transform their organisations. This paper provides a conceptual framework to understand and reduce variation in product quality through the use of system thinking and statistical methods.

Keywords and phrases: Process control, quality function deployment, sampling inspection, statistical thinking, Taguchi methods, variation

4.1 Statistical Thinking

The development of the concept of total quality management (TQM) has brought with it a renewed emphasis on statistical thinking and its role in the management of organisations [see Snee (1990)]. The word 'statistics' or 'statistical' might invoke fear in many managers but statistical thinking is not difficult to comprehend. It is about analysing a business process and recognising the difference between when it is 'business as usual' and when it has changed. Statistical thinking encompasses three key components:

1. Systems thinking.

2. Understanding variation.

35

3. Using data to guide decision making processes.

The concept of systems thinking, developed over the last fifty years, 'takes the idea of a whole entity which may exhibit properties as a single whole, properties which have no meaning in terms of the parts of the whole' [Checkland and Scholes, (1990, p. 25)]. A number of authors [e.g. Forrester (1961), Checkland (1981), Goldratt (1990), Senge (1992), and Deming (1993)] have contributed towards the development of a body of knowledge in systems thinking. Some important aspects of systems thinking include:

- All work is a system

- The majority of the problems are in the system

- It is a must that the appropriate processes/systems are in place.

No two things are exactly the same. Variation exists in all aspects of our lives - variation among people, among raw materials, and among equipment. The important thing here is to understand the nature of variation and recognise that the existence of variation represent an opportunity for improvement. The ability to distinguish between patterns of variation is the key to reducing variation effectively and minimise losses resulting from the misinterpretation of patterns. Typical losses resulting from misinterpretation are [Nolan and Provost (1990)]:

- Wasting time looking for explanations of a perceived trend when nothing has changed.

- Investing resources for new equipment when it is not necessary.

- Blaming people for problems beyond their control.

In order to be able to distinguish between patterns of variation in a system, it is vital to understand the causes of variation - special causes and common causes. Deming (1986) pointed out that variation can be reduced and quality improved in two ways:

1. Eliminate the special causes, causes that are not part of the system but arise because of special circumstances, and bring the system into a state of statistical control. A system under statistical control is neither good nor bad, it simply means that the outcome of the system is predictable. Some of the benefits of a system which is under statistical control are [Deming (1986, p. 340)]:

   - The system has an identity.
   - Costs are predictable.
   - Productivity is at a maximum and costs at a minimum under the present system.

- The effect of changes can be measured with greater speed and reliability.

2. Improve the system by reducing the common causes - those causes of variation that still exist when a system is in a state of statistical control. Here, the improvement of the system is possible only by changing the system.

The concept of statistical thinking in quality management is shown schematically in Figure 4.1.

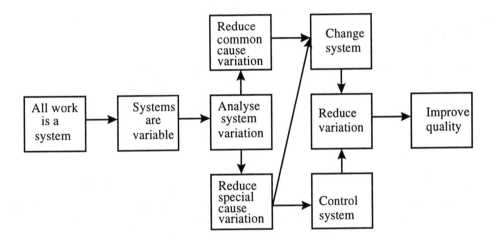

Figure 4.1: Statistical thinking and quality improvement
[adapted from Snee (1990)]

The first two blocks - all work is a system and systems are variable, represent 'system thinking' component whereas the next four blocks represent 'use of data', and 'understanding variation' component of statistical thinking. The next two blocks - change system and control system, are related to management decisions to reduce variation and improve quality.

4.2 Variation Reduction

The 'understanding variation' component of statistical thinking can be used to reduce variation and improve quality through the use of statistical concepts, methods, and tools.

The development process of a product/service follows the sequence of idea generation and design formulation, product production and after-production packaging, storing and distribution (Figure 4.2). Interestingly, the quality management approaches developed and applied to assess and improve product quality have followed the opposite sequence (Figure 4.3). Firstly, the inspection-based quality control approach was introduced. The purpose of this approach is to screen out inferior products at the after-production stage. The approach became very popular and applied widely as a sole approach to the control of product quality.

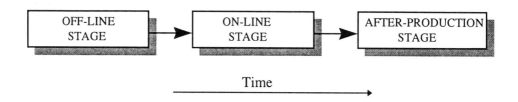

Figure 4.2: The stages of product development

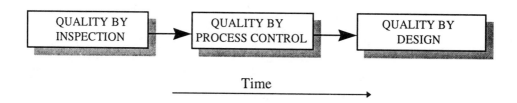

Figure 4.3: Development, application, and emphasis of quality management approaches over time

Realizing that it is too late to be able to improve quality by simply inspecting the finished product, the emphasis of quality management then shifted from inspection to process control, i.e., from the after-production stage to the on-line stage. The current belief about quality is that it is a 'virtue of design', i.e., the robustness of products is more a function of good design than of on-line control. In order to produce robust products which reflect customer needs the emphasis of quality management is gradually being shifted to the design phase, i.e., the off-line stage of product development. We discuss below three approaches to quality management.

4.2.1 Quality by inspection

The inspection-based system was possibly the first scientifically designed quality control system introduced to evaluate quality. The system is applied to incoming goods and components which would be used as input for production processes and/ or finished products. Under this system one or more quality characteristics are examined, measured and compared with required specifications to asses conformity. It is, therefore, a screening process which merely isolates conforming from non-conforming products without having any direct mechanism to reduce defects. Moreover, no matter how scientific a basis sampling inspection may have, the method rests on the notion that a certain level of defect is inevitable. The theory of the (acceptance) sampling inspection system was developed by Dodge in the late 1920s. The system was widely applied in the 50s, 60s and 70s as a sole methodology to control product quality. Sampling plans, operating characteristic curves, MIL STD Tables, and Dodge-R oming Inspection Tables are some of the tools used under this system. A quality control system based on sampling inspection still remains a good technique, but does not directly help attain the goal of continuous improvement and zero defects.

4.2.2   Quality by process control

The concept of quality control based on inspection has been replaced by the philosophy of continual process improvement through a consistency of purpose. This concept, introduced by Deming, is considered a necessary ingredient for the long term survival of an organisation [Deming (1982)]. Japanese industries have benefited immensely by using the concept. The concept is based on a method popularly known as the Deming wheel. It consists of four stages: plan, do, study, and act (PDSA). Initially a plan is developed based on objectives, goals and measures. The plan is then tested for a period of time, formal reviews are conducted to check progress toward goals, and corrective actions are taken. These appropriate corrective actions may, in turn, lead to a new cycle beginning with a revised detailed plan, so the wheel continues upwards in the path of never ending improvement.

4.2.3   Quality by design

If the aim of the Deming approach is to shift quality emphasis a step back from inspection to process control, the approach of quality by design takes a further step back from process to design. There is a consensus that product robustness is more a function of good design than of on-line control. Robust means that products 'perform their intended functions regardless of customer-imposed usage environments, manufacturing variations, variation imposed by suppliers, and degradation over the useful product life' [Blake et al. (1994, p. 100)].

Two important techniques for producing robust products are quality function deployment (QFD) and the Taguchi methods of experimental design. In the 90s, many believe that the method of experimental design should be considered a strategic weapon to compete globally [Blake et al. (1994)]. It can be applied to produce robust products with high reliability, reduced time to market, and reduced life-cycle cost.

Taguchi Methods. The Taguchi methods of experimental design are a set of effective and efficient techniques for designing robust products and processes. The heart of the Taguchi method is the concept of 'quality loss function'. Traditionally, manufacturers have felt that, as long as a product falls within predetermined specifications, the quality is as good as required. Taguchi pointed out that two products that are designed to perform the same function may both meet specifications, but can impart different losses to society. This was illustrated by the performance of a Sony TV produced in Tokyo and San Diego [see Taguchi and Clausing (1990)].

Quality loss is defined as deviation from a target value; the farther from target, the greater the loss. Taguchi expressed such a loss as a simple quadratic function. Process A in Figure 4.4 imparts greater loss than process B since deviation from the target value in process A is greater although both the processes fall within specification limits. Minimising losses for processes A and B is equivalent to minimising the mean square deviations from target. An emphasis on minimising mean square errors from the target, rather than simply being any where within the specification limits, makes more sense and is far more in line with the idea of continuous and never-ending improvement.

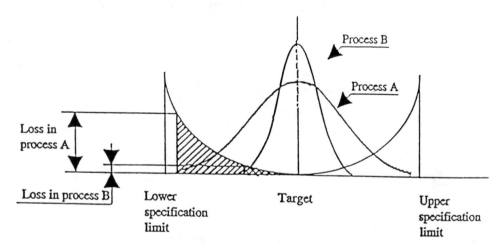

Figure 4.4: Conformance to specification and quality loss

The minimisation of mean square errors about a target value is one of the strategies of what Taguchi calls *parameter design*. Other strategies of parameter design are making products (1) robust to environmental factors, and (2) insensitive to component variation. Environmental factors are conditions in which the product will be used by customer. Such factors include human variations in operating the product, low and high temperature, temperature fluctuation, shock, vibration, and humidity.

Component variation is piece-to-piece variation. Most industrial products are assembled from a number of components. The performance of a final assembly varies if the individual components vary. The purpose of parameter design is to separate controllable factors and uncontrollable factors responsible for variation and identify the relationship between them. The identification of interactions between controllable factors and uncontrollable factors through experimental design is the key to achieving product robustness.

Quality Function Deployment.   Another technique used for robust design is known as quality function deployment (QFD). It is a kind of conceptual map which can systematically translate customers' needs into operational requirements for designing products and processes. These operational requirements are:

- Customer requirements - *what* is important to the customer?

- Technical requirements - *how* is it provided?

- The relationship between the *whats* and *hows*.

- Target design characteristics - how much must be provided by the hows to satisfy customers?

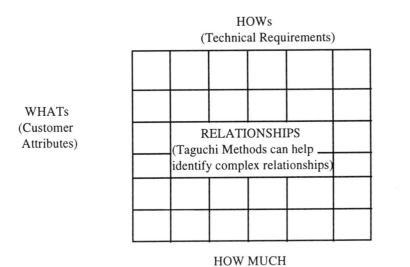

Figure 4.5: The role of Taguchi methods of experimental design in QFD

To determine appropriate design characteristics of products or processes, an understanding of the nature of the relationships between whats and hows is crucial. Simple relationships are quite obvious. Complex relationships, however, are often difficult to understand. In these cases, the Taguchi methods of experimental design can help identify the nature of relationships between whats and hows and their relative strengths (Figure 4.5). Hence QFD and Taguchi methods of experimental design are complementary techniques that can be used to design robust products.

4.3   Conceptual Framework

A framework integrating the methods and tools in three approaches is presented in Figure 4.6. This can be used to understand and reduce variation, and improve quality in the entire value chain of customer needs identification, translation of needs into measurable characteristics, and design and production of robust products and services.

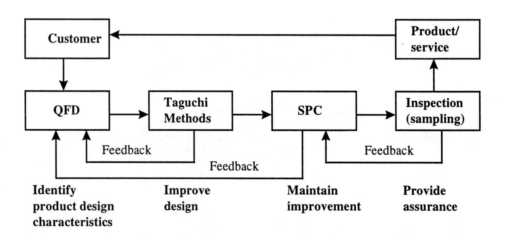

Figure 4.6: Framework for variability reduction

In this framework improvement of quality begins with identifying customer needs. The QFD process allows the design/innovation team to translate customer needs into appropriate product and process characteristics. Taguchi methods are used to make the product robust to input and environmental factors, whereas statistical process control (SPC) helps to maintain what has been achieved in the design stage. The process of sampling inspection provides assurance about the quality of the product. The feedback loops provide a mechanism for continuous improvement.

Generally speaking, all three approaches are important in order to maintain and improve quality. However it is the degree of emphasis amongst the approaches that is important. A product or service developed and produced with greater emphasis on design and process control can be expected to be of better quality.

4.4   Conclusion

With the introduction of mass production systems, the need for inspection was recognised and widely applied as a sole approach to quality management. Realising that it is not possible to improve quality only by inspecting the end product, the emphasis gradually shifted from inspection to process control. Today, good quality is considered more a function of good design than of process control. There is evidence that by better understanding customer needs and carefully incorporating these needs into product design, companies can reduce significantly the number of design changes in the innovation process, and reduce start -up costs and lead times for product development [Sullivan (1986)].

The techniques such as Taguchi methods and QFD for developing robust designs and SPC for process control are becoming increasingly popular in Australia. Companies like BHP, Ford, and Alcoa are not only successfully applying these techniques but also extending these ideas incorporating local needs [Gilmour and Hunt (1995)]. Currently, the author is carrying out a survey to assess the impact of the application of such a framework on the product quality in manufacturing organisations in Australia.

References

1. Blake, S., Launsby R. G. and Weese, D. L. (1994). Experimental design meets the realities of the 1990s, *Quality Progress*, October, 99-101.

2. Checkland, P. (1981). *Systems Thinking, Systems Practice*, John Wiley, Chichester, UK.

3. Checkland, P. and Scholes, J. (1990). *Soft Systems Methodology in Action*, John Wiley, Chichester, UK.

4. Deming, W. E. (1982). *Quality, Productivity and Competitive Position,* MIT - Center for Advanced Engineering Study, Cambridge, Mass., USA.

5. Deming, W. E. (1986). *Out of the Crisis,* MIT- Center for Advanced Engineering Study, Cambridge, Mass., USA

6. Deming, W. E. (1993). *The New Economics for Industry, Government, Education,* MIT, Cambridge, Mass., USA.

7. Forrester, J. W. (1961). *Industrial Dynamics,* MIT Press, Cambridge, Mass., USA.

8. Gilmour, P. and Hunt, R. A. (1995). *Total Quality Management,* Longman, Melbourne, Australia.

9. Goldratt, E. M. (1990). *The Theory of Constraints,* North River Press, Croton-on-Hudson, NY, USA.

10. Nolan, T. W. and Provost, L. P. (1990). Understanding variation, *Quality progress,* May, 70–78.

11. Senge, P. M. (1992). *The Fifth Discipline - The Art & Practice of the learning Organisation,* Random House, NSW, Australia.

12. Snee, R. D. (1990). *Statistical thinking and its contribution to total quality,* The American Statistician, **44**, 116–121.

13. Sullivan, L. P. (1986). Quality function deployment, *Quality Progress,* June, 39–50.

14. Taguchi, G. and Clausing, D. (1990). Robust design, *Harvard Business Review,* January-February, 65–75.

5

The Role of Academic Statisticians in Quality Improvement Strategies of Small and Medium Sized Companies

John F. Brewster, Smiley W. Cheng, Brian D. Macpherson and Fred A. Spiring

University of Manitoba, Winnipeg, Manitoba, Canada

Abstract: This paper will describe the relationships and interactions that have developed between academic statisticians in the Department of Statistics at the University of Manitoba and small to medium sized companies. The size of the local community and the limited resources available to many of these small companies mean that few knowledgeable commercial consultants are readily available. As a result, these companies have turned to the statisticians at the University of Manitoba for assistance. The paper will provide some details on the type and range of activities that the Department has undertaken. The benefits of these activities to the local industrial community as well as the benefits to the academic statisticians and their students are described.

Keywords and phrases: Consulting, industry liason, quality improvement, training, workshop

5.1   Background

The activities that are to be described in this paper are shaped in no small way by the physical location of the individuals who are involved. The initial impetus and the ongoing approach has evolved because of the characteristics of the province, city, university, and department within which the participants work.

The Province of Manitoba, with a population of just over one million people, is located in the geographical centre of Canada. The city of Winnipeg is the capital city of the Province and is home to some 650,000 people. An important aspect of its location is its relative isolation from other major centres of population. The closest major cities are more than 600 kilometers distant. There is a very healthy and active manufacturing industry in Winnipeg with over 1300 companies involved. Of these companies, 85 per cent have fewer than 50 employees and 93 per cent have fewer than 100 employees.

The Department of Statistics of the University of Manitoba offers a wide range of courses directed towards students in many disciplines as well as honours, masters and Ph.D. programs in statistics. The Department has a staff of fourteen statisticians, and the Statistical Advisory Service has one full-time statistical Consultant. Currently the Department has 7 Masters and 10 Ph.D. students. Because of its origin and evolution, the department might well be classified as having a leaning towards the application side of statistics. In that regard, within the Department there are two initiatives through which interaction with individuals and companies outside the Department has been accomplished. The Statistical Quality Control Research and Applications Group (SQCRAG) provides a wide range of Workshops to industry, and the Statistical Advisory Service (SAS) carries out consultation and collaborative research with individual and industrial clients.

5.2 Workshop Training for Industry

In the mid 1980's several faculty members in the department became interested in statistical process control and its role in quality improvement. Some modest amount of research and publishing had been done and it was felt that this might well be an area in which a coordinated approach would be profitable. The Department adopted this as a priority area in which to devote time and energy. The faculty members interested in looking at opportunities with a statistical process control focus formed an informal grouping called the Statistical Quality Control Research and Applications Group within the Department. This group had no particular structure and no source of funding so that its initial efforts were directed towards the development of awareness and the sharing of ideas. An attempt was made, given very limited resources, to collect materials, articles, books and the like on the subject matter. The initiative gained some measure of impetus from the thesis work of some of its graduate students. Over the intervening years, the academic staff have become very active, producing widely recognized research in the area, supervising graduate student dissertations, serving on national and international quality related organizations, actively contributing to the editorial boards of journals with a statistical process control (SPC) focus, and providing workshops and advice to industrial, service, and governmental clients throughout Manitoba.

Shortly after the formation of the SQCRAG, a major manufacturing company in Winnipeg contacted the Department with a request for assistance with the development of a statistical process control strategy in their operation. This company was a supplier of a large international concern that was demanding conformance to a supplier standard that emphasized the need for statistical process control. The local company had searched extensively for help and discovered that nothing satisfactory seemed to be available from local consulting organizations and individuals. They had reached the conclusion that they would have to look somewhat further afield for such help involving either importing consulting assistance from the United States or sending large numbers of their employees outside of Manitoba for training. This company had learned that, for some period of time, several of the larger Winnipeg companies had been sending their employees to training facilities in the United States and also had been bringing consultants in to their establishments for training and consultation sessions. In addition the company noted that a typical strategy adopted by some of the smaller companies had been to send a few of their key individuals away to seminars or workshops given perhaps by Edward Deming, or other such well known individuals, consulting institutes, or companies. Their personnel were expected to absorb sufficient knowledge in a short period of time to be able to spearhead not only the implementation of SPC on their shop floor but also to train the employees who would be closely involved. Upon returning to their company these individuals often found their task to be intimidating indeed, even if their enthusiasm for the task was very high.

Because of the distances involved and the logistics of such strategies the results were often very unsatisfactory and the costs involved were extensive. Somewhat as a last resort, this particular company telephoned the Department of Statistics at the University to see if there was any way that assistance could be obtained from members of the Department. As a result of this call, a workshop was planned and delivered to the senior members of the company with enthusiastic response. It was from this serendipitous beginning that a very active program of workshops and consulting activities have grown and service to a large number of local companies has resulted.

5.3   Workshop Activities

Since that first workshop, the Department has received many requests from companies for advice and assistance. These requests have come from companies with a variety of motivations for seeking the help of the statisticians in the Department. Some of these companies have found themselves in the position of being required to implement statistical process control procedures by virtue of having to satisfy a quality standard of one or more of its customers in order to achieve a recognized supplier status. The widely recognized and adopted ISO 9000 and QS 9000 standards have raised the issue of the use of statistical process control procedures without providing a great deal of direction about the how's, what's, and why's of the tools and techniques involved. Companies faced with such requirements may well simply want someone to provide them with the requisite amount of training to satisfy the various clauses of the "standard". On the other hand, management of some companies have recognized that statistical process control procedures are fundamental to a quality improvement strategy and want to develop and implement such a strategy for the continued well being of the company, regardless of explicit external pressures. In some cases, this has led senior management to recognize their need for awareness and understanding of the nature and roles of the various statistical process control tools. As a result, many companies have requested that a programme of education and training for its workforce be developed so that statistical process control tools can be developed, implemented, and maintained on the shop floor.

Over the years, the reputation of the Department has developed and the number of requests for assistance has steadily increased. As a result, the SQCRAG has developed and delivered workshop training to companies involved with the manufacture of discrete parts such as tool and die concerns, sheet metal fabricators, and the like. Workshops have been given to companies in the aerospace, electronics, and mining industries, in blood product processing, in commercial printing operations, in agricultural food product processing, in service industries such as life insurance companies, government departments and agencies, school administrations, and academic offices and departments. Within these organizations, the education and training have been directed to senior executives and management personnel, engineers, and shop-floor workers.

The workshops that have been delivered have been predominantly associated with statistical process control tools and techniques. The content of these workshops has been quite traditional with emphasis on the seven tools of SPC. Providing the participants with sufficient background to understand the workings of control charts of various kinds, and the allied tools needed to improve their shop-floor and business processes has been the prime focus of the SPC workshops. In several cases follow-up workshops have been developed to extend these basic tools with content developed to introduce other process monitoring tools such as CUSUM and EWMA charts, process capability analyses, and to introduce the idea of improving processes through scientific experimentation. In addition, special workshops have been given on design of experiments, basic statistical methods, total quality management, reengineering concepts, and SPC implementation issues.

The SQCRAG has conducted workshops in a variety of locations, typically on the premises of the company involved or in an external location such as a hotel. Except for the workshops directed specifically towards University of Manitoba departments, none of the workshops given by the SQCRAG have been given on the campus of the University. This has been done for a variety of reasons, including problems with visitor parking at the University campus, wanting to avoid an "academic" atmosphere being associated with the workshops, and cost to the company. In addition, several workshops have been given for general audiences, in association with consortia of companies, and with government funded training initiatives. Such workshops have involved large numbers of participants and have included external invited presenters with international reputation and status.

The workshop format and length are developed around the particular needs of the company involved. Workshops involving two, three, and five days of training have been given. The longer workshops typically directed towards more senior people who want a broader range of topics and in some cases greater depth, and the shorter workshops for shop floor workers stressing the application of tools and techniques. The workshops have been given on consecutive days, on alternate days, once a week, and in a series of half days depending upon the availability of the company personnel for training, as well as the staff of the Department for conducting the training. During the academic year, September through April, teaching duties of staff within the Department have been arranged such that staff may well be free on Tuesday and Thursday, or on Monday, Wednesday and Friday. The workshops can then be organized around these non-teaching days. In one case, a series of workshops were given on Saturdays over a period of several months during the winter. During May through August however, workshops can be given in a somewhat more flexible format.

Companies for which workshops are developed and delivered are charged a standard per diem fee for workshops with up to fifteen participants. Additional participants are charged for on a per person fee basis. These fees cover the cost of all disposable supplies and materials used in the workshop, including a manual developed specifically for the company, as well as any other expenses associated with the development of the specific workshop content and delivery. The client company is responsible for providing the location and facility for the workshop and any incidental costs such as provision of meals to the participants as appropriate. The Department typically provides at least two of its academic staff to present the workshop and often includes one or more graduate students as assistants.

While companies are charged for the services provided, one rather unique aspect of the operation is that the academic staff involved with the workshops receive no remuneration for their time and effort. Student assistants are paid, however, at a per diem rate consistent with pay scales at the University. All revenues gained from the workshop activities go directly to the Department and maintained in a special "income account". The funds generated are then available to be used to support graduate students, to buy books and equipment for the students and staff of the Department, to pay for travel to conferences that have a "quality" focus, and to support visitors to the Department.

5.4   Workshop Preparation and Delivery

When the SQCRAG is first approached, extensive discussions are held to determine exactly what expectations the company officials have for the workshop training they are requesting. At this point, SQCRAG tries to make its particular role and function as clear as is possible. It is emphasized to the potential client that the departmental member's first priority is their regular assigned academic duties. It is also emphasized that the staff members are involved in doing only those aspects of quality that are squarely within their field of expertise, namely statistics. Companies are cautioned that our involvement is not as general consultants in management principles, or as motivational gurus in the area of quality. Our efforts are directed to the educational aspects of topics primarily in statistical process control.

Early on in our discussions, the level of the participants is determined, both with respect to their job function within the company (management, operator/technician) as well as their facility with English and their formal education. In addition, the SQCRAG arranges for several members to have a tour of the company. The purpose of this is many faceted but it primarily affords the opportunity for us to learn about the company involved, its needs, its wants, its motivation for requesting the workshop, the nature of its operation, and to obtain a feeling for the level of current SPC activity in the company's operations. During the tour, an attempt is made to identify typical processes which are felt to be candidates for the application of SPC, and to determine the availability of data, or to identify places where data which would be useful in the workshop presentation could be gathered. A set of topics is agreed upon, and dates for delivering the workshop are determined. A formal proposal is then developed which includes an outline of the topics that are to be included, dates, materials, and costs.

Materials are developed which reflect the specific needs of the company as well as the nature of the business activity of the company. Data obtained from each client are used wherever and whenever possible, and processes unique to the client company are discussed and used for illustration purposes. This requires extensive consultation with the client company in the development stage of the workshop. Staff and graduate students from the Department may well spend time at the company gathering useful data and gaining an understanding of the company activities. A manual is prepared which is placed in a binder and given to each participant together with additional materials such as calculation worksheets, model control charts, etc. Participants are provided with simple calculators to use during the workshops.

Care has been taken to develop course materials through which such concepts as brainstorming, flowcharting, collecting and graphing data, sampling, and experimentation can effectively be introduced. Various projects and participatory exercises are used throughout the workshop. Because of the challenging nature of some of the material, especially for an older group of participants and for those whose first language may not be English, an emphasis is placed on making the workshop fun and entertaining. Props of various kinds are used which will reinforce the relevance of the material to the particular company and its operation. Use has been made of such items as wheel hubs, engine parts such as stator vanes and pistons, metal washers, printed circuit boards, forged metal pieces, peas and beans, specific measuring devices, for illustration purposes.

Participants are involved in making and flying paper helicopters, shooting coloured balls from small catapults, examining boxes of animal crackers and packages of M & M's or Smarties, drawing samples from a bowl containing coloured beads, drawing numbered ping pong balls from a box, selecting samples of measurements from various populations contained in hats, using water pistols, and the like. As much as is possible the participants are actively involved in the workshop. Many of the concepts are presented in a manner so as to lead to the participants discovering the fundamental principles involved rather than simply having those principles given to them.

For a basic introductory workshop on statistical process control, the objective is to bring the participants to a level whereby they have an understanding of some of the fundamental aspects of the statistics encountered in the construction and use of the basic tools of SPC, sampling, control charts, process capability, etc. It is our view that the teaching of techniques alone is not only insufficient but also unwise if successful implementation and continued use of the learned material is to be achieved.

In workshops such as Design of Experiments, the participants are given exercises and projects which involve hands-on experience with the use of computer packages. The helicopters and catapults are used extensively to explain the fundamental ideas of measurement and variability, factorial and fractional factorial experimentation, and the use of response surface methodology to find optimal settings. They are also encouraged to identify real situations within their operations which have the potential to benefit from a designed experiment. They are assisted with developing an experimentation strategy for the problem they have identified.

Typically two or more academic staff from the department are involved in development and delivery of the workshop, and graduate students are often included as assistants during the presentation phase. Participants are provided with as much one-on-one attention as they require in order to gain an understanding of the tools and techniques presented.

5.5   The Pros and Cons of Our Workshop Involvement

One of the perhaps unique aspects of the program that has developed is that the workshops that are prepared and delivered by the SQCRAG are done using academic staff of the Department on a purely volunteer basis with no direct remuneration provided to them for such service. There are several reasons why the staff have agreed to do this. There is a recognition that we are located in an area where expertise in SPC and allied techniques in the private sector is not readily available and yet there is a genuine need on the part of industry for training in the use of these statistical tools. Few of the companies doing business in the Winnipeg area are large enough to justify bringing in consulting firms from other areas of Canada or the United States to conduct the required training. The academic staff have recognized that they have the requisite knowledge and skill to fill the need. There is also the feeling that because we are academic staff at a publicly funded university there is a certain degree of obligation to assist local industry when called upon to do so.

The decision to volunteer our services is not altogether altruistic as there are certainly tangible and intangible benefits flowing from our involvement. Not the least of these benefits is the new challenges and opportunities that working with industry presents.

The staff are exposed to real life problems requiring the application of some of the basic statistical knowledge they possess. The need to teach rather complex ideas and techniques even of a basic nature to adult learners with extremely diverse educational backgrounds has resulted in the development of teaching tools and aids that have found their way into our regular classroom activities.

In addition, there has been a direct benefit to the research effort of the Department as staff have encountered problems which have suggested new areas of interesting research. This has lead to publications in learned journals and to dissertation topics for our graduate students.

We have also succeeded in raising the visibility of the field of statistics within local industry as well as of the Department of Statistics itself. Such increased awareness of what is available in the Winnipeg area has the added benefit of improving the prospect of local companies hiring graduates from our programmes on a full or part-time basis. Bringing some of our graduate students to the workshop sessions to assist the participants has introduced our client companies to our students and has the added benefit of exposing our students to the real world of industry.

Of course, the giving of these workshops generates funds and these funds are an invaluable asset as they are used to enhance departmental functions and activities. At a time of drastically reduced funding from government we have been able to use these workshop funds to provide some additional graduate student support. We have used the funds to purchase much needed equipment such as computers, as well as for obtaining books and other such materials. We have been able to encourage staff and students to travel to conferences related to quality and statistics through use of the moneys generated. We have also been able to support the visits of distinguished colleagues to our department to give seminars and to participate in special workshops.

It is also felt that the interaction we have developed with our client companies has provided a direct benefit to industry as well. We have made affordable training and consultation services of very high quality available to a wide range of companies. The training is done by individuals who are professional teachers and competent researchers in the area of statistics in general, and as applied to quality issues in particular. In addition, because they are availing themselves of a local resource, there is ample opportunity for post-training support, something which may not necessarily be as readily available from consultants brought to Winnipeg from other areas.

Following any training that is done by the SQCRAG the client is encouraged to contact us if they encounter problems or need any follow-up support. This provides the small local company with somewhat of a pseudo in-house resource with which to consult when the need arises. Visits to client companies have been arranged within a matter of hours when problems have arisen. Additionally there is a supply of students which may be available to the company as the need arises on a part-time basis.

There remains the problem of the smallness of many of the local businesses. The SQCRAG prefers to give its workshops to groups of twelve to fifteen people. Many companies simply do not have that many employees that they want to train or can release for training at one time. In such situations we have encouraged groups of companies to join together, perhaps a company and one or more of its suppliers. Additionally consortia of local companies with similar focus, such as in the aerospace industry or electronics, have requested the help of the SQCRAG in training and consulting across the member companies. This has enabled the smaller companies to send one or two employees for training locally at a reasonable cost.

There are some significant difficulties that have been identified within the approach that the Department has chosen for this interaction with industry. As has been noted, this workshop activity is done using academic staff who volunteer their time and expertise. The staff are not compensated through remuneration or through release time from their regular duties. Anyone involved with this programme of activities is doing so over and above all other duties expected of a staff member with faculty rank. This results in a very heavy workload associated with workshop activities; visiting clients, preparing materials, and delivering the workshop presentations. As a result there is a very limited amount of time available to devote to these activities, which leads to scheduling difficulties often during the academic year.

The Department of Statistics has not developed any form of infrastructure to support the work of the SQCRAG. There are no paid employees devoting time directly to its work. Annually a faculty member is designated by the Head of the Department as the Coordinator, again purely on a volunteer basis. The primary consequence of this structure is a recognition that we have not entered into these activities as a commercial enterprise. Our clients are informed at the point of first contact that these activities of necessity are over and above our regular academic duties and as such must be made to fit in and around these responsibilities. On occasion there have been clients who wanted training to commence almost immediately, often because of an impending visit from a supplier auditor, and we simply are unable to respond that rapidly. Additionally, we have never undertaken to advertise the fact that we do undertake workshop training in SPC, design of experiments, quality improvement tools, and the like. Were the SQCRAG to advertise its services there is ample evidence to suggest that we would be unable to respond to the volume of requests that we would receive. Virtually all contacts with new industrial clients have resulted from referrals from companies for which we have done training in the past. In spite of these limitations, the SQCRAG has conducted over 70 workshops over the past ten years.

One source of some frustration to those involved with giving these workshops is the lack of direct support for these activities from the University itself, and the absence of recognition of the value to the University community that our service to industry has provided. This is in spite of the fact that our activities can be seen to support directly one of the explicit roles of the University to foster improved relationships with industry in Manitoba.

5.6   Statistical Consultation Services

One of the main services that has been provided for many years by the Department of Statistics has been consultation with students, faculty members, and others on the application of statistical methods. In the earliest days of the Department its academic staff would spend significant amounts of their time meeting with graduate students and researchers in a variety of fields. This often involved serving on graduate committees for students in other disciplines or assisting with the design and analysis of research projects of many kinds. As the awareness of the availability of a large number of statisticians to provide assistance spread the frequency and diversity of contacts increased dramatically. In addition, requests from people outside of the academic community began to appear in greater numbers.

In order to better focus the consulting work that was being requested across the expertise in the Department and to provide better service to researchers the Statistical Advisory Service (SAS) was formed within the Department. Over the years the Service has grown and now is formally structured with a Board of Directors, a Director, and a Coordinator. These individuals are members of the academic staff of the Department who have volunteered their time to organize this Service. Through the support of the National Sciences and Engineering Research Council (NSERC) the Department has been able to hire a full-time Consultant who is responsible for the day-to-day operation of SAS and serves as the first point of contact with individuals seeking help from the SAS. The Consultant runs a permanent facility which contains office space for the Consultant, sufficient space for meeting with small groups of researchers, as well as a smaller office for consulting sessions between individual clients and a statistical advisor. The SAS has been provided with access to various types of computing facilities to assist with its consulting activities.

Whenever possible, the consultations provided are done so on a fee-for-service basis, typically based on an hourly rate. Researchers are encouraged to include consultation fee requests in their applications for research funding. Students and other researchers without funding are provided with assistance in spite of not being able to pay for such help. In many cases, under the direction of the Consultant, senior students are assigned to specific projects to assist researchers with their statistical analyses. A specific course on Statistical Consulting attracts a number of students each year and the SAS provides a vehicle for providing the practical experience for these students.

The extent of the statistical consultation provided is extremely broad, ranging from fairly straight forward advice to graduate students on statistical analysis for their thesis work, to very extensive data handling and computer analyses for senior researchers. The Statistical Advisory Service has from time to time advertised in various university publications and has run a drop-in service staffed by graduate students. As a general policy, the SAS has tried to restrict its activities to assistance and advice concerning the planning, and conducting of research projects, the handling and analysis of large amounts of data which may include data entry, and the presentation of the statistical results. It normally does not itself conduct the research project such as hiring individuals to conduct interviews or make phone calls for a sample survey.

One of the very important benefits deriving from this structure has been the development of research projects involving close collaboration of researchers from many disciplines with statisticians from the Department of Statistics. Strong ties have been forged across the University and many research papers have appeared in learned journals as a result of the collaborative research work. In addition, exposure of statisticians to statistical problems in these research projects has often given rise to the development of research topics in statistics itself. Many of our graduate students have used their experience with the consulting clients to develop thesis topics of their own.

Over the years, a steady increase in requests for assistance from clients outside of the university research community has been noted. Some of these contacts have resulted in consulting projects of major proportions, often requiring the development of formal contracts to cover the extent of the work to be undertaken. Some of these contracts have been with large governmental agencies such as the Canadian Grain Commission, and others with private sector industry such as a major nickel mining company and an aerospace company. The projects have required a great deal of time and effort but have proved to be very rewarding to everyone involved. In several cases the interesting statistical problems have been developed into Masters and Ph.D. thesis work, and have spawned numerous research papers.

5.7   Quality Resource Centre

The University of Manitoba has undertaken to examine much of its operation with a view to making all of its administrative and academic processes work more effectively and efficiently. To that end the University established a Quality Resource Centre through which all quality improvement activities of the University could be coordinated and facilitated. One of the academic staff members of the Department of Statistics was seconded to a half-time position attached to the President's Office of the University to develop the structure of what was called the "Quality First" project of the University. It was this individual's responsibility to develop materials and training programmes, to facilitate teams of administrative staff engaged in quality improvement projects, and to coordinate the activities of the Quality Resource Centre. During the two years of tenure in this position great progress was made and many interesting projects were completed.

Although the secondment is now completed, the staff person involved was able to develop some interesting statistical research ideas which are currently being pursued.

5.8   Recent Initiatives

The Department of Statistics has had a difficult time over the years in convincing local industry of the usefulness of having a properly trained statistician available on their staff. Many companies have indicated an interest but seem to feel that a statistician on staff is a luxury that a company of their small size cannot afford. They may feel that there simply isn't enough statistical work generated to occupy a statistician full-time.

The Department has recently placed one of our Masters graduates in a local company essentially as a full-time on-site consultant for a six month trial period. An academic staff member from the Department worked closely with this individual over this period of time and provided back-up and support. At the end of the six month period the company placed the individual on its permanent staff.

The Statistical Advisory Service has also undertaken to provide its Consultant to a consortium of aerospace companies as an on-site consultant two full days per week. This Consultant has been provided with office space and spends time with staff from several of the member companies at their work sites. It is hoped that the long term benefit of such an initiative will be that these companies will come to realize the extent of the statistical work that they in fact are involved with. In the future it is our hope that one or more of our graduates will find employment in industry through this project.

In addition, in the current year, two academic staff members of the Department are on Research/Study Leave of Absence from the University. As part of their research activities during their leaves these faculty members have included time to be spent at local manufacturing businesses. They have spent time or intend to spend time working with these companies, sharing their statistical knowledge with them, and learning more about the operation of these businesses. Again it is anticipated that the visibility of the Department of Statistics will be increased leading to more cooperative work in the future and perhaps employment opportunities for graduates of our programmes.

The Statistical Quality Control Research and Applications Group has recently participated with several Manitoba organizations to produce a special workshop on the topic of Implementing Statistical Process Control. Several individuals from outside of the Manitoba area with particularly extensive experience in implementing SPC strategies in companies both large and small were invited to participate in the presentation of this workshop. Given the rather unique focus of this workshop, it was broadly advertised and a large group of participants was attracted. Through the participation of a Government of Manitoba agency involved with training initiatives, and the support of SQCRAG, the cost of this workshop was underwritten to the extent that the fee to participants for the two day workshop was held to a very low amount. This enabled even the smallest of our local businesses to send one or more individuals. The response to this initiative was most positive and future activities of a similar nature may well be contemplated.

5.9   Conclusion

The Department of Statistics has over the years made a conscious effort to reach out to our local business community. The response to these initiatives has been rewarding both from a financial as well as an academic point of view. The particular model chosen by the members of the Department to provide the service involves volunteer effort, little infrastructure, and a great deal of very rewarding work. There is a general feeling amongst those individuals involved with these activities that the benefits to the Department and to themselves have been substantial. Improved facilities, increased graduate student support, enriched teaching methods, practical industrial experience, and increased research opportunities are some of the results that are seen.

With some degree of confidence it can be stated that there must be many places in which local industry is desperately in need of help with statistical problems as basic as using statistical process control procedures to more complex problems of process optimization. Departments of statistics around the world can have a major impact on their local business community by sharing their expert knowledge and their teaching skills. It takes time and effort but the results are particularly rewarding to all parties concerned.

PART II
STATISTICAL PROCESS CONTROL

6

Developing Optimal Regulation and Monitoring Strategies to Control a Continuous Petrochemical Process

A. J. Ferrer, C. Capilla, R. Romero and J. Martin

Polytechnic University of Valencia, Valencia, Spain

Abstract: The purpose of this paper is to discuss the development of optimal strategies for regulating and monitoring a continuous petrochemical process using the Algorithmic Statistical Process Control (ASPC) methodology. The Single-Input/Single-Output (SISO) case is discussed, using the polymer viscosity as controlled variable (output), and reactor temperature as manipulated variable (input). Alternative regulation strategies are considered. Robustness properties and stability conditions of the controllers are discussed. The performance and the adequacy of the regulation schemes in a simulation of realistic assignable causes affecting the process are studied. In this context the benefits of implementing an integrated EPC/SPC system (ASPC system) are compared with those of using an EPC system alone.

Keywords and phrases: ASPC, constrained controller, engineering process control, process monitoring, process regulation, statistical process control

6.1   Introduction

Engineering Process Control (EPC) and Statistical Process Control (SPC) are two complementary strategies for quality improvement that until recently have developed independently. SPC monitoring procedures seek to reduce output variability by detecting and eliminating assignable causes of variation. On the other hand, EPC is usually applied to minimize output variability in the presence of dynamically related observations by making regular adjustments to one or more compensatory processing variables. However, ideas from both fields can be used together in an integrated EPC/SPC system to secure both optimization and improvement. Notion of superimposing statistical process control on a closed-loop system is quite recent and has opened new lines of research in the area of quality improvement; see, Box and Kramer (1992), MacGregor (1988), Montgomery, Keats, Runger and Messina (1994), Tucker, Faltin and Vander Wiel (1993) and Vander Wiel, Tucker, Faltin and Doganaksoy (1992).

This paper describes a case study of integrating the EPC and SPC approaches in a polymerization process. Several control strategies to reduce polymer viscosity (output variable) deviations from target are developed and compared with the actual control done by process operators. Their performance is also analyzed under assignable causes of variability. Monitoring procedures to implement the SPC component of a combined system are proposed and studied.

6.2   Polymerization Process

6.2.1   Process description

The present case study involves a commercial scale polymerization process that produces large volumes of a polymer (high density polyethylene) of a certain grade used in many familiar consumer products. Processing is performed continuously. Observations of the output polymer properties and opportunities for adjustment occur at discrete equispaced intervals of time. Samples of reactor effluent are taken every two hours and analyzed off-line for polymer processability. The key quality characteristic is polymer viscosity, which is measured by Melt Index (MI). The objective is to minimize MI variation around a target level of .8 viscosity units. Adjustments to viscosity can be made by varying the temperature of the reactor, which is a readily compensatory variable.

Control strategies have been developed independently by the Engineering Process Department and the Quality Control Department. Traditionally, experienced operators determine the adjustments of the temperature based on MI laboratory results and some general guidelines provided by the responsible process engineer. On the other hand, the Quality Control Department have tried to apply Shewhart charts to MI measurements. Therefore, a control action was taken only when the chart provided an out-of-control signal. However, as the MI values are correlated over time, SPC charts were very inefficient. In fact, when this study began control charts were no longer being applied.

An analysis of the problem shows that the control of the process is not optimized. This indicates the need for a consistent control rule to reduce the effect of predictable MI deviations from target and the cost incurred by off-specifications, and to detect efficiently assignable causes of variability.

6.2.2   Model formulation and fitting

The dynamics of the process and the disturbance model have been identified and estimated using plant data from various manufacturing periods. For comparative purposes, we have focused on manufacturing period five (January 1993). Given the consistency of the dynamics of the process, the results can be easily extrapolated to other manufacturing periods. The assumed model is:

$$\nabla MI_t = w_0 \nabla T_t + w_1 \nabla T_{t-1} + a_t \tag{6.1}$$

where MI_t is the observed melt index at time t (every two hours); T_t is the average reactor temperature during the two hours before t; $w_0 = .185(.087)$ and $w_1 = .274(.086)$; $\{a_t\} \sim$ white noise ($\sigma_a^2 = .004$). Note here that the same notation is used for parameters and their estimates, and that the quantities in parenthesis are the standard errors.

The inertial properties of model (6.1) can be appreciated from the consideration that two time periods after a unit step change is made in T the estimated change in MI will be .46 units ($w_0 + w_1$).

6.3   EPC Component: Developing the Control Rule

Control algorithms can be designed to minimize the mean square deviation of viscosity from its target value. These Minimum Mean Square Error controllers (MMSEC) optimize the performance index

$$\min E\left[MI_{t+1} - Target\right]^2 = \min\left[\sigma^2\left(MI_{t+1}\right) + bias^2\right]. \tag{6.2}$$

Although such a strategy has an appealing motivation, the MMSEC may not work properly in practice. These controllers may have undesirable properties, such as requiring excessive control action, and having performance and stability characteristics that are sensitive to the accuracy of the process model [see Box and Jenkins (1970), Bergh and MacGregor (1987), and Harris and MacGregor (1987)].

Modified control schemes can be employed in which reduced control action can be achieved at a cost of small increases in the mean square error (MSE) at the output. So, in these modified schemes the MSE of the output will be minimized subject to a constraint on the variance of the temperature adjustments ∇T_{t+1}. From these constrained schemes, a simple approach has been proposed by Clarke and Hasting-James (1971) and Clarke and Gawthrop (1975). They treat the simpler problem of minimizing an instantaneous performance index

$$\min\left\{\left(MI^*_{t+1/t} - Target\right)^2 + r\left(\nabla T_{t+1}\right)^2\right\}, \tag{6.3}$$

where $MI^*_{t+1/t}$ is the minimum variance forecast of MI_{t+1} made at time t; the constrained parameter r is like a Lagrangian multiplier. Using the performance indices (6.2) and (6.3), the following controllers have been designed.

6.3.1 The minimum mean square error controller (MMSEC)

Given the process model (6.2), the control action at time t that produces MMSE of viscosity about its target is obtained by setting T_{t+1} so that the one-step ahead minimum variance forecast of MI taken at time t equals the target value .8 [see Box and Jenkins (1970)]. So, the control rule of this MMSE controller is given by

$$\nabla T_{t+1} = \frac{(0.8 - MI_t)}{(w_0 + w_1 B)} = \frac{-e_t}{U(B)}; \qquad (6.4)$$

where $U(B) = w_0 + w_1 B$. Substituting equation (6.4) into the estimated AR-MAX model (6.1) results in the output error - that is, the MI deviation from the target, - of the adjusted process being white noise, $e_{t+1} = MI_{t+1} - .8 = a_{t+1}$. So, under model (6.1) the temperature adjustment series would follow an autoregressive process of order 1 (AR(1)) and the fact that $w_0 < w_1$ violates the stationarity condition that would ensure that the ∇T series have a positive and finite variance. This makes the MMSE controller unstable and, thus, is of no practical use. This led us to develop a viable control rule following the MMSE criterion, but focusing on MI_{t+2} when setting T_{t+1}.

6.3.2 The two-step ahead forecasting controller (TSAFC)

Assuming that T will be constant for the next four hours ($T_{t+2} = T_{t+1}$), the control action at time t should be such that the two-step ahead minimum variance forecast of MI taken at time t, $MI^*_{t+2/t}$, equals the target value .8. This leads to the control rule of the TSAF controller

$$\nabla T_{t+1} = \frac{-e_t}{(w_0 + w_1) + w_1 B} = \frac{-e_t}{U'(B)} \qquad (6.5)$$

where $U'(B) = (w_0 + w_1) + w_1 B$.
Using the parameter estimates, the root of $U'(B)$ is $B = -(w_0 + w_1)/w_1 = -1.68$ and lies outside the unit circle in the B plane. So, this controller is stable, as the gain of the adjustments in equation (6.5) is bounded. This means that to cancel an output error the adjustments sequence will converge to zero. It can be shown that equation (6.5) can be expressed as

$$\nabla T_{t+1} = -a_t/(w_0 + w_1). \qquad (6.6)$$

Therefore, under model (6.1) the temperature adjustment series resulting from this controller follows a white noise process. This result will be used for monitoring purposes in Section 6.7. After some manipulation, one obtains that in this case the output error of the adjusted process

$$MI_t - 0.8 = e_t = \left(1 + \frac{w_1}{(w_0 + w_1)}B\right)a_t \qquad (6.7)$$

follows a moving average process of order 1 that satisfies the invertibility condition $((w_0 + w_1) > w_1)$, which ensures that the current output error e_t depends on past output errors $e_{t-1}, ..., e_{t-k}$, with weights which decrease as k increases.

6.3.3   Clarke's constrained controller (CCC(r))

Under model (6.1), the minimization of the cost function (6.3) proposed by Clarke and Hasting-James (1971) and Clarke and Gawthrop (1975) leads to select the change in temperature adjustment ∇T_{t+1} which, at each instant of time t, attempts to drive the one-step ahead minimum variance forecast of the MI, $MI^*_{t+1/t}$, to the target value .8, subject to a constraint on the magnitude of the present adjustment. It can be shown that the temperature adjustment ∇T_{t+1} that will be set by this controller at time t will be

$$\nabla T_{t+1} = \frac{-e_t}{\left(w_0 + \frac{r}{w_0}\right) + w_1 B}. \qquad (6.8)$$

If $r = 0$ the algorithm of the MMSEC is obtained. On the other hand, if $r = w_0 w_1$, equation (6.8) yields the control rule of the TSAFC. Therefore, the TSAFC is a particular case of a CCC. In this process, the TSAFC is the CCC(.05). Equation (6.8) can be expressed as

$$\nabla T_{t+1} = \frac{-a_t}{\left(w_0 + \frac{r}{w_0}\right) + \left(w_1 - \frac{r}{w_0}\right)B} = \frac{1}{(1 - \phi B)}\frac{-a_t}{\delta},$$

$$\text{where } \phi = -\frac{w_1 - (r/w_1)}{w_0 + (r/w_0)} \text{ and } \delta = w_0 + (r/w_0). \qquad (6.9)$$

So, assuming that the process model (6.1) is running under the CCC, the temperature adjustment series follows an AR(1). The variance of this process can be expressed as

$$\frac{\sigma^2(\nabla T)}{\sigma_a^2} = \frac{1}{(1 - \phi^2)\delta^2}. \qquad (6.10)$$

It follows from this equation that the stationarity condition for this series (positive and bounded variance) will be satisfied if and only if $r > w_0(w_1 - w_0)/2$ ($r > .008$ in this case). It can be shown that the resulting output error of the adjusted process under this controller follows an autoregressive moving average ARMA(1,1) process

$$(1 - \phi B)e_t = (1 - \theta B)a_t,$$

$$\text{where } \phi = -\frac{w_1 - (r/w_0)}{w_0 + (r/w_0)} \text{ and } \theta = \frac{-w_1}{w_0 + (r/w_0)}. \qquad (6.11)$$

For the ARMA(1,1) process in this case,

$$\frac{\sigma_e^2}{\sigma_a^2} = \frac{1 + \theta^2 - 2\phi\theta}{1 - \phi^2}, \qquad r > w_0 \left(w_1 - w_0\right)/2 = .008$$

$$\frac{\sigma_e^2}{\sigma_a^2} = \infty, \qquad \text{otherwise.}$$

It is easy to verify that the condition for this ARMA(1,1) process to be both stationary and invertible is $r > w_0 \left(w_1 - w_0\right) = .016$.

A compromise between good control performance and small manipulating effort can be obtained for $r/g^2 = .1, ..., .25$ [Isermann (1981)], where g is the process gain. In our process $g = w_0 + w_1 = .459$, and therefore, following the previous rule, the constraint parameter r should be set at values ranging from .02 to .05.

6.4   Comparison of Control Strategies

Assuming that the dynamics of the process is given by equation (6.1) and there are no assignable causes affecting the process, the residuals from manufacturing period five have been used to obtain by simulation the MI and T values that would have resulted had the different control rules MMSEC, TSAFC and CCC been applied.

To determine suitable values for the constraint parameter r of the CCC, different r values have been simulated ($r \geq .016$ for the controller to be stable). The resulting MSE of the MI and the maximum T adjustment that would have resulted from each controller under different r values are plotted in Figure 6.1. A lower bound for the MSE of the MI, obtained with the unstable MMSEC, is also drawn for purposes of comparison.

As shown in Figure 6.1, for small values of r, moving from $r = .02$ to .05, the MSE of the MI increases only slightly, but the maximum temperature adjustment, ∇T, decreases a great deal. A suitable compromise between the variances of the manipulated and controlled variables can be found by setting $r = .05$, which yields the TSAFC and leads to a maximum temperature adjustment of .35°C, causing no problem in practice. For purposes of comparison we also consider in our simulations the CCC with $r = .02$. In this case, a lower MSE of the simulated MI can be obtained at the expense of increasing the maximum temperature adjustment to .8°C.

The performances of the MMSE, TSAF and CC(.02) controllers have been compared with the actual control done by process operators (MANUAL) and the simulated situation where no Engineering Process Control would have resulted from setting T fixed (NO EPC). The performance measurement is the MSE of the MI under each control strategy. This is shown in Figure 6.2.

These results indicate that, although operators were doing a good job, the feedback algorithms can reduce variability even more and give a control strategy independent of the particular rules of each process operator. Anyway, the experience of each operator can be used to improve the performance of the process.

6.5   Robustness Properties of the Controllers

Controllers designed following performance indices (6.2) and (6.3) are optimal under the assumption that the process model is correct. Nevertheless, when uncertainties exist in the process model, optimal performance cannot be maintained. A controller is said to be robust if the stability and performance of the closed-loop system are not severely compromised by moderate deviations from the true process model. If the deterioration in performance is small for the type of modeling errors that might feasibly be expected, this will be referred to as *performance robustness*. Nevertheless, when modeling errors exist, the closed-loop system can become unstable. If the stability region is wide enough to cover the modeling errors which might occur in practice, this is referred to as *stability robustness*. The size of the stability region in the model parameter space is a measure of the relative robustness of a controller.

In this section we are going to study the stability and performance robustness of the controllers designed in Section 6.3 to process/model mismatch errors. These controllers are designed under the assumption that the dynamics of the process is given by equation (6.1). Nevertheless, let us assume that the dynamics of the real process is given by

$$\nabla M I_t = \left(w_0' + w_1' B\right) \nabla T_t + a_t. \tag{6.12}$$

This equation is derived from the estimated model (6.1). Note that the model mismatch involves only in the parameters of the process transfer function model.

Stability Robustness.   It can be shown that the output error of a closed-loop system under the CCC in this case would follow an ARMA(2,1) model

$$(1 - \phi_1 B - \phi_2 B)\, e_t = (1 - \theta B)a_t,$$

where $\phi_1 = -\dfrac{(w_0' + w_1 - \delta)}{\delta}$, $\phi_2 = -\dfrac{(w_1' - w_1)}{\delta}$, $\theta = -\dfrac{w_1}{\delta}$,   and $\delta = w_0 + \dfrac{r}{w_0}$.

The stationarity condition implies that parameters ϕ_1 and ϕ_2 must lie in the triangular region [Box and Jenkins (1970)]:

$$\phi_1 + \phi_2 < 1, \ \phi_2 - \phi_1 < 1, \ -1 < \phi_2 < 1. \tag{6.13}$$

This condition ensures the stability of the process controlled with the CCC in the sense that the variance of the output error is bounded. Therefore, equation (6.13) defines the closed-loop stability region of the system. Figure 6.3 shows the contours of stability for the closed-loop system when using the TSAFC and the CCC(.02) designed by using the estimated model (6.1), but when there are different mismatch errors between the real (w') and estimated (w) parameters of the transfer function model. The stability region for each controller is the area inside the corresponding contour line. Nevertheless, given the dynamics of this process, parameters w_0' and w_1' are not expected to take negative values and, thus, in practice the fractional error will not be lower than -1. One might expect that as the r value increases, the CC controllers would also increase their robustness properties to process modeling errors [see Harris and MacGregor (1987)]. This is illustrated in Figure 6.3. So, as the constraining parameter r increases, the stability region of the controller becomes wider and stability robustness also increases. Note that robustness of TSAFC has been achieved with very little sacrifice in performance (ECM of MI) with respect to CCC(.02), (see Figures 6.1 and 6.2).

Performance Robustness.   Control engineers were interested in knowing the consequences of moderate changes in the transfer function parameters in the system regulated using the studied controllers. This question was adressed by simulation, considering changes of up to 50% in the w parameters. For purposes of discussion, suppose that the process is given by (6.1) with known parameter values equal to the ones estimated in Section 6.2; however, suppose that in a realization of the process, beginning with period 25, a sudden change in the w parameters occurs and the correct model follows equation (6.12). In spite of this, the closed-loop system is not modified. This change in the process dynamics can be due to any of several sources, including effects of raw materials feeding the reactor, factors affecting heat-exchange effectiveness or a change in the properties of the catalyst. Plant engineers would be interested in knowing how the quality of the output produced during the day following the change could be affected. The performance of a closed-loop system under TSAFC and CCC(.02), and the hypothetical case of no control in an open-loop system (NO EPC) have been compared. Table 6.1 presents the results based on 50 simulation runs for each simulated condition.

The first row in Table 6.1 gives the results in the open-loop system. In this case, as the control system is deactivated, no action is taken from the model. This explains why, in spite of the mismatch errors, all the figures are the same. Comparing the first column (no change) with the others, it is clear that the controller performance, and mainly the TSAFC, is seen to be practically un-affected by moderate changes in the transfer function parameters, except for the case of $(w_0' - w_0)/w_0 = .5$, $(w_1' - w_1)/w_1 = -.5$, where the CCC(.02) be-comes unstable, (see also Figure 6.3). Note that even with moderate mismatch errors the closed-loop systems studied are noticeably better strategies than the no-control situation.

6.6 EPC Component: Performance Under Assignable Causes

The performance of the EPC component alone when assignable causes occur has been investigated through simulation. We have considered disturbances which may be consequences of realistic problems in the process and, therefore, take different forms. The performance of the MMSEC, TSAFC and CCC(.02) has been compared with the situation where there is no Engineering Process Control. Due to limited space, this paper is only going to present one of the several assignable causes studied.

MI Measurement Error in Laboratory: We assume that at time period $t = 25$ a very extreme viscosity value is reported by the laboratory, because the measurement process introduces an analytical error of magnitude $\delta = .3$ units ($\simeq 5\sigma_a$). The assignable cause takes the form of an observed/actual MI values mismatch.

$$
\begin{aligned}
MI_{t\ observed} &= MI_{t\ actual} + \delta I\,[t = 25] \\
&= MI_{t-1} + w_0 \nabla T_t + w_1 \nabla T_{t-1} + a_t + \delta I[t = 25]
\end{aligned}
$$

where

$$
\begin{aligned}
I[t = 25] &= 0, \quad \text{if } t \neq 25 \\
&= 1, \quad \text{if } t = 25.
\end{aligned}
$$

In this section we have considered the use of only the EPC component, which may reduce and compensate for the effect of any process disturbance but does not detect or remove them. Process engineers were interested in evaluating how the output quality would be affected during the following day after the disturbance occurs. On the other hand, EPC components actively compensate for the effect of process disturbances by transferring variability in the output variable MI to the input control variable T. From a practical point of view, T adjustments up to an upper bound are tolerable, so process engineers were also interested in the maximum input adjustment that would have resulted from each controller under the special cause of variation.

Table 6.2 presents the average values of the mean square error (MSE) of MI and the averages of the maximum T adjustment during the following 24 hours after the assignable cause has been introduced in the process ($t = 25$), based on 50 simulation runs for each condition. In each closed-loop simulation run the resulting MSE of MI and the maximum T adjustment have been recorded. Although the MMSEC controller is unstable and will lead to non-viable performance, it has been included as a reference.

The first column gives the results under the open-loop system (NO EPC). Note that even when control actions are based on wrong information (MI analytical error), situations which result in overcompensation, the EPC rules give superior performance to no-control.

Comparing the first row with the second one it is clear that the controllers' performance is affected by assignable causes. TSAFC and CCC(.02) compensate for the assignable causes, as we can see in the MSE of MI values under special causes, which are close to the minimum attainable (MMSEC).

6.7 SPC Component: Developing Monitoring Procedures

The development of a monitoring scheme leads to various questions, including what to monitor and what is an appropriate monitoring procedure. As commented by Faltin, Hahn, Tucker and Vander Wiel (1993), monitoring process performance (MI deviations from target, $\{e_t\}$), allows us to evaluate whether the control scheme continues to operate satisfactorily, but it is generally insufficient. Process performance already includes the compensating impact of the control variable and fundamental changes that may be compensated for by ever-increasing control action may remain undetected for a considerable time. Some *special causes* producing a temporary deviation from the underlying system model could be evidenced by outliers or unusual patterns of points in the sequence of deviations from predicted performance $\{a_t\}$. Therefore, the monitoring of series $\{a_t\}$ could be useful to detect this kind of problems and to lead to remedial action. The control actions (temperature adjustments, ∇T), as well as the input series (Temperature, T) could also be usefully charted as a means of detecting these kinds of process changes. The predicted output deviations from target $\{e_t\}$ when no temperature adjustments are made can be computed using the observed sequence $\{a_t\}$ to evaluate process performance without including the compensating impact of manipulated (control) variable.

In considering the appropriate monitoring scheme to apply, a Shewhart control chart for individuals with run rules is a good candidate because of its simplicity. More sophisticated schemes, such as CUSUM and EWMA charts are also good candidates, because they are more effective in detecting changes of small magnitude; in fact we have also considered an EWMA procedure with $\lambda = .25$ as an alternative to the Shewhart chart.

It is well known that the assumption of uncorrelated data is critical to the performance of control charts, which are very inefficient when observations are dependent; for example, see Johnson and Bagshaw (1974), Bagshaw and Johnson (1975), Vasilopoulos and Stamboulis (1978) and MacGregor and Harris (1990). Some of the quantities we have mentioned previously are autocorrelated. This may lead to more frequent false alarms and to slower detection of out-of-control situations, and, therefore, to bad performance of control charts.

Consequently, in this paper, Shewhart and EWMA control charts are applied to deviations from predicted performance $\{a_t\}$ which are due to random perturbations in the process and which under control are uncorrelated and $N(0, \sigma_a^2)$. With the transfer function model (6.1), the TSAFC yields a series of temperature adjustments ∇T which under control are independent and $N(0, \sigma_a^2/(w_0 + w_1)^2)$ (see Section 6.3.2). Therefore, standard monitoring charts can be applied to ∇T without departures from the assumptions of these procedures. TSAF control rule also yields a process output error $\{e_t\}$ which under a common cause system follows an MA(1) process, $e_t = (1 - \theta B)a_t$, where $\theta = -w_1/(w_0 + w_1)$ (see Section 6.3.2). This result will be considered to calculate the control limits of a Shewhart chart for $\{e_t\}$ series. But the chart will be used without the run rules because of the autocorrelation. Finally, a run chart without control limits will be applied to the two other quantities which may be useful to monitor the input series T and the predicted deviations from target when no control action is performed. Note that this latter chart (simulated $\{e_t\}$ series under NO-EPC) would be the same for any other control algorithm and would show the whole special cause effect, as it does not include the compensating action.

In this section, we analyze the performance of a TSAFC/SPC integrated system in the polymerization process studied, in comparison to applying the TSAFC alone, when the out-of-control situation considered in Section 6.6 has been introduced into the process at time period $t = 25$.

Assuming process model (6.1), the sequence of white noise residuals from manufacturing period five has been used to calculate the MI and T values that would have resulted had the TSAFC and the *monitoring* procedures been applied. We use the same performance measure as in Sections 6.5 and 6.6: The MSE of MI during the first 24 hours following the out-of-control situation, but assuming that the assignable cause is eliminated as soon as it is detected by any of the SPC schemes.

Figure 6.4 shows the monitoring charts which may be integrated with the TSAFC for a process realization including a MI measurement error in the laboratory as a special cause. In these charts, observation number $t = 0$ corresponds to the third value of the process realization. Therefore, the assignable cause occurs at charted observation $t = 23$. Shewhart limits are shown on the graphs for the sequences of deviations from predicted performance $\{a_t\}$ (Chart a), deviations from process performance $\{e_t\}$ (Chart c) and T adjustments (Chart e). An EWMA chart for $\{a_t\}$ series with $\lambda = .25$ and $L = 2.2$ is also included (Chart b). The two run-charts without control limits are for predicted performance series when no adjustments are made (Chart d, $\{e_t\}$ NO-EPC) and for T series (Chart f).

As shown in Figure 6.4, there is a false out-of-control signal in the control charts for deviations from target $\{e_t\}$ and deviations from predicted performance $\{a_t\}$ (above the upper control limit) almost as soon as the laboratory measurement of MI_{25} is reported. In the Shewhart chart for $\{e_t\}$ (Fig.4c), a run of two observations beyond 2 sigma appears after the point which falls out of the control limits. It occurs because the analytical error has not been eliminated and the feedback control algorithm overcompensates, as the points below and above the control limits in the chart for T adjustments show. The error affects only the level of T at $t = 26$ because the control action for that period is based on MI_{25}, and this leads to two control adjustments greater than expected. In the run-chart for predicted deviations from target under the NO EPC situation, the analytical error affects all predictions beyond $t = 25$ with a level shift.

When no action is adopted from the SPC schemes, the MSE of MI during 24 hours after the error, under TSAF rule alone, is .0103. However, assuming that the assignable cause is eliminated as soon as it is detected, the MSE of MI under the combined TSAFC/SPC procedure is .0039, which is equal to the performance measure for the same period when no assignable cause is present. Thus, the combined procedure provides a substantial reduction in variability.

6.8   Concluding Remarks

From this study, it appears that the TSAFC is a good choice. On one hand, the robustness-performance combination achieved by this controller is quite good. On the other hand, this is quite effective in compensating for the assignable causes as simulated, with moderate increases in compensatory adjustments ∇T.

Nevertheless, although TSAFC has good performance in the simulated contexts, it does not detect or remove the assignable causes; it uses continuous adjustments to keep MI on target. So, the implementation of a system which allows the elimination of special causes will result in additional reduction of output variability. That is the role of the SPC component in an integrated EPC/SPC scheme.

The results in this paper show that integrating monitoring procedures and the TSAF controller can lead to a substantial reduction in output deviation from target when certain *unexpected* forms of disturbance occur. The appropriate monitoring scheme depends on the out-of-control situation and to effectively detect and identify the source of disturbance it is advisable to use several procedures simultaneously.

References

1. Bagshaw, M. and Johnson, R. A. (1975). The effect of serial correlation on the performance of CUSUM Tests II, *Technometrics*, **17**, 73–80.

2. Bergh, L. G. and MacGregor, J. F. (1987). Constrained minimum variance controllers: internal model structure and robustness properties, *Industrial and Engineering Chemistry Research*, **26**, 1558–1564.

3. Box, G. E. P. and Jenkins, G. M. (1970). *Time Series Analysis: Forecasting and Control*, San Francisco: Holden Day.

4. Box, G. E. P. and Kramer, T. (1992). Statistical process monitoring and feedback adjustment - A discussion, *Technometrics*, **34**, 251–285.

5. Clarke, D. W. and Gawthrop, B. A. (1975). Self-tuning controller, *Proceedings of the Institution of Electrical Engineers*, **122**, 929–934.

6. Clarke, D. W. and Hasting-James, R. (1971). Design of digital controllers for randomly disturbed systems, *Proceedings of the Institution of Electrical Engineers*, **118**, 1503–1506.

7. Harris, T. J. and MacGregor, J. F. (1987). Design of multivariable linear-quadratic controllers using transfer functions, *American Institute of Chemical Engineers Journal* (AIChE Journal), **33**, 1481–1495.

8. Isermann, R. (1981). *Digital Control Systems*, Berlin: Springer-Verlag.

9. Johnson, R. A. and Bagshaw, M. (1974). The effect of serial correlation on the performance of CUSUM Tests, *Technometrics*, **16**, 103–112.

10. MacGregor, J. F. (1988). On-line statistical process control, *Chemical Engineering Progress*, **84**, 21–31.

11. MacGregor, J. F. and Harris, T. J. (1990). Discussion of "Exponentially Weighted Moving Average Control Schemes: Properties and Enhacements," by J. M. Lucas and M. S. Saccucci, *Technometrics*, **32**, 1–29.

12. Montgomery, D. C., Keats, B. J., Runger, G. C. and Messina, W. S. (1994). Integrating statistical process control and engineering process control, *Journal of Quality Technology*, **26**, 79–87.

13. Tucker, W. T., Faltin, F. W. and Vander Wiel, S. A. (1993). Algorithmic statistical process control: An elaboration, *Technometrics*, **35**, 363–375.

14. Vander Wiel, S. A., Tucker, W. T., Faltin, F. W. and Doganaksoy, N. (1992). Algorithmic statistical process control: concepts and an application, *Technometrics*, **34**, 286–297.

15. Vasilopoulos, A. V. and Stamboulis, A. P. (1978) Modification of control chart limits in the presence of data correlation, *Journal of Quality Technology*, **10**, 20–30.

Appendix - Tables and Figures

Table 6.1: Averages of the closed-loop mean square error (MSE × 1E-3) of the MI during one day of production following a moderate process/model fractional mismatch error using the TSAF controller and the CC controller with $r = .02$. The open-loop case (NO EPC) is also included for comparative purposes. Standard errors of the averages (× 1E-3) (in brackets) are estimated for each simulated condition. The change occurs at $t = 25$

$(w_0' - w_0)/w_0$	0	−.5		.5	
$(w_1' - w_1)/w_1$	0	−.5	.5	−.5	.5
NO EPC	24.7 (3.4)	24.7 (3.4)	24.7 (3.4)	24.7 (3.4)	24.7 (3.4)
TSAFC	5.4 (.3)	7.2 (.6)*	7.0 (.4)*	4.8 (.3)	5.4 (.3)
CCC(.02)	4.9 (.3)	6.1 (.4)*	8.3 (.5)*	Unstable	6.0 (.4)*

* (significantly different from the no change case, $p < .05$)

Table 6.2: Averages of the mean square error (MSE) of MI (× 1E-3) and averages of the maximum T adjustment for EPC rules based on 50 simulations, during one day of production after the assignable cause occurs. Standard deviations of the averages are given in parentheses. The open-loop case (NO EPC) is also included for purposes of comparison. The special causes occur at period $t = 25$

	NO EPC	MMSEC	TSAFC		CCC(.02)	
	MSE	MSE	MSE	Max.Adj.	MSE	Max.Adj.
No special causes	24.7	4.2	5.4	.26	4.9	.46
	(.0034)	(.00021)	(.00031)	(.0084)	(.00025)	(.0180)
MI analytical error	24.7	11.3	9.1	.73	11.2	1.55
	(.0034)	(.00059)	(.00064)	(.0177)	(.00064)	(.0418)

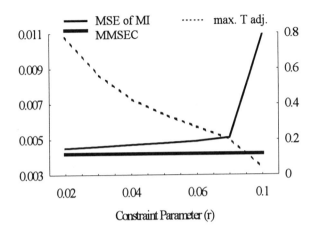

Figure 6.1: Illustration of the resulting MSE of the output MI ($\times$ 1E-4) and the maximum adjustment ∇T under different constraint parameter values r of the constrained controller

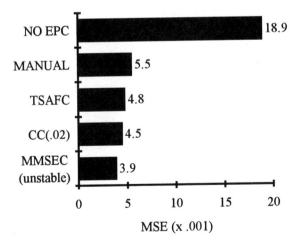

Figure 6.2: MSE of the output MI ($\times$ 1E-3) under different control strategies (normal conditions)

Fractional Mismatch Error in w_1

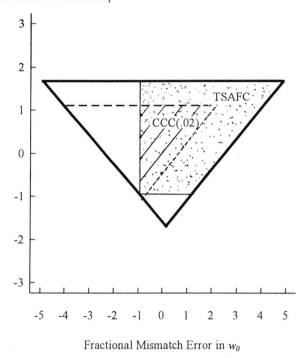

Fractional Mismatch Error in w_0

Figure 6.3: Closed-loop stability regions for fractional mismatch errors in transfer function model parameters ($w_i' - w_i/w_i$ for $i = 0, 1$) using the CC(.02) and TSAF controllers

(a)   (b)

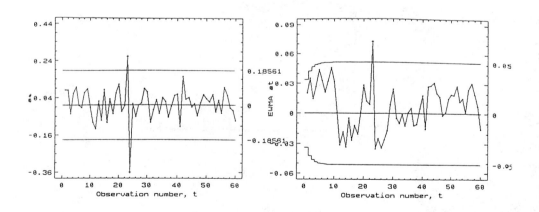

Figure 6.4: MI measurement error: (a) Shewhart chart for deviations from predicted performance $\{a_t\}$; (b) EWMA chart for $\{a_t\}$

(c) (d)

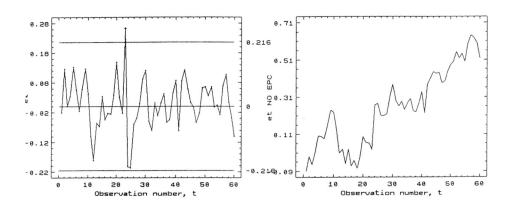

(e) (f)

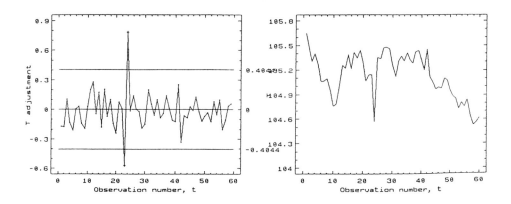

Figure 6.4: (continued) MI measurement error: (c) Shewhart chart for deviations from target $\{e_t\}$; (d) predicted deviations from target when no temperature adjustments are applied; (e) Feedback-control actions ∇T; (f) Manipulated variable series T

7

Capability Indices When Tolerances Are Asymmetric

Kerstin Vännman

Luleå University of Technology, Luleå, Sweden

Abstract: Different process capability indices, which have been suggested for asymmetric tolerances, are discussed and, furthermore, two new classes of appropriate indices are suggested. We study the properties of these new classes of indices. We provide, under the assumption of normality, explicit forms of the distributions of these new classes of the estimated indices. We suggest a decision rule to be used to determine if the process can be considered capable. Using the results regarding the distributions it is possible to calculate the probability that the process is to be considered capable. Based on these results we suggest criteria for choosing an index from the class under investigation.

Keywords and phrases: Asymmetric tolerances, estimation, distribution, process capability indices

7.1 Introduction

The tool most frequently used to measure the capability of a manufacturing process is some form of process capability index, designed to quantify the relation between the actual performance of the process and its specified requirements. The three most widely used capability indices in industry today are

$$C_p = \frac{USL - LSL}{6\sigma}, \quad C_{pk} = \frac{\min(USL - \mu, \mu - LSL)}{3\sigma},$$

$$\text{and } C_{pm} = \frac{USL - LSL}{6\sqrt{\sigma^2 + (\mu - T)^2}}, \tag{7.1}$$

where $[LSL, USL]$ is the specification interval, T is the target value, μ is the process mean, and σ is the process standard deviation under stationary controlled conditions. To obtain a capability index which is more sensitive than

C_{pk} and C_{pm} with regard to departures of the process mean from the target value Pearn et al. (1993) introduced C_{pmk} as

$$C_{pmk} = \frac{\min{(USL - \mu, \ \mu - LSL)}}{3\sqrt{\sigma^2 + (\mu - T)^2}}. \tag{7.2}$$

The process parameters μ and σ in (7.1) and (7.2) are usually unknown and are estimated by appropriate sample statistics.

Capability indices have received much interest in the statistical literature during recent years. For thorough discussions of the indices in (7.1) and (7.2) and several related indices and their properties see, e.g., Kane (1986), Chan et al. (1988), Gunter (1989), Boyles (1991), Pearn et al. (1992), Rodriguez (1992), Kotz and Johnson (1993), Pignatiello and Ramberg (1993), Vännman (1995, 1997a), and Vännman and Kotz (1995a, b).

Many published articles focus on the case when the specification interval is two-sided with the target value T at the mid-point M of the specification interval, i.e., when the tolerances are symmetric. There have been relatively few articles published dealing specifically with the case when the tolerances are asymmetric, i.e., when $T \neq M$. When Kane (1986) and Chan et al. (1988) introduced the indices C_{pk} and C_{pm}, respectively, in (7.1) these were intended for the case $T = M$. But they also suggested corresponding indices for asymmetric tolerances, namely

$$C_{pk}^* = \frac{d - |R - M| - |\mu - T|}{3\sigma} \quad \text{and} \quad C_{pm}^* = \frac{d - |T - M|}{3\sqrt{\sigma^2 + (\mu - T)^2}}. \tag{7.3}$$

Boyles (1994) makes a comprehensive study of six proposed indices for asymmetric tolerances, not including C_{pk}^* and C_{pm}^* in (7.3), and compares them with respect to process yield and process centering. Based on this study Boyles (1994) recommends the index

$$S_{pn} = \frac{1}{3}\Phi^{-1}\left(\frac{1}{2}\Phi\left(\frac{\mu - LSL}{\tau}\right) + \Phi\left(\frac{USL - \mu}{\tau}\right)\right), \quad \text{with} \quad \tau^2 = \sigma^2 + (\mu - T)^2, \tag{7.4}$$

where Φ denotes the standard normal cumulative distribution function. He also considers the index C_{pmk} in (7.2) for asymmetric tolerances (but denotes it by C_{pn}) and shows that C_{pmk} may be considered as an approximation to S_{pn}. As commented by Boyles (1994), C_{pmk} is easier than S_{pn} in (7.4) to calculate by hand and in many ways easier to work with analytically. Boyles (1994) refers to Choi and Owen (1990) as the ones who proposed C_{pn} for asymmetric tolerances.

As pointed out by Boyles (1994) there are many instances where asymmetric tolerances appear, although the case of symmetric tolerances is more common. Asymmetric tolerances simply reflect the customer's view that deviations from the target are less tolerable in one direction than in the other. An interesting example showing such a reasoning is given by Boyles (1994). Asymmetric tolerances can also arise from a situation where the tolerances are symmetric to begin with but the process distribution is skewed and the data are transformed to achieve approximate normality.

Unfortunately, the estimator of S_{pn} has not been studied by Boyles (1994), and, with the properties of the estimator unknown, the practical use of the index is limited. In the present paper we propose two new classes of indices, one which contains C_{pmk}, and one which contains C_{pk}^* and C_{pm}^*, when the tolerances are asymmetric. We study the behavior of the indices in these classes. We also consider estimators of the proposed classes and provide, under the assumption of normality, explicit forms of the distributions of the families of the estimated indices. Furthermore, numerical investigations are made to explore the behavior of these estimators for different values of the parameters.

7.2 Capability Indices for Asymmetric Tolerances

In Vännman (1995) a new family of capability indices, depending on two non-negative parameters, u and v, is defined as

$$C_p(u,v) = \frac{d - u|\mu - M|}{3\sqrt{\sigma^2 + v(\mu - T)^2}}, \qquad (7.5)$$

where $d = (USL - LSL)/2$, i.e., half the length of the specification interval, $M = (USL + LSL)/2$, i.e., the mid-point of the specification interval. We obtain the four indices C_p, C_{pk}, C_{pm}, and C_{pmk} in (7.1) and (7.2) by setting $u = 0$ or 1 and $v = 0$ or 1 in (7.5), i.e., $C_p(0,0) = C_p, C_p(1,0) = C_{pk}, C_p(0,1) = C_{pm}, C_p(1,1) = C_{pmk}$.

In the case of symmetric tolerances it was shown by Vännman (1995) and Vännman and Kotz (1995a,b) that the case $u = 0$ gives rise to indices with suitable properties. However, when $T \neq M$ we will exclude the case $u = 0$ since this u-value will render certain drawbacks. It has been pointed out by, e.g., Kotz and Johnson (1993), and Boyles (1994), that when the process shifts away from target then the index C_{pm}, i.e., the case $u = 0, v = 1$, is evaluated without respect to direction. The same conclusion can of course be drawn for $u = 0$ and $v > 0$. If the characteristic of the process is symmetrically distributed then a shift, when σ is fixed, towards the specification limit which is closest to the target value will give rise to a larger expected percentage of nonconforming than the corresponding shift towards the middle of the specification interval. Hence, a shift towards the specification limit which is closest to the target value ought to be considered more serious and give rise to a lower index value than the corresponding shift towards the middle of the specification interval. An additional problem with C_{pm} shown by Boyles (1994) is that C_{pm} clearly can overstate the process capability with respect to expected percentage noncon-forming when the process is on target, and the same is true for $C_p(u,v)$ when

$u = 0$. To avoid this last problem with C_{pm} Chan et al. (1988) suggested the index C^*_{pm} in (7.3).

Indices which might attain negative values when the process is on target are not of interest. Hence we will consider $C_p(u,v)$ in (7.5) only when $u < d/|T - M|$. We will assume that the target value T is strictly within the specification interval and hence $d/|T - M| > 1$. To avoid that the u-value will depend on the location of T within the specification interval we will restrict our attention to $u \le 1$. When $T = M$ it has been shown by Vännman (1995) and Vännman and Kotz (1995b) that $u > 1$ is not desirable since in that case the bias and the mean square error of the estimator could be large. Furthermore, it was shown that not very much is gained by using non-integer values of u and v. We can expect that the same will be true in the case $T \ne M$. Hence we will consider $C_p(u,v)$ in (7.5) only when $u = 1$ and denote this index by $C_p(v)$, i.e.,

$$C_p(v) = \frac{d - |\mu - M|}{3\sqrt{\sigma^2 + v(\mu - T)^2}}. \tag{7.6}$$

The numerator in (7.6) is the same as in C_{pk} and C_{pmk}. Furthermore, we have $C_p(0) = C_{pk}$ and $C_p(1) = C_{pmk}$. In this paper we will consider $C_p(v)$ only when $v > 0$ to achieve sensitivity for departures of the process mean μ from the target value T, which was one reason for introducing indices like $C_p(u,v)$, with $v > 0$.

A capability index should not only decrease when σ increases, as all suggested indices do, but also decrease when the process mean, μ, shifts away from the target value T. Furthermore, according to the discussion above the index should decrease more steeply when μ shifts towards the closest specification limit than when μ shifts towards the specification limit furthest away from T.

To study the behavior of $C_p(v)$ when μ shifts away from the target it is convenient to express $C_p(v)$ in (7.6) as the function

$$G(\delta) = C_p(v) = \frac{1 - |\delta + \beta sgn(T - M)|}{3\sqrt{\gamma^2 + v\delta^2}}, |\delta + \beta sgn(T - M)| < 1, \tag{7.7}$$

where

$$\beta = \frac{|T - M|}{d}, \quad \gamma = \frac{\sigma}{d}, \quad \delta = \frac{\mu - T}{d},$$

$$\text{and } sgn(T - M) = \begin{cases} -1, & T < M, \\ +1, & T > M. \end{cases} \tag{7.8}$$

Taking the derivative of $G(\delta)$ in (7.7) with respect to δ we find (see Vännman, 1997a) that $G(\delta)$ is not symmetric around $\delta = 0$ and attains its maximum not for $\delta = 0$ but for $\delta = -sgn(T - M) \min (\beta, \gamma^2 (v(1 - \beta)))$. Hence when μ shifts away from T the index $C_p(v)$ in (1.6) will decrease if μ shifts towards the closest specification limit, but first increase, and then decrease if μ shifts towards the middle of the specification interval and the maximum will be attained when μ is between T and M or when $\mu = M$. When γ is small or v is chosen to be large then the maximum will be attained for μ close to T.

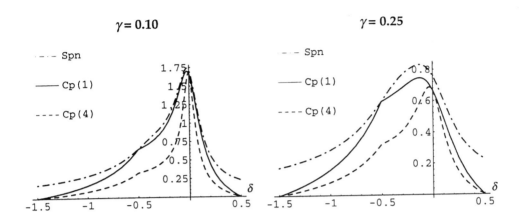

Figure 7.1: Each of the indices S_{pn} in (7.4) and $C_p(v)$ in (7.7) as a function of δ for $T > M$, $\beta = 0.5$, $\gamma = 0.10$, 0.25 and $v = 1.4$. $S_{pn} \geq C_p(1) \geq C_p(4)$ for $-1.5 \leq \delta \leq 0.5$

In Figure 7.1 we illustrate, for $v = 1, 4$ and $\gamma = 0.1, 0.25$, the behavior of $C_p(v)$ together with S_{pn} in (7.4). We have used the same example as Boyles (1994), i.e., $LSL = 26$, $USL = 58$, and $T = 50$. Hence $T > M$, $d = 16$ and $\beta = 0.5$. We can see from Figure 7.1 that when σ is small, i.e., γ is small, and $v = 4$, then the maximum is attained for μ close to T. Hence by choosing v large we can move the maximum close to T.

A capability index should indicate with a small value both if σ is too large and if μ is shifted away from the target value T. It is clear that $C_p(v)$ is decreasing as a function of σ. The drawback that $C_p(v)$ does not attain its maximum when $\mu = T$ and does not decrease to begin with, when μ shifts towards the middle of the specification interval, may be considered not to be too serious, since a shift towards the middle of the specification interval is less serious with regard to expected percentage nonconforming than is a shift in the opposite direction. Furthermore, when σ is small and v chosen not to be too small, the maximum is attained for μ close to T.

We can see from Figure 7.1 that S_{pn} has the same drawback as $C_p(v)$ of not attaining its maximum when $\mu = T$ and, furthermore, $C_p(v)$ is more sensitive for departures of the process mean μ from the target value T compared to S_{pn}, especially when v is large. Boyles (1994) showed that $S_{pn} \geq C_{pmk}$ and hence $S_{pn} \geq C_p(v)$ for $v \geq 1$. We can also see that $C_p(v)$ decreases more steeply when

μ shifts towards the closest specification limit than when μ shifts towards the specification limit furthest away from T. This is an advantage since the index will indicate more clearly a shift towards "the wrong side" of T than towards the middle of the specification interval. Furthermore, $C_p(v)$ will not overstate the process capability with respect to expected percentage nonconforming when the process is on target. Hence $C_p(v)$ avoids the drawback mentioned above regarding the case $u = 0$.

Another possible class of indices for asymmetric tolerances is obtained by generalizing the indices in (7.3) and introducing the family

$$C_{pv}(u,v) = \frac{d - |T - M| - u|\mu - T|}{3\sqrt{\sigma^2 + v(\mu - T)^2}}, \tag{7.9}$$

where u and v are non-negative parameters. The index $C_{pv}(u,v)$ has the advantage of having its maximum when $\mu = T$, and it decreases as μ shifts away from T in either direction. However, it has the drawback of being symmetric around $\mu = T$.

In order to obtain an index which will decrease more steeply when μ shifts away from T towards the closest specification limit than when μ shifts towards the specification limit furthest away from T, we combine $C_p(v)$ and $C_{pv}(u,v)$ to introduce a new class

$$C_{pa}(u,v) = \frac{d - |\mu - M| - u|\mu - T|}{3\sqrt{\sigma^2 + v(\mu - T)^2}}, \tag{7.10}$$

where u and v are non-negative parameters. For the same reasons as when we decided not to multiply $|\mu - M|$ in (7.6) by a constant we do not consider it of interest to multiply $|\mu - M|$ in (7.10) by a constant. Note that $C_{pa}(0,v) = C_p(v)$ and that $C_{pa}(u,v) = C_{pv}(u,v) = C_p(v)$ when the process is on target. Furthermore, when $T = M$, we have that $C_{pa}(u,v)$ in (7.10) is equal to $C_p(u + 1, v)$ in (7.5) and that $C_{pv}(u,v) = C_p(u,v)$. We will, as before in this paper, consider $C_{pa}(u,v)$ only when $v > 0$ to achieve sensitivity for departures of the process mean μ from the target value T.

First we can conclude from (7.10) that $C_{pa}(u,v)$ is not symmetric around $\mu = T$. To study the behavior of $C_{pa}(u,v)$ when μ shifts away from the target it is convenient to express $C_{pa}(u,v)$ in (7.10) as the function

$$H(\delta) = C_{pa}(u,v) = \frac{1 = |\delta + \beta sgn(T - M)| - u|\delta|}{3\sqrt{\gamma^2 + v\delta^2}}, |\delta + \beta sgn(T - M)| < 1 \tag{7.11}$$

where γ, β and δ are given in (7.8). Taking the derivative of $H(\delta)$ in (7.11) with respect to δ we find [see Vännman (1996)], by choosing $u \geq 1$ that the index $C_{pa}(u,v)$ will decrease when μ shifts away from T in either direction and, furthermore, will decrease more steeply when μ shifts away from T towards the closest specification limit than when μ shifts towards the specification limit furthest away from T. Hence by choosing $u \geq 1$ and $v > 0$ in (7.10) we have obtained an index with the desired properties. In Figure 7.2 we illustrate, for the same parameter values as in Figure 7.1, the index $C_{pa}(1,v)$ together with the index $C_{pv}(1,v)$ in (7.9) and S_{pn} in (7.4). We can see from Figure 7.2 that both $C_{pa}(1,v)$ and $C_{pv}(1,v)$ are more sensitive for departures of the process mean μ from the target value T compared to S_{pn}, especially when v is large.

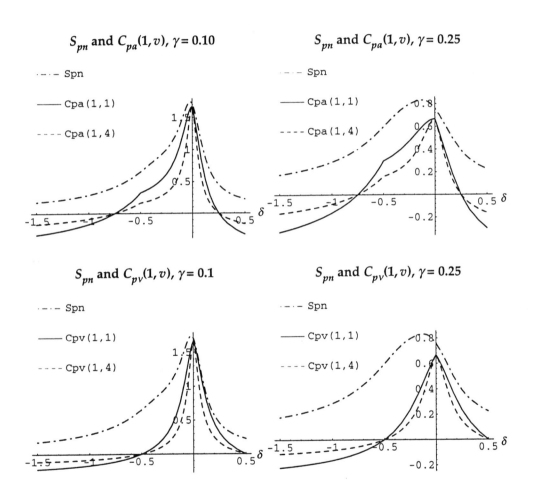

Figure 7.2: Each of the indices S_{pn} in (7.4), $C_{pa}(1,v)$ in (7.11), and $C_{pv}(1,v)$ in (7.9) as a function of δ for $T > M$, $\beta = 0.5$, $\gamma = 0.10$, 0.25, and $v = 1.4$

In the rest of this paper we will study $C_{pa}(u,v)$ for $u = 0$, $u \geq 1$, and $v > 0$

introducing indices like $C_{pa}(u, v)$ is to achieve sensitivity for departures of the process mean from the target value. It is easily seen [see Vännman (1996)] that by selecting large values of u and v we obtain an index with a larger sensitivity. An example of the effect that v has on the sensitivity can be seen graphically in Figure 7.2.

In order to gain more understanding regarding $C_{pa}(u, v)$ we will study what Boyles (1994) calls process centering, i.e., we will derive bounds on $\mu - T$ for a given value of the index. We find that [see Vännman (1996)] for a given value $C_{pa}(u, v) = c > 0$ if $u \geq 1$ or $c > 1/(3\sqrt{v})$ if $u = 0$ we get

$$-\min\left(\frac{(1-\beta)d}{\mu - 1 + 3c\sqrt{v}}, \frac{(1+\beta)d}{u + 1 + 3c\sqrt{v}}\right) < (\mu - T)sgn(T - M) < \frac{(1-\beta)d}{u + 1 + 3c\sqrt{v}},$$
$$(7.12)$$

where β is given in (7.8). When $u = 0$ and $v = 1$ the result in (1.12) is the same as in Boyles (1994) for the index that he denotes by C_{pn}, which is the same index as $C_{pa}(0, 1) = C_{pmk}$ in (7.2). We can see from (7.12) that the bounds will be tighter around $\mu - T$ for large values of u and v. The bounds in (7.12) are tighter than the corresponding bounds for S_{pn}, which according to Boyles (1994) are very close but slightly larger than those for $C_{pn} = C_{pa}(0, 1)$. One interpretation of the inequality in (7.12) is that if we have, e.g., $T < M$, $\beta = 0.25$, $u = 1$, and $v = 4$ then the $C_{pa}(u, v)$-value 1 implies that the process mean μ lies between $T - 0.093d$ and $T + 0.125d$.

The original reasons for introducing capability indices seem to be associated with the expected percentage of non-conforming items or the probability of non-conformance, that is, the probability of obtaining a value outside the specification limits. According to today's modern quality improvement theories the focus should not be on the probability of non-conformance only. It is also very important to keep the process on target. See, e.g., Bergman and Klefsjo (1994). Hence if μ is far away from the target T the process should not be considered capable even if σ is so small that the probability of non-conformance is small. Indices like $C_p(u, v)$, when $T = M$, and $C_{pa}(u, v)$, when $T \neq M$, take both the closeness to target and the variation into account and hence adhere to modern quality improvement theories. However, it is important to know that when the capability index is large enough to make the process considered capable then the probability of non-conformance is bounded above by a small known bound.

The indices $C_{pa}(u, v)$ in (7.10) for $u = 0$, $u \leq 1$, and $v > 0$ have this last-mentioned property. If $C_{pa}(u, v) > 0$, and the studied characteristic of the process is normally distributed, then we have the probability of non-conformance is never more than $\Phi\left(-3C_{pa}(u, v)\right) + \Phi\left(-3C_{pa}(u, v)(1 + \beta)/(1 - \beta)\right)$ [see Vännman (1997b)].

One important reason when Boyles (1994) recommended S_{pn} was its property regarding process yield. He showed that the probability of non-conformance is equal to $2\Phi\left(-3S_{pn}\right)$ when $\mu = T$ and approximately the same for μ in a neighborhood of T. However, the more appealing properties for the class $C_{pa}(u,v)$ are regarding process centering and sensitivity for departures of the process mean from the target value, together with its simplicity and the fact that there is an easily calculated estimator of $C_{pa}(u,v)$, whose distribution is given in the next section; we recommend $C_{pa}(u,v)$ to be used instead of S_{pn}.

If, for some reason, it is as important to protect against a shift of μ from T towards the specification limit furthest away from T as towards the closest specification limit, the index $C_{pv}(u,v)$ in (7.9) may be used. From (7.9) and (7.11) we easily see that, for fixed values of u and v, $C_{pv}(u,v) > C_{pa}(u,v)$ when μ is not equal to T and closest to the specification limit closest to T, and $C_{pv}(u,v) < C_{pv}(u,v)$ when μ is not equal to T and closest to the specification limit furthest away from T. We also see that $C_{pv}(1,v) = C_{pa}(0,v) = C_p(v)$ when $\mu \neq T$ and closest to the specification limit closest to T.

In the next sections we will study the distributions of the estimators of $C_{pa}(u,v)$ in (7.10) and $C_{pv}(u,v)$ in (7.9) and then, based on the results obtained, we will suggest suitable values of u and v.

7.3 The Estimator of $C_{pa}(u,v)$ and Its Distribution

We treat the case when the studied characteristic of the process is normally distributed. Let $X_1, X_2, \ldots, X_n$ be a random sample from a normal distribution with mean μ and variance σ^2 measuring the characteristic under investigation. In accordance with Vännman (1995) we consider two estimators of $C_{pa}(u,v)$, differing in the way the variance σ^2 is estimated. The first estimator is defined as

$$C_{pa}^*(u,v) = \frac{d - |\overline{X} - M| - u|\overline{X} - T|}{3\sqrt{\sigma^2 + v(\overline{X} - T)^2}} \quad \text{with } \sigma^{2*} = \frac{1}{n}\sum_{i=1}^{n}\left(X_i - \overline{X}\right)^2, \quad (7.13)$$

where the mean μ is estimated by the sample mean and the variance σ^2 by the maximum likehood estimator. The second estimator is obtained by estimating the variance σ^2 by the sample variance s^2. Since there is a simple relationship between the two estimators, cf. Vännman (1995), the statistical properties of the second estimator can easily be derived once the properties of $C_{pa}^*(u,v)$ are derived.

The cumulative distribution function of $C_{pa}^*(u,v)$ is derived by Vännman (1997b) and we will only state the result here. Let F_ξ denote the cdf of ξ, where ξ is distributed according to a central χ^2-distribution with $n-1$ degrees of freedom, and let f_η denote the probability density function of η, where η is distributed according to the standardized normal distribution $N(0,1)$. Then we have the following theorem.

Theorem 7.3.1 *If the characteristic of the process is normally distributed, $0 \leq u < 1/\beta$, and $v > 0$, then the cumulative distribution function of $C_{pa}^*(u,v)$ is as follows:*

$$
F_{C_{pa}^*(u,v)}(X) = \begin{cases}
0, & x \leq -\frac{u+1}{3\sqrt{v}}, \\[2ex]
\int_{-\infty}^{\frac{-D(1+\beta)}{u+1+3x\sqrt{v}}} L(x,t)dt + \int_{\frac{D(1-\beta)}{u+1+3x\sqrt{v}}}^{\infty} L(x,t)dt, & -\frac{u+1}{3\sqrt{v}} < x \leq 0, \\[2ex]
1 - \int_{\frac{-D(1+\beta)}{u+1+3x\sqrt{v}}}^{\frac{D(1+\beta)}{u+1+3x\sqrt{v}}} L(x,t)dt, & 0 < x \leq \frac{1-\beta u}{3\beta\sqrt{v}}, \\[2ex]
1 - \int_{\frac{D(1+\beta)}{u+1+3x\sqrt{v}}}^{\frac{D(1+\beta)}{u+1+3x\sqrt{v}}} L(x,t)dt, & x > \frac{1-\beta u}{3\beta\sqrt{v}},
\end{cases}
$$

$$(7.14)$$

where

$$
g = \frac{(\mu - T)\sqrt{n}}{\sigma}, \quad D = \frac{d\sqrt{n}}{\sigma}, \quad \text{and} \tag{7.15}
$$

$$
L(x,t) = \begin{cases}
f_\xi\left(\frac{(D - |t + \beta D| - u|t|)^2}{9x^2} - vt^2\right) f_\eta(t - g\, sgn(T - M)), & x \neq 0, \\[1ex]
f_\eta(t - g\, sgn(T - M)), & x = 0.
\end{cases}
$$

$$(7.16)$$

The result for the case when $u \geq 1/\beta$, and $v > 0$ is given in Vännman (1997b) but not presented here.

7.4   The Estimator of $C_{pv}(u,v)$ and Its Distribution

We study the following estimator of $C_{pv}(u,v)$

$$
C_{pv}^*(u,v) = \frac{d - |T - M| - u|\overline{X} - T|}{3\sqrt{\sigma^{2*} + v(\overline{X} - T)^2}}, \quad \text{with } \sigma^{2*} = \frac{1}{n}\sum_{i=1}^{n}\left(X_i - \overline{X}\right)^2, \tag{7.17}
$$

where the mean μ is estimated by the sample mean and the variance σ^2 by the maximum likehood estimator. As for the case of $C_{pa}^*(u,v)$ there is a simple relationship between the estimator in (7.17) and the one obtained by estimating the variance σ^2 by the sample variance s^2.

The distribution of $C_{pv}^*(u,v)$ can easily be derived from the distribution of the estimator $C_{p,n}(u,v)$ of $C_p(u,v)$ in (7.5) in the case when $T = M$. From the proof of Theorem 1 in Vännman (1997a) it is easily seen that the distribution of $C_{pv}^*(u,v)$ is obtained by replacing d in Theorem 1 in Vännman (1997a) by $d - |T - M|$. Hence we have the following result.

Theorem 7.4.1 *If the characteristic of the process is normally distributed the cumulative distribution function of $C_{pv}^*(u,v)$ is as follows:*

$$
F_{C_{pv}^*(u,v)}(x) =
\begin{cases}
0 & x \le -\frac{u}{3\sqrt{v}}, \\[2mm]
\int_{\frac{B}{u+3x\sqrt{v}}}^{\infty} F_\xi\left(\frac{(B-ut)^2}{9x^2} - vt^2\right) h(t)dt, & -\frac{u}{3\sqrt{v}} < x < 0, \\[2mm]
1 - \Phi\left(\frac{B}{u} - g\right) - \Phi\left(-\frac{B}{u} - g\right), & x = 0 \\[2mm]
1 - \int_0^{\frac{B}{u+3x\sqrt{v}}} F_\xi\left(\frac{(B-ut)^2}{9x^2} - vt^2\right) h(t)dt, & x > 0
\end{cases}
\tag{7.18}
$$

where g is defined in (7.15), $B = \frac{(d-|T-M|)\sqrt{n}}{\sigma}$, and

$$
h(t) = f_\eta(-t-g) + f_\eta(t-g) = \frac{1}{\sqrt{2\pi}}\left(\exp\left(-\frac{(t+g)^2}{2}\right) + \exp\left(-\frac{(t-g)^2}{2}\right)\right).
\tag{7.19}
$$

When $v = 0$ the inequality $-u/(3\sqrt{v}) < x < 0$ in (7.18) should be interpreted as $x < 0$. When $u = 0$ then (7.18) should be interpreted as $F_{C_{pv}^*(0,v)}(x) = 0$ for $x \le 0$.

7.5 Some Implications

We can use Theorems 7.3.1 and 7.4.1 to explore the behavior of the estimators of the indices for different values of u and v. We will follow the same kind of reasoning as in Vännman and Kotz (1995a) and first consider $C_{pa}^*(u,v)$. Here we will assume that a process is capable when the process is on target and $\sigma < d(1-\beta)/3$, i.e., when $\mu = T$ and $C_{pa}(u,v) > 1$. As is noted in Section 7.2 the probability of non-conformance is never more than $\Phi(-3)+\Phi(-3(1+\beta)/(1-\beta))$ when $C_{pa}(u,v) = 1$. If $\sigma > d(1-\beta)/3$ then the process is assumed to be non-capable whether the process is on target or not. We should also consider the process capable if the process is shifted only a short distance away from the target value, if at the same time we have that σ is smaller than $d(1-\beta)/3$. Furthermore, the process will be considered non-capable when σ is smaller than $d(1-\beta)/3$ and if at the same time the process is shifted further away from the target value.

To obtain an appropriate decision rule we consider a hypothesis test with the null hypothesis $\mu = T$ and $\sigma = d(1-\beta)/3$ and the alternative hypothesis $\mu = T$ and $\sigma < d(1-\beta)/3$. Under the null hypothesis we have $C_{pa}(u,v) = 1$ regardless of the values of u and v. The null hypothesis will be rejected whenever

$C_{pa}^*(u,v) > c_\alpha$, where the constant c_α is determined so that the significance level of the test is α. The decision rule to be used is then that, for given values of α and n, the process will be considered capable if $C_{pa}^*(u,v) > c_\alpha$ and non-capable if $C_{pa}^*(u,v) \leq c_\alpha$.

Using Theorem 7.3.1 we can, for given values of α and n, determine c_α and calculate the probability that the decision rule will consider the process to be capable.

Of special interest is the power of the test, i.e., the probability that the process is considered capable when $\mu = T$ and $\sigma < d(1-\beta)/3$. When $\mu = T$ and $\sigma < d(1-\beta)/3$ we want the decision rule to detect that the process is capable with a high probability. Hence indices with large power for given values of σ, where $\sigma < d(1-\beta)/3$, will be suitable. It is also of importance that the decision rule will detect departures from the target value. Hence, when μ shifts away from T then the probability that the process is considered capable should decrease, the faster the better. We can conclude that an appropriate index is one with the following two properties:

(i) The index should result in a decision rule with large power, i.e., a large probability that the process is considered capable when $\mu = T$ and $\sigma < d(1-\beta)/3$.

(ii) At the same time the index should be sensitive to departures from the target value, in the sense that the probability that the process is considered capable should be sensitive to departures from the target value.

Vännman (1997b) investigated numerically the probability that the process is considered to be capable as a function of $a = (\mu - T)/\sigma$ with respect to the power as well as to the sensitivity to departures from the target value. It was found that the probability that the process is considered capable decreases more steeply when m shifts away from T towards the closest specification limit than when μ shifts towards the specification limit furthest away from T. However, when $u \geq 1$, the probability that the process is considered capable does not decrease to begin with, when μ shifts towards the middle of the specification interval. Hence, although $C_{pa}(u,v)$, with $u \geq 1$, decreases, when μ shifts away from T in either direction, the probability that the process is considered capable when using the corresponding estimator does not. Here we have yet another example showing that studying the properties of the class of indices alone, without taking the properties of its estimators into account, might be misleading, Vännman (1995) and Vännman and Kotz (1995a,b).

It can easily be seen, using Theorem 7.3.1, that for a fixed value of v the probability that the process is considered capable when $\mu = T$, i.e., the power, decreases when u increases, and that for a fixed value of u the power decreases when v increases. Using numerical investigations Vännman (1997b) showed that for a fixed value of v the sensitivity increases with u, and for a fixed value of u the sensitivity increases with v. Hence large values of u and v will give us the most sensitive index. At the same time with large values of u and v we obtain an index with small power. Hence by increasing u and v we will increase the sensitivity but at the same time decrease the power when $\mu = T$ and $\sigma < d(1\beta)/3$, which is an unfavourable situation. In particular, large values of u will decrease the power.

Since large values of u decrease the power it does not seem of interest to consider $u > 1$, and we will consider only indices with $u = 0$ and $u = 1$ and choose u and v according to the properties (i) and (ii) above. In Figure 7.3 the probability that the process is considered capable using $C^*_{pa}(u, v)$ is plotted for some values of u and v when $T > M$, $\beta = 0.5$, $\gamma = 0.1$ (corresponding to $C_{pa}(u, v) = 5/3$ when $\mu = T$), $\alpha = 0.05$, and $n = 20$. More plots are given in Vännman (1997b).

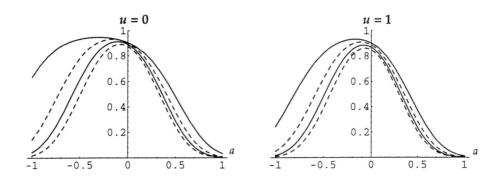

Figure 7.3: The probability that the process is considered capable as a function of $a = (\mu - T)/\sigma$, i.e., $P(C^*_{pa}(u, v) > c_{0.05})$, for $-1 \leq a \leq 1$, given that $T > M$, $\beta = 0.5$, $\gamma = 0.1$, $n = 20$, and $\alpha = 0.05$, where $c_{0.05}$ is determined by $P(C^*_{pa}(u, v) > c_{0.05}) = 0.05$, given that $a = 0$ and $\gamma = 1/6$. In each case for a fixed value of u the curves correspond to $v = 1, 2, 3, 4$ counted from the top

We see from Figure 7.3 that when $u = 0$ the power is approximately the same for $v = 1, 2, 3$, and 4, but $v = 3$ and 4 will give more sensitive indices. Approximately the same power as when (u, v) is equal to $(0,3)$ and $(0,4)$ is obtained when (u, v) is equal to $(1,2)$ and $(1,3)$. The same pattern is found for other values of β, γ, and n. Among these (u, v)−values, having approximately the same power, (u, v) equal to $(0,4)$ or $(1,3)$ are the most sensitive ones to departures from the target value. Since $u = 0$ will give an index which is somewhat simpler than when $u = 1$, we recommend, based on the numerical calculations done, the index with $u = 0$, $v = 4$, i.e., $C_{pa}(0, 4) = C_p(4)$.

We can conclude that, although the index $C_{pa}(u, v)$ in (7.10), with $u \geq 1$, has more attractive properties than the index $C_p(v)$ in (7.6), by studying the statistical properties of the estimators we find that we might as well restrict our attention to the somewhat simpler index $C_p(v)$.

Investigating $C_{pv}^*(u, v)$ in the same way as $C_{pa}^*(u, v)$ above we come to the same conclusion as earlier that large values of u and v decrease the power and it does not seem of interest to consider $u > 1$, and so we will consider only indices with $u = 0$ and $u = 1$. Using $C_{pv}^*(u, v)$ the probability that the process is considered capable as a function of $a = (\mu - T)/\sigma$ is symmetric around 0, as is the index $C_{pv}(u, v)$. In Figure 7.4 the probability that the process is considered capable using $C_{pv}^*(u, v)$ is plotted for $u = 1$ and some values of v when $T > M$, $\beta = 0.5$, $\gamma = 0.1$, $\alpha = 0.05$, and $n = 20$. We also see in Figure 7.4 the probability that the process is considered capable using $C_{pa}^*(0, 4)$ compared to using $C_{pv}^*(1, 3)$. These two estimated indices have approximately the same

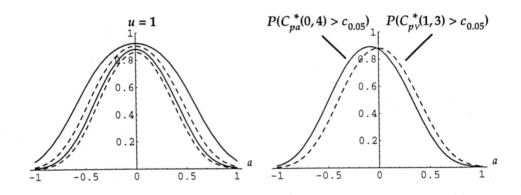

Figure 7.4: To the left the probability that the process is considered capable as a function of $a = (\mu - T)/\sigma$, i.e., $P(C_{pv}^*(u, v) > C_{0.05})$, for $-1 \leq a \leq 1$, given that $T > M$, $\beta = 0.5$, $\gamma = 0.1$, $n = 20$, and $\alpha = 0.05$, where $C_{0.05}$ is determined by $P(C_{pv}^*(u, v) > C_{0.05}) = 0.05$, given that $a = 0$ and $\gamma = 1/6$. For $u = 1$ the curves correspond to $v = 1, 2, 3, 4$ counted from the top. To the right the probability that the process is considered capable as a function of a using $C_{pa}^*(0, 4)$ and $C_{pv}^*(1, 3)$

power but $C_{pa}^*(0,4)$ is more sensitive to departures from the target value when μ shifts away from T towards the closest specification limit. If it is as important to protect against a shift of μ from T towards the specification limit furthest away from T as towards the closest specification limit then $C_{pv}(u,v)$ is to be preferred, otherwise $C_{pa}(0,4) = C_p(4)$ is still recommended.

One advantage with the results in Theorems 7.3.1 and 7.4.1 is that they can be used to determine an index with the properties, concerning both power and sensitivity, that are of importance in each particular case under consideration. When searching for a suitable index we do not recommend using a criterion based solely on power or one based solely on sensitivity. The two properties (i) and (ii) mentioned above ought to be combined, e.g., in the following way. Determine, for fixed values of α and n, a smallest acceptable power for a given value of γ and then, among the indices obtained, select values of u and v which will render the index to be the most sensitive one to departures from the target value. When a decision has been made about which index to use, the results of Theorems 7.3.1 or 7.4.1 are easy to apply for concluding whether the process is capable or not, by calculating the so called p-value, i.e., the probability that $C_{pa}^*(u,v)$ or $C_{pv}^*(u,v)$ exceeds the observed estimated index when $\mu = T$ and $\sigma = d(1-\beta)/3$, and then compare this probability with the significance level α.

7.6   Concluding Remarks

By considering the classes of indices discussed in this paper together with the distributions of the estimated indices we can choose values of u and v to obtain a suitable index and its estimator with properties of interest like those described in (i) and (ii) in Section 7.5. The explicit form of the distribution of the estimated index, based only on the central χ^2-distribution and the normal distribution, increases the utility of the classes of indices under investigation. When a decision has been made about which index to use then the results of Theorems 7.3.1 and 7.4.1 are easy to apply to conclude whether the process is capable or not.

The classes of indices described by $C_{pa}(u,v)$ and $C_{pv}(u,v)$ can be generalized to be used for one-sided specification intervals with a target value in analogy with the one-sided versions of C_{pk}, see Vännman (1996).

Acknowledgement. The author is grateful for the financial support provided by the Volvo Research Foundation, project number 96:08.

References

1. Bergman, B. and Klefsjö, B. (1994). *Quality from customer needs to customer satisfaction*, McGraw-Hill, New York and Studentlitteratur, Lund.

2. Boyles, R. A. (1991). The Taguchi capability index, *Journal of Quality Technology*, **23**, 17–26.

3. Boyles, R. A. (1994). Process capability with asymmetric tolerances, *Communications in Statististics-Simulations*, **23**, 615–643.

4. Chan, L. K., Cheng, S. W. and Spiring, F. A. (1988). A new measure of process capability: Cpm, *Journal of Quality Technology*, **20**, 162–175.

5. Choi, B. C. and Owen, D.B. (1990). A study of a new process capability index, *Communications in Statististics-Theory and Methods*, **19**, 1231–1245.

6. Gunter, B. H. (1989). The use and abuse of C_{pk}: Parts 1-4, *Quality Progress*, **22**; January, 72–73; March, 108–109; May, 79–80; July, 86–87.

7. Kane, V. E. (1986). Process capability indices, *Journal of Quality Technology*, **18**, 41–52.

8. Kotz, S. and Johnson, N. L. (1993). *Process capability indices*, Chapman and Hall: London.

9. Pearn, W. L., Kotz, S. and Johnson, N. L. (1992). Distributional and inferential properties of process capability indices, *Journal of Quality Technology*, **24**, 216–231.

10. Pignatiello, Jr, J. J. and Ramberg, J. S. (1993). Process capability indices: just say "No!", *ASQC Quality Congress Transactions 1993*, Boston, 92–104.

11. Rodriguez, R. N. (1992). Recent developments in process capability analysis, *Journal of Quality Technology*, **24**, 176–186.

12. Vännman, K. (1995). A unified approach to capability indices, *Statistica Sinica*, **5**, 805–820.

13. Vännman, K. (1996). Families of capability indices for one-sided specification limits, Research Report, Division of Quality Technology & Statistics, Lulee University, 1996:6.

14. Vännman, K. (1997a). Distribution and moments in simplified form for a general class of capability indices, *Communications in Statististics-Theory and Methods*, **26**, 159–180.

15. Vännman, K. (1997b). A general class of capability indices in the case of asymmetric tolerances, *Communications in Statististics-Theory and Methods*, **26**, 2049–2072.

16. Vännman, K. and Kotz, S. (1995a). A superstructure of capability indices-distributional properties and implications, *Scandinavian Journal of Statistics*, **22**, 477–491.

17. Vännman, K. and Kotz, S. (1995b). A superstructure of capability indices-asymptotics and its implications, *International Journal of Reliability, Quality, and Safety Engineering*, **2**, 343–360.

8
___

Robustness of On-line Control Procedures
___

M. S. Srivastava

University of Toronto, Toronto, Ontario, Canada

Abstract: In this paper, robustness of on-line control procedures for process adjustment using feedback is considered. Most commonly used models are random walk models and integrated moving average (IMA) process of order one in which the disturbances are assumed to be normally distributed. When the cost of inspection is zero, there is no sampling interval. For this case, the optimum value of the action limit, where the adjustment needs to be made, is obtained for any disturbance that is distributed symmetrically around zero. Numerical values of average adjustment interval and mean squared deviations are given for symmetric uniform distribution and symmetric mixture normal distributions. However, when the cost of inspection is not zero, and sampling interval is present, optimum values of the control parameters, sampling interval and action limit, are given only for the symmetric random walk model. It is shown that when there is no sampling interval, the optimum action limit is robust for moderate departure form normality. However, when sampling interval is present, the optimum values of the control parameters are not as robust. Since the random walk model is a special case of the IMA process, the same conclusion applies to this case as well.

Keywords and phrases: Integrated moving average process, optimal control parameters, process adjustment, random walk, sampling interval
___

8.1 Introduction

In many situations in quality control monitoring, the usual assumption that the observations are independent and identically distributed cannot be assumed since several factors affecting the quality may not be controlled at a specified level. For example, it is usually difficult to control the quality of the feed material, temperature and humidity etc without substantially increasing the cost of the production. In such cases, it is assumed that the quality process is of wandering kind. For such a process, an on-line control procedure is needed so that an adjustment can be made whenever the deviations from the target value exceeds a specified action limit. The adjustment can be effected automatically or manually and it can be applied directly to the process or by compensating an input variable employing feedback control to produce the desired adjustment.

In order to obtain an optimum value of the action limit, a reasonable model for the deviations from the target value is desired. Since the deviations are of the wandering kind, a non-stationary model such as a random walk model or an IMA process of order one will be more appropriate. We shall consider both these models. The cost of production will include the costs of inspection and adjustment, and the cost of deviating from the target value. When there is no cost of inspection, each manufactured item will be inspected and an adjustment will be made whenever the deviations of the quality characteristic exceeds the action limit. In this case, we would need only to find the optimum value of the action limit that minimizes the long term average cost rate. However, when the cost of inspection is present, sampling is done at m units of interval and we would need to find the optimum value of the action limit as well as the sampling interval. When the disturbance in the above models are normally distributed, optimum values of the control parameters have been obtained by Srivastava and Wu (1991a,b) for the normal random walk model and by Srivastava (1996) for the IMA model. These models have been previously considered by Box and Jenkins (1963), Box and Kramer (1992), Srivastava and Wu (1992,1995) Taguchi et.al (1989).

In this paper, we consider the case when the disturbance is symmetrically distributed but not necessarily normal. Numerical evaluations are, however, carried out for the uniform and a mixture of normal distributions. It is shown that when there is no sampling interval, the optimum action limit is robust. However, when sampling interval is present, the optimal control parametes are robust only for slight departure from normality. The organization of the paper is as follows.

In Section 8.2, we consider the case when there is no inspection cost and each manufactured item is inspected. The effect of departure from normality on the optimum action limit is investigated when the distrubance of the IMA model for deviations of the optimum process from the target value is symmerically distributed around zero. In Section 8.3, we consider the case when the cost of inspection is present. However, we, consider only a random walk model.

8.2 Robustness Study When Inspection Cost Is Zero

In this section, we shall assume that there are no inspection cost and hence each manufactured item is inspected. However, there is an adjustment cost of C_A of adjusting the process whenever the deviation of the quality process from the target value goes out of the action limit. There is also a cost of being off target. It is assumed that the cost of being off target is proportional to Z_t^2, the square of the deviation from the target value and C_0 is the constant of proportionality. Taguchi et.al (1989) chooses $C_0 = \frac{C}{\Delta^2}$, where C is the cost of producing a defective item and Δ is the tolerance limit, while Box and Luceño (1994) chooses it as $\frac{C_T}{\sigma_a^2}$, where C_T is the cost of being off target for one time interval by an amount σ_a^2 where σ_a^2 is the variance of the disturbance in the process Z_t. We shall assume that the sequence of observations $\{Z_t\}$ follow an IMA process which can be represented as

$$Z_t = \hat{Z}_t + a_t , \tag{8.1}$$

where a_t is a sequence of iid random variables with mean 0 and variance σ_a^2 and $\hat{Z}_t$ is independent of a_t and is an exponentially weighted moving average (EWMA) of the past data defined by

$$
\begin{aligned}
\hat{Z}_t &= \lambda[Z_{t-1} + \Theta Z_{t-2} + \Theta^2 Z_{t-3} + \ldots] \\
&= \lambda Z_{t-1} + \Theta \hat{Z}_{t-1} , \tag{8.2}
\end{aligned}
$$

where $\Theta = 1 - \lambda, 0 < \lambda \leq 1$ and is the predicted or forecasted value of the tth observation having observed $Z_1, \ldots, Z_{t-1}$. Since $\hat{Z}_{t-1}$ is the predicted value of $(t-1)^{th}$ observation, it is clear that as soon as we obtain the $(t-1)^{th}$ observation, we can easily update it to find the predicted value of tth observation using equation (8.2). For example, given the predicted value of the tth observation $\hat{Z}_t$ and observed value Z_t, we get the predicted value of the $(t+1)^{th}$ observation as

$$\hat{Z}_{t+1} = \lambda Z_t + \Theta \hat{Z}_t ,$$

Thus, when $\lambda = 1$ (random walk model), the predicted value of the next future observation is the current observation.

From (8.1), we can also write

$$\hat{Z}_{t+1} = \lambda[\hat{Z}_t + a_t] + (1 - \lambda)\hat{Z}_t ,$$

giving

$$\hat{Z}_{t+1} - \hat{Z}_t = \lambda a_t ,$$

Thus, the first difference of Z_t is the first order moving average model

$$
\begin{aligned}
Z_{t+1} - Z_t &= \hat{Z}_{t+1} + a_{t+1} - \hat{Z}_t - a_t \\
&= a_{t+1} - a_t + \lambda a_t \ .
\end{aligned}
$$

which, by summing and noting that $Z_1 - a_1 = \hat{Z}_1$ from (8.2), gives

$$
Z_t = \hat{Z}_1 + a_t + \lambda \sum_{i=1}^{t-1} a_i \ ,
$$

$$
\hat{Z}_{t+1} = \hat{Z}_1 + \lambda \sum_{j=1}^{t} a_j \ ,
$$

If at time $t = 1$, the adjustment is prefect, then $\hat{Z}_1 = 0$ and from (8.2), $Z_1 = a_1$. We shall assume that the adjustment is prefect at $t = 1$, and the integrated moving average process of order one, denoted by $IMA_1(\lambda, \sigma_a^2)$, is given by

$$
Z_t = a_t + \lambda \sum_{i=1}^{t-1} a_i \ , \ Z_1 = a_1, 0 \le \lambda \le 1 \ , \tag{8.3}
$$

where $a_i's$ are iid with the mean 0 and variance σ_a^2. We shall assume that $a_i's$ are also symmetrically distributed around zero. The parameter λ is called the nonstationarity parameter. Most of the derivations of the optimum action limit is based on the assumption that $a_i's$ are iid $N(0, \sigma_a^2)$, see Box and Jenkins (1963), Srivastava and Wu (1991a,b) and Srivastava (1996). The optimum action limit is the one that minimizes the long-run average cost rate given by

$$
C(L) = \frac{C_A}{(AAI)} + C_0[MSD] \ , \tag{8.4}
$$

where,

$$
MSD = \frac{E[\Sigma_{j=1}^{N} Z_j^2]}{(AAI)} \ , \tag{8.5}
$$

and AAI is the average adjustment interval, where an adjustment is made as soon as the forecasted value $\hat{Z}_{n+1}$ of the $(n+1)^{st}$ observation exceeds an action limit L, that is, at

$$
N = min\{n \ge 1 : |\hat{Z}_{n+1}| > L\} \tag{8.6}
$$

Thus,

$$
AAI = E(N) \ . \tag{8.7}
$$

When $a_j's$ are iid $N(0, \sigma_a^2)$, the cost function $C(L)$ was evaluated by Box and Jenkins (1963) and for $\lambda = 1$ by Srivastava and Wu (1991a,b). They also obtained the optimum value of L, the one that minimizes the average cost rate $C(L)$. It may be noted that from (8.3)

$$\hat{Z}_{n+1} = \lambda \sum_{i=1}^{n} a_i \ , \quad since \ \hat{Z}_1 = 0 \tag{8.8}$$

Thus, $\hat{Z}_{n+1}$ is a random walk which is a sum of n iid random variables with mean 0 and variance $\lambda^2 \sigma_a^2$.

Let

$$w_j = \frac{a_j}{\sigma_a} \ , \quad j = 1, 2, 3, \ldots \tag{8.9}$$

Then $\{w_j\}$ is a sequence of iid symmetric random variables with mean 0 and variance 1 and pdf f. Let

$$h(\lambda) = E[e^{i\lambda w}] = \int e^{i\lambda w} f(w) dw \tag{8.10}$$

be its characteristic function. Since, it has been assumed that w_j is symmetrically distributed around zero, $h(\lambda)$ will be real. Let

$$\rho_1 = -\frac{1}{\pi} \int_0^\infty \lambda^{-2} \log_e \left[\frac{1 - h(\lambda)}{\lambda^2/2} \right] d\lambda \ , \tag{8.11}$$

We shall denote by $f(0)$ and $f''(0)$ as the pdf and its second derivative evaluated at zero respectively Then, it has been shown by Srivastava (1995) that

$$AAI \simeq \begin{cases} u^2 + 2\rho_1 u + (\rho_1^2 + \frac{b}{12}) \ , & u \geq 1, \\ 1 + \xi_1 u + \xi_1^2 u^2 + \xi_3 u^3 \ , & u < 1. \end{cases} \tag{8.12}$$

and

$$(MSD/\lambda^2\sigma^2) \simeq \begin{cases} \frac{1}{6}(u^2 + 2\rho_1 u + \rho_1^2 - \frac{7b}{12}) + \frac{1}{\lambda^2} \ , & u \geq 1, \\ \frac{1}{3}\xi_1 u^3 + \frac{1}{\lambda^2} \ , & u < 1, \end{cases} \tag{8.13}$$

where,

$$u = (L/\lambda\sigma_a) \tag{8.14}$$

$$b = E(w_j^4) \ , \tag{8.15}$$

the fourth moment of the random variable w_j,

$$C_1 = \left(1 - \frac{\lambda^2 b}{6} \right) \tag{8.16}$$

and

$$\xi_1 = 2f(0) \ , \quad \xi_3 = \frac{1}{3}f''(0) + \xi_1^3 \tag{8.17}$$

The optimum value of L or u denoted by L_0 and u_0 respectively, is given by

$$u_0 = (L_0/\lambda\sigma_a) = \begin{cases} \left(\frac{6C_A}{\lambda^2\sigma_a^2 C_0} \right)^{\frac{1}{4}} - \rho_1 \ , & u_0 \geq 1 \ , \\ \frac{2C_A f(0)}{2\lambda^2\sigma_a^2 C_0 f(0) - f''(0) + 24 f^3(0)} \ , & u_0 \geq 1 \ . \end{cases} \tag{8.18}$$

8.2.1   Effect of non-normality

In this section, we consider some non-normal distributions and compare it with the results of normal distribution. Specifically, we consider uniform distribution over $(-\sqrt{3}, \sqrt{3})$ and a mixture of normal distributions

$$(1-\alpha)\phi(;0,\sigma_1^2) + \alpha\phi(;0,\sigma_2^2)$$

where $\sigma_1^2 = 1 - \varepsilon$, and $\sigma_2^2 = 1 + \varepsilon\frac{1-\alpha}{\alpha}$, $0 < \alpha \leq 1$, and $\phi(;,0,\sigma^2)$ denotes a normal pdf with mean 0 and variance σ^2. Both these distributions have mean 0 and variance one. The comparison will be in terms of AAI and a function of MSD,

$$1 + g() = \frac{MSD}{\lambda^2\sigma_a^2} - \frac{1}{\lambda^2} + 1 , \tag{8.19}$$

since it does not depend on λ or σ_a. Also, we shall compare u_0 or L_0.

Table 8.1: Comparsion of AAI and $1 + g$, the entries BL are from Box and Luceño's Table 2. (No entries available where dashes appear)

	Normal				Uniform			
	AAI		1+g		AAI		1+g	
u	BL	S	BL	S	BL	S	BL	S
0.25	1.245	1.245	1.004	1.004	1.169	1.168	1.003	1.003
0.50	1.607	1.605	1.031	1.033	1.406	1.396	1.024	1.024
0.75	2.115	2.115	1.092	1.112	1.764	1.702	1.081	1.081
2.00	—	6.922	—	1.820	—	6.481	—	1.880
3.00	—	13.09	—	2.848	—	12.51	—	2.885
4.00	—	21.25	—	4.209	—	20.55	—	4.224
6.00	—	43.59	—	7.931	—	42.61	—	7.902

In order to compare these quantities, we need the values of b, ρ_1, ξ_1 and ξ_3. For the three distributions, these quantities are obtained as follows:

(a) Standard Normal Distribution.
    For standard normal distribution, we have $b = 3$, $\rho_{1\phi} = 0.583$, $\xi_{1\phi} = 0.798$, $\xi_{3\phi} = 0.376$

(b) Uniform Over $(-\sqrt{3}, \sqrt{3})$.
    For the standard uniform distribution, $b_u = \frac{9}{5}$, $\rho_{1u} = 0.5161$, $\xi_{1u} = 0.289$, $\xi_{3u} = 0.024$

(c) Mixture Normal Distribution.
    For standard mixture normal distribution described above $b_m = 3\left[1 + \frac{\varepsilon^2(1-\alpha)}{\alpha}\right]$, $\rho_{1m} \simeq 0.5830 + 0.1303\frac{\varepsilon^2(1-\alpha)}{\alpha} + o(\varepsilon^2)$

$$\xi_{1m} \simeq 0.798[\tfrac{1-\alpha}{\sigma_1} + \tfrac{\alpha}{\sigma_2}] \ , \ \xi_{3m} \simeq 0.508[\tfrac{1-\alpha}{\sigma_1} + \tfrac{\alpha}{\sigma_2}]^3 - 0.133[\tfrac{1-\alpha}{\sigma_1^3} + \tfrac{\alpha}{\sigma_2^3}] \ ,$$

where $\sigma_1^2 = 1 - \varepsilon$ and $\sigma_2^2 = 1 + \varepsilon\tfrac{1-\alpha}{\alpha}$.

The comparsion of AAI and $1 + g$ for uniform distribution with normal distribution is given in Table 8.1 and that of mixture normal is given in Tables 2 and 3. Also, from the calculation of ρ's, it is clear that when there is no sampling interval, the optimum value of u_0 or L_0 will not be affected for the uniform distribution and for small to moderate departure from normality in the case of mixture normal distribution.

8.3 Robustness Study When Inspection Cost Is Not Zero

We shall now consider the case when the inspection cost is C_I, and thus we inspect every m^{th} manufactured item $m \geq 1$, instead of every item as was done in the previous section. However, we consider only the random walk model since the correspomding result for the IMA model is not yet available. Although, it will be of great interest to have these theoretical results for the IMA model also, it is not as important for the robustness study since it is shown that unless the departure from normality is slight, the optimal control parameters are not robust. The random walk model is given by

$$Z_n = Z_{n-1} + a_n$$

where $\{a_i\}$ is a sequence of iid random variables distributed symmetrically around zero with variance σ_a^2. Since every m^{th} item is inspected, the adjustment will take place at the Nth inspection where N is a random variable defined by

$$N = min\{n \geq 1 : |Z_{mn}| \geq L\}$$

and thus Nm is the total number of items manufactured before the adjustment was found necessary. Thus, the average adjustment interval is given by

$$AAI = mE(N) \ ,$$

and the average mean squared deviation is given by

$$MSD = \frac{1}{mE(N)} \ E\left[\sum_{i=1}^{mN} Z_i^2\right]$$

The long run average cost rate is given by

$$C(L,m) = \frac{C_I}{m} + \frac{C_A}{AAI} + C_0(MSD)$$

Let $u = \left[\frac{L}{\sigma_a}\right]$ and $b = E\left[\left(\frac{a_j}{\sigma_a}\right)^4\right]$. In this section, we shall confine our attention to the case when $\left(\frac{u}{\sqrt{m}}\right) \geq 1$; the other case can be handled on the same line as the results of Section 8.2. Srivastava and Wu (1996) have shown that

$$C(L,m) \simeq \frac{C_I}{m} + \frac{C_A}{u^2 + 2\rho_1 u m^{\frac{1}{2}} + (\rho_1^2 + \frac{b}{12})m}$$

$$+ \frac{C_0\sigma_a^2}{6}\left[u^2 + 2\rho_1 u m^{\frac{1}{2}} + (\rho_1^2 + \frac{b}{12})m\right] + C_0 m \sigma_a^2 \left(\frac{9-2b}{18}\right) + \frac{C_0\sigma_a^2}{2},$$

$$AAI \simeq u^2 + 2\rho_1 u m^{\frac{1}{2}} + \left(\rho_1^2 + \frac{b}{12}\right)m,$$

and

$$\left(\frac{MSD}{\sigma_a^2}\right) \simeq \frac{1}{6}\left[u^2 + 2\rho_1 u m^{\frac{1}{2}} + \left(\rho_1^2 + \frac{b}{12}\right)m\right] + \frac{(9-2b)m}{18} + \frac{1}{2}$$

Minimizing the loss function $C(L,m)$, we find that the optimum value of L and m are respectively given by

$$L^* \simeq \sigma_a\left[\left(\frac{6C_A}{\sigma_a^2 C_0}\right)^{\frac{1}{4}} - \rho_1 m^{*\frac{1}{2}}\right],$$

and

$$m^* \simeq max\left\{1, \left(\frac{18C_I}{C_0\sigma_a^2(9-2b)}\right)^{\frac{1}{2}}\right\}, \quad b < \frac{9}{2}.$$

We shall assume that $b < \frac{9}{2}$.

8.3.1 Effect of non-normality

As in Section 8.2, we shall consider uniform and mixture normal distributions. However, while for uniform distribution $b = \frac{9}{5} < \frac{9}{2}$, for mixture normal distribution, in order that $b < \frac{9}{2}$, we shall require that

$$1 + \left[\epsilon^2(1-\alpha)/\alpha\right] < 3/2$$

That is,

$$\epsilon < \left[\alpha/2(1-\alpha)\right]^{1/2}.$$

For example, if $\epsilon = 0.5, \alpha = 0.6$, then $b_m = 3.5 < 9/2$. For this value of ϵ and α, $\rho_1 \simeq 0.607$. If m_0 denotes the optimum value of m under the normal distribution, then

$$m^* = \left(3/(9-2b)\right)^{1/2} m_0,$$

which for the above mixture normal with $\epsilon = 0.5$ and $\alpha = 0.6$, becomes

$$m^* = \left(3/2\right)^{\frac{1}{2}} m_0$$

Thus, the inspection has to be carried out after approximately twenty-three percent more items produced than in the normal distribution case. Similarly, the optimum action limit has to be decreased considerably. Thus, unless the departure from normality is slight, the optimum control parameters are not robust.

Table 8.2: Comparison of average adjustment interval
Normal vs Mixture Normal

u	ϵ	Normal	Mixture Normal α							
			0.95	0.90	0.85	0.80	0.75	0.70	0.60	0.50
0.5	0.1		1.605	1.605	1.606	1.606	1.606	1.606	1.607	1.608
	0.2		1.606	1.607	1.608	1.609	1.610	1.611	1.614	1.618
	0.3	1.605	1.607	1.609	1.611	1.614	1.616	1.619	1.626	1.635
	0.4		1.609	1.613	1.617	1.622	1.627	1.633	1.646	1.662
	0.5		1.612	1.619	1.627	1.635	1.644	1.654	1.676	1.703
1.0	0.1		2.756	2.757	2.757	2.758	2.758	2.759	2.760	2.763
	0.2		2.757	2.759	2.761	2.763	2.765	2.768	2.774	2.783
	0.3	2.756	2.759	2.763	2.767	2.771	2.777	2.782	2.797	2.816
	0.4		2.762	2.769	2.776	2.784	2.793	2.804	2.830	2.865
	0.5		2.766	2.776	2.788	2.801	2.816	2.833	2.873	2.929
2.0	0.1		6.922	6.923	6.924	6.924	6.925	6.926	6.928	6.931
	0.2		6.924	6.926	6.929	6.931	6.935	6.938	6.947	6.959
	0.3	6.922	6.927	6.932	6.937	6.944	6.951	6.959	6.979	7.006
	0.4		6.930	6.940	6.950	6.962	6.975	6.989	7.025	7.074
	0.5		6.936	6.951	6.967	6.986	7.006	7.030	7.087	7.164
4.0	0.1		21.25	21.26	21.26	21.26	21.26	21.26	21.26	21.27
	0.2		21.26	21.26	21.26	21.27	21.27	21.28	21.29	21.31
	0.3	21.25	21.26	21.27	21.28	21.29	21.30	21.31	21.34	21.39
	0.4		21.27	21.28	21.30	21.32	21.34	21.36	21.42	21.49
	0.5		21.28	21.30	21.33	21.36	21.39	21.42	21.51	21.63

Table 8.3: Comparison of $1 + g = (\text{MSD}/\lambda^2\sigma_a^2) + (1 - 1/\lambda^2)$
Normal vs Mixture Normal

| | | Normal | Mixture Normal | | | | | | | |
| | | | | | | α | | | | |
u	ϵ		0.95	0.90	0.85	0.80	0.75	0.70	0.60	0.50
0.25	0.2	1.003	1.003	1.003	1.003	1.003	1.003	1.003	1.003	1.003
	0.4		1.003	1.003	1.003	1.003	1.003	1.003	1.003	1.004
	0.6		1.003	1.003	1.003	1.003	1.004	1.004	1.004	1.004
	0.8		1.003	1.004	1.004	1.004	1.004	1.004	1.004	1.004
0.50	0.2	1.021	1.021	1.021	1.021	1.021	1.021	1.021	1.021	1.021
	0.4		1.021	1.021	1.021	1.021	1.021	1.021	1.021	1.021
	0.6		1.021	1.021	1.021	1.021	1.021	1.022	1.022	1.022
	0.8		1.021	1.022	1.022	1.022	1.023	1.023	1.024	1.024
0.75	0.2	1.053	1.053	1.053	1.053	1.053	1.053	1.053	1.053	1.053
	0.4		1.053	1.053	1.053	1.053	1.053	1.053	1.054	1.054
	0.6		1.053	1.053	1.054	1.054	1.054	1.054	1.054	1.054
	0.8		1.054	1.054	1.055	1.055	1.055	1.056	1.055	1.054
2.00	0.2	1.636	1.634	1.634	1.634	1.634	1.633	1.633	1.632	1.631
	0.4		1.634	1.633	1.632	1.631	1.630	1.628	1.625	1.621
	0.6		1.633	1.631	1.629	1.626	1.624	1.620	1.613	1.603
	0.8		1.631	1.628	1.624	1.620	1.615	1.610	1.596	1.577
3.00	0.2	2.343	2.342	2.341	2.341	2.341	2.340	2.340	2.339	2.337
	0.4		2.341	2.340	2.338	2.337	2.335	2.333	2.329	2.322
	0.6		2.340	2.337	2.334	2.331	2.327	2.323	2.312	2.298
	0.8		2.338	2.333	2.328	2.323	2.316	2.309	2.291	2.266
4.00	0.2	3.354	3.353	3.353	3.352	3.352	3.351	3.351	3.349	3.347
	0.4		3.352	3.351	3.349	3.347	3.345	3.342	3.336	3.328
	0.6		3.350	3.347	3.343	3.339	3.334	3.328	3.315	3.296
	0.8		3.348	3.342	3.335	3.328	3.319	3.310	3.286	3.254
6.00	0.2	6.342	6.341	6.340	6.340	6.339	6.338	6.337	6.335	6.332
	0.4		6.339	6.337	6.334	6.331	6.328	6.324	6.315	6.302
	0.6		6.337	6.331	6.325	6.319	6.311	6.303	6.282	6.253
	0.8		6.333	6.323	6.312	6.300	6.287	6.272	6.235	6.186

8.4   Conclusion

In this paper we first considered the IMA model when there is no cost of inspection and the disturbance term is symmetrically distributed. We obtain the optimum value of the action limit and showed that it is robust against the departure from normality, at least in two cases when the disturbance is symmetrically distributed (round the target value) as a mixture of normal or uniform. When the cost of inspection is also present, we consider only the random walk model and showed that the optimum control parameters are not robust. This contradicts the findings of Box and Luceño (1994) who claim it to be robust.

Acknowledgments. Research supported by Natural Sciences and Engineering Research Council of Canada.

References

1. Box, G. E. P. and Jenkins, G. M. (1963). Further contributions to adaptive quality control: simulations estimation of dynamics: Non-zero costs, *Bulletin of the International Statistical Institute*, **34**, 943–974.

2. Box, G. E. P. and Kramer, T. (1992). Statistical process monitoring and feedback adjustment - A Discussion, *Technometrics*, **34**, 251–267.

3. Box, G. and Luceño, A. (1994). Selection of sampling interval and action limit for discrete feedback adjustment, *Technometrics*, **36**, 369–378.

4. Srivastava, M. S. (1995). Robustness of control procedures for integrated moving average process of order one, *Techical Report No. 9501*, University of Toronto.

5. Srivastava, M. S. (1996). Economical process adjustment with sampling interval, *Communications in Statistics Theory & Methods*, **25(10)**, 2403–2430.

6. Srivastava, M. S. and Wu, Y. (1991a). A second order approximation on Taguchi's on-line control procedure, *Communications in Statistics Theory & Methods*, **20(7)**, 2149–2168.

7. Srivastava, M. S. and Wu, Y. (1991b). Taguchi's on-line control procedures and some improvements, *Technical Report No. 9121*, University of Toronto.

8. Srivastava, M. S. and Wu, Y. (1992). On-line quality control procedures based on random walk model and integrated moving average model of order (0,1,1), *Technical Report No. 9203*, University of Toronto.

9. Srivastava, M. S. and Wu, Y. (1995). An improved version of Taguchi's on-line control procedure, *Journal of Statistical Planning & Inference*, **43**, 133–145.

10. Srivastava, M. S. and Wu, Y. (1996). Economical quality control procedures based on symmetric random walk model, *Statistica Sinica*, **6**, 389–402.

11. Taguchi, G., Elsayed, E. A. and Hsiang, T. (1989). *Quality Engineering in Production System*, New York: MacGraw-Hill.

9

An Application of Filtering to Statistical Process Control

A. Thavaneswaran, B. D. Macpherson and B. Abraham

University of Manitoba, Winnipeg, Manitoba, Canada,
University of Manitoba, Winnipeg, Manitoba, Canada,
University of Waterloo, Waterloo, Ontario, Canada

Abstract: There has been growing interest in the Kalman filter as an estimation technique in statistical process control. In cases where prior information about the process is available, procedures based on the 'optimal' [Godambe (1985)] smoother can be superior to the classical procedures like Shewhart and CUSUM control charts. We also discuss the relationship among EWMA, Kalman filtering and the 'optimal' smoother. This smoother and its applications are also illustrated through an example.

Keywords and phrases: ARIMA processes, estimating function, EWMA, Kalman filter, optimal filter, optimal smoother

9.1 Introduction

Recently there has been growing interest in the general theory of statistical process control through the use of exponentially weighted moving average (EWMA) charts [Hunter (1986), Montgomery and Mastrangelo (1991)], for autocorrelated data. In this paper an 'optimal' smoother is proposed; it is optimal in the sense that the smoother is a solution of the optimal estimating function, [see for example, Thavaneswaran and Abraham (1988), Godambe, (1985)]. This smoother essentially incorporates the correlation structure of the underlying process, and leads to a control chart with better properties than the EWMA chart.

In the literature two different methods for constructing control charts are proposed for correlated data. In the first method, the basic idea is to model the autocorrelation structure in the original process using an autoregressive integrated moving average (ARIMA) model and apply control charts to the residuals. The second method uses a control chart based on the EWMA statistic, a function of the one step ahead forecast errors. The exponentially weighted moving average statistic gives a procedure which is optimal (in the minimum mean square error (MSE) sense) for a limited class of ARIMA (p,d,q) processes with $(p, d, q) = (0, 1, 1)$ [see for example, Abraham and Ledolter (1986)].

Shewhart Control Charts and other Statistical Process Control (SPC) techniques are very useful in industry for process improvement, estimation of process parameters and determination of process capabilities. The assumption of uncorrelated observations is fundamental to the use of the Shewhart control charts [Hunter (1986)]. In this situation a simple model that is used for the observations is:

$$X_t = \mu + \epsilon_t$$

where μ is the process mean and ϵ_t $(t = 1, 2, ...)$ are independent identically distributed (iid) random variables with mean zero and variance σ^2.

The existence of autocorrelated errors violates the conditions of this model and failure to detect, or ignoring autocorrelations can lead to misleading results. Detection of autocorrelation can be accomplished through diagnostic plots or through a formal test. A simple plot of the residuals from the model can be helpful. If the residuals are plotted against time, and unusually large numbers of residuals with the same sign are observed clustered together, then this is an indication that the errors are governed by positive autocorrelation. On the other hand, rapid changes in sign may indicate the presence of negative auto-correlation. Positively correlated errors can lead to substantial underestimation of σ^2 and an increase in the frequency of false alarms; in other words, the in control Average Run Length (ARL) is much shorter than it would be for a process with uncorrelated observations. Thus often, the state of control of the process can not be determined from the usual control charts.

In Section 9.2, the ARIMA (p,d,q) modeling approach to quality control and EWMA are briefly discussed. Section 9.3 provides the filtering and prediction algorithms based on estimating functions. It is also shown there that Kalman filtering algorithm does not take into account the autocorrelation structure of the observed process . In addition we provide control charts for correlated observations using a smoother, the form of which depends on the autocorrelation structure of the observed process of interest. Section 9.4 provides some special cases and an application. Section 9.5 gives some concluding remarks.

9.2 Control Charts for Autocorrelated Data

9.2.1 Modeling the autocorrelations using ARIMA (p,d,q) models

The primary approach is to fit an appropriate time series model to the observations and apply a Shewhart control chart to the residuals from this model. A commonly used time series model is the Autoregressive Integrated Moving Average model which is given by

$$\phi_p(B)(1-B)^d X_t = \theta_q(B)\epsilon_t$$

where

$$\phi_p(B) = 1 - \phi_1 B - \theta_2 B^2 - \cdots - \phi_p B^p$$

$$\theta_q(B) = 1 - \theta_1 B - \theta_2 B^2 - \cdots - \theta_q B^q$$

and the ϵ_t's are iid random variables with mean zero and variance σ^2.

If $\hat{X}_t$'s are the predicted values from a fitted ARIMA model, then

$$e_t = X_t - \hat{X}_t , \quad t = 1, 2, \ldots, n$$

are the residuals which are considered to be approximately identically distributed independent random variables. The process X_t will be declared 'out of control' if a mean shift is detected on the control chart applied to the residuals.

9.2.2 An application of the EWMA statistic to autocorrelated data

The EWMA approach was first suggested by Roberts (1959) and has been discussed by several authors [for example, Abraham and Kartha (1978,1979)]. The EWMA statistic Z_t is defined as

$$Z_t = \lambda X_t + (1-\lambda)Z_{t-1}, \quad 0 < \lambda < 1 .$$

Montgomery and Mastrangelo (1991) and Hunter (1986) have shown that if the observations are uncorrelated, the control limits for the EWMA control chart under steady state conditions are given by :

$$LCL = \overline{X} - 3\sigma\sqrt{\frac{\lambda}{n(2-\lambda)}}$$

$$UCL = \overline{X} + 3\sigma\sqrt{\frac{\lambda}{n(2-\lambda)}}$$

where $\overline{X}$ is the overall average and σ is the process standard deviation. The EWMA can also be used for autocorrelated data. As an illustration, consider a situation where the data can be modelled as an ARIMA (0,1,1) process:

$$X_t = X_{t-1} + \epsilon_t - \theta\epsilon_{t-1}.$$

When $\lambda = 1 - \theta$, the EWMA is the optimal one step ahead forecast (prediction) for this process (i.e. $Z_t = \hat{X}_{t+1/t}$, the one step ahead forecast of X_{t+1} made at time t). In this case, the one step ahead forecast errors $\hat{X}_t - \hat{X}_{t/t-1}, t = 2, 3, \ldots$ are independent with mean zero and standard deviation σ if the fitted ARIMA (0,1,1) model is correct. Thus, we could set up control charts for the one step ahead forecast errors.

Montgomery and Mastrangelo (1991) argue that generally the EWMA, with a suitable λ, will give an "excellent one step ahead forecast" even if the observations from the process are positively autocorrelated or the process mean does not drift too quickly. In addition they indicate that the EWMA provides good forecasts for models which are not exactly ARIMA (0,1,1) and that some processes which follow a slow random walk, can be well represented by the ARIMA(0,1,1) model. We show however that the use of the EWMA in these situations results in a loss of almost 50% efficiency compared to the smoother based on the optimal estimating function.

9.3   Optimal Filter and Smoother

Let us consider a simple time series model,

$$X_t = \phi X_{t-1} + \epsilon_t$$

where $\{\epsilon_t\}$ is an iid sequence with mean zero and variance σ^2 and $|\phi| < 1$. We can define two different means and variances. The unconditional mean of X_t is zero and the unconditional variance of X_t is $\frac{\sigma^2}{1-\phi^2}$. In addition, the conditional mean and variance of X_t given $X_{t-1}, X_{t-2}, \cdots, X_1$ are given by

$$E\left[X_t | X_{t-1}, X_{t-2}, \cdots, X_1\right] = \phi X_{t-1}$$

and

$$\mathrm{Var}\ \left[X_t | X_{t-1}, X_{t-2}, \cdots, X_1\right] = \sigma^2$$

respectively.

Most of the recent inferences for non-linear time series models are based on conditional moments. In the next section we use the conditional moments to obtain an optimal filter and compare this with the Kalman filter.

9.3.1   Optimal filter

Consider a process of the form

$$X_t = \theta_t + N_t \quad t = 1, \cdots, n, \cdots$$

where $\{N_t\}$ is an autocorrelated sequence with zero mean and known covariance structure. Estimation of $\theta_t, t = 1, 2, \cdots, n$ from n-observations on $\{X_t\}$ without making any restrictions on the parameter sequence is an ill-posed problem. In order to obtain a filtered estimate of θ_t we assume that θ_t follows a random walk

$$\theta_t = \theta_{t-1} + a_t$$

where $\{a_t\}$ is a white noise sequence with mean zero and variance σ_a^2.

The following theorem on optimal estimation of θ_t obtained by identifying the sources of variation and optimally combining elementary estimating functions [see Heyde (1987)] from each of these sources, gives the filtering formula for θ_t.

Theorem 9.3.1 *Let F_{t-1}^x be the σ-field generated by $X_1, \cdots, X_{t-1}$. The recursive form of the estimator of θ_t based on $X_1, \cdots, X_t$ is given by:*

$$\hat{\theta}_t = \hat{\theta}_{t/t-1} + P_{t/t-1} \left(X_t - \hat{X}_{t/t-1} \right) \Big/ r_t$$

where $P_{t/t-1} = E\left[\left(\theta_t - \hat{\theta}_{t/t-1} \right)^2 | F_{t-1}^x \right]$, $\hat{\theta}_{t/t-1}$ is an estimate of θ_t based on $X_{t-1}, \cdots, X_1$, $\hat{X}_{t/t-1}$ is the predictor of X_t based on $X_{t-1}, \cdots, X_1$, and $r_t = E\left[\left(X_t - \hat{X}_{t/t-1} \right)^2 | F_{t-1}^x \right]$ its mean square error. Moreover,

$$P_{t/t-1} = P_{t-1/t-1} + \sigma_a^2$$

and

$$\frac{1}{P_{t/t}} = \frac{1}{P_{t/t-1}} + \frac{1}{r_t}.$$

PROOF. The elementary estimating function for fixed θ_t based on the tth observation is

$$h_{1t} = \left(X_t - \hat{X}_{t/t-1} \right) \Big/ r_t$$

and the corresponding information [see for example, Abraham et al. (1997)] associated with h_{1t} is $1/r_t$.

The elementary estimating function for the parameter of interest, say θ, is

$$\left(X_t - \hat{X}_t \right) \frac{\partial \hat{X}_t}{\partial \theta} \Big/ r_t.$$

Now h_{2t}, the corresponding estimating function for θ_t based on prior information is

$$h_{2t} = \left(\theta - \hat{\theta}_{t/t-1} \right) \Big/ P_{t/t-1}.$$

Combining the estimating functions h_{1t} and h_{2t} the Theorem follows. ∎

Under normality assumption on the errors, the Kalman filtering algorithm turns out to be a special case of Theorem 9.3.1.

It is of interest to note that the above filtering algorithm, which could be used in a wider context (for non-Gaussian linear processes as well), provides the estimator in terms of the first two conditional moments of the observed process. The same argument is also true for Kalman filtering. However, for many industrial situations the observed processes are autocorrelated and the Kalman filtering formula does not incorporate the autocorrelation structure. In Section 9.3.2 we present a smoother which incorporates the autocorrelation structure and is different from the Kalman filter.

9.3.2   Optimal smoother

In Abraham et al. (1997), an optimal smoother based on conditional moments has been proposed for nonlinear time series models. In this section an optimal smoother which is a solution of an optimal estimating function (based on moments of the observed process) is proposed for linear processes. In this context, the control limits of the chart for monitoring the process level (conditional mean) with the optimal smoother are functions of the autocorrelation of the observed process of interest.

For an industrial process $\{X_t\}$ having mean zero and covariances $r(t,s)$ the following theorem provides the form of the optimal smoother.

Theorem 9.3.2 *The optimal predictor of X_t, based on $X_{t-1}, X_{t-2}, \cdots, X_1$ is given by*

$$\hat{X}_{t/t-1} = \sum_{s=1}^{t-1} \frac{r(t,s)}{r(s,s)} X_s \qquad (9.1)$$

and the corresponding mean square error, $v_t = E\left[X_t - \hat{X}_{t/t-1}\right]^2$, is given by

$$\nu_t = r(t,t) - \sum_{s=1}^{t-1} \frac{r^2(t,s)}{r(s,s)} \qquad (9.2)$$

We can also give the predictions in terms of predication errors $X_t - \hat{X}_{t/t-1}$ having covariances $r^*(t,s)$:

$$\hat{X}_{t/t-1} = \sum_{s=1}^{t-1} \frac{r^*(t,s)}{r^*(s,s)} \left(X_s - \hat{X}_{t/t-1}\right) \qquad (9.3)$$

and the corresponding mean square error, $V_t = E\left[X_t - \hat{X}_{t/t-1}\right]^2$, is given by

$$V_t = r^*(t,t) - \sum_{s=1}^{t-1} \frac{r^{*2}(t,s)}{r^*(s,s)}. \qquad (9.4)$$

PROOF. When the normality assumption is made, the proof follows from a theorem on normal correlations as in Abraham and Thavaneswaran (1990) or as in Brockwell and Davis (1991) using projection arguments. For the nonnormal case, the result follows by applying the unconditional version of the optimality criterion given in Thavaneswaran and Thompson (1988). ∎

9.4 Applications

9.4.1 ARMA(0,0,1) process or MA(1) process

Consider the recursive smoothing algorithm for an MA(1) process

$$X_t = \epsilon_t - \theta \epsilon_{t-1} \tag{9.5}$$

where $\{\epsilon_t\}$ is a white noise process having mean zero and variance σ^2.

Then it can be easily shown that the one step ahead smoother of X_{t+1} and its mean square error are given by

$$\hat{X}_{t+1/t} = -\theta \left(X_t - \hat{X}_{t/t-1} \right) / V_t$$

and

$$V_t = \sigma_a^2 \left(1 + \theta^2 \right) - \theta^2 / V_{t-1}, \quad \text{respectively}$$

where

$$V_0 = \sigma_a^2 \left(1 + \theta^2 \right).$$

Based on n observations,

$$V_n = E \left(X_{n+1} - \hat{X}_{n+1/n} \right)^2 = \sigma^2 \left(1 - \theta^{2n+4} \right) / \left(1 - \theta^{2n+2} \right)$$

which converges to σ^2 as $n \to \infty$, i.e. MSE of the smoother converges to σ^2 for an MA(1) process. Now we look at the asymptotic properties of the EWMA statistic for the MA(1) process and compare its mean square error with that of the smoother.

Consider the EWMA,

$$\hat{Z}_t = \lambda X_t + (1 - \lambda) Z_{t-1}.$$

For the constant mean model with iid errors given in Section 9.1, Hunter (1986) showed that the mean of the EWMA,

$$E[Z_t] \to \mu$$

and the variance

$$\text{Var}[Z_t] \to \frac{\lambda}{2 - \lambda} \sigma^2.$$

or in terms of the discount coefficient $\omega = 1 - \lambda$,

$$\text{Var}\,[Z_t] \cong \sigma^2 \frac{(1-\omega)}{(1+\omega)}.$$

For the MA(1) process (9.5) it can be shown that the asymptotic variance (MSE) of the EWMA statistic is given by

$$\text{Var}\,[Z_t] = \sigma^2 \frac{(1-\omega)}{(1-\omega)}\left[(1+\theta^2) - 2\omega\theta\right].$$

Note that for small values of ω, and θ close to 1 this variance is as large as $2\sigma^2$ and we lose about 50% of the efficiency by using EWMA. For different values of λ and θ, Table 9.1 provides the MSE for the EWMA statistic when $\sigma^2 = 1$.

Table 9.1: MSE for the EWMA

θ \ λ	0	.25	.5	.75	1.0
-1	1.85	2.38	3.10	4.12	5.7
-0.9	1.68	2.15	2.80	3.71	5.14
-0.8	1.51	1.94	2.52	3.34	4.62
-0.7	1.37	1.75	2.27	3.00	4.14
-0.6	1.25	1.59	2.04	2.69	3.69
-0.5	1.14	1.44	1.84	2.40	3.29
-0.4	1.05	1.32	1.66	2.15	2.92
-0.3	0.98	1.22	1.51	1.93	2.59
-0.2	0.93	1.14	1.38	1.74	2.30
-0.1	0.90	1.08	1.28	1.58	2.04
0	0.88	1.04	1.21	1.45	1.83
.1	0.88	1.03	1.16	1.34	1.65
.2	0.90	1.03	1.13	1.27	1.51
.3	0.94	1.06	1.13	1.23	1.41
.4	1.00	1.11	1.15	1.22	1.34
.5	1.08	1.18	1.20	1.23	1.31
.6	1.17	1.27	1.27	1.28	1.32
.7	1.28	1.39	1.37	1.36	1.36
.8	1.41	1.52	1.49	1.46	1.45
.9	1.56	1.68	1.64	1.59	1.57
1.0	1.73	1.85	1.81	1.76	1.72

It follows from Table 9.1 that the MSE of EWMA is larger than one when $|\theta|$ is large. For instance when $\theta = -.8$ and $\lambda = .25$ the mean square error is 1.94.

9.4.2 ARMA(1,1) process

Now we consider an ARMA(1,1) process of the form

$$X_t - \phi X_{t-1} = \epsilon_t - \theta \epsilon_{t-1}.$$

Then it can be shown that the one step ahead smoother of X_{t+1} and its mean square error are given by

$$\hat{X}_{t+1/t} = \sigma \phi X_{t-1} - \theta \left(X_t - \hat{X}_{t/t-1} \right) / V_t$$

and

$$V_t = \sigma^2 \left(1 + \theta^2 \right) - \sigma^2 \theta^2 / V_{t-1}$$

respectively, where

$$V_0 = \sigma^2 (1 + 2\theta\phi - \theta)/(1 - \phi^2).$$

As in the previous case it can be shown that V_t converges to σ^2 while the mean square error of the EWMA depends on ϕ, θ and λ. It is of interest to note here that for purely autoregressive processes the optimal linear smoother turns out to be the conditional expectation.

9.4.3 A numerical example

Montgomery and Mastrangelo (1991) used a control chart based on EWMA for a set of data containing 197 observations from a chemical process. This chart (not presented here to save space) identifies observations 4, 5, 9, 33, 44, 59, 65, 108,148, 173, 174, 183, 192 and 193 as out-of-control points with the optimal value of the smoothing constant. Moreover, examination of the autocorrelation structure leads to an AR(2) model

$$\hat{X}_t = 54.5973 + .4260 X_{t-1} + .2538 X_{t-2}$$

for these data. The residuals from this model are uncorrelated.

Figure 9.1 presents the 197 observations from the same process, along with the optimal smoother with two sigma control limits. This chart identifies observations 6, 33, 44, 45, 61, 65, 108, 174, 192 and 193 as out-of-control points but does not detect any out-of-control conditions between observations 108 and 174. This is because different methods are used to estimate σ^2 (one step ahead forecast error variance) on the control charts; this can result in observations being classified differently.

As discussed in the paper by Montgomery and Mastrangelo (1991) and Lowry and Montgomery (1995), the EWMA control chart is a very useful procedure that can be used when the data from a process are autocorrelated. However, it should be noted that the EWMA smoother

(i) is just an approximation for any process other than an ARIMA (0,1,1) and does not have any optimality property if used for other models,

(ii) has the difficulty of choosing the smoothing constant using an iterative procedure.

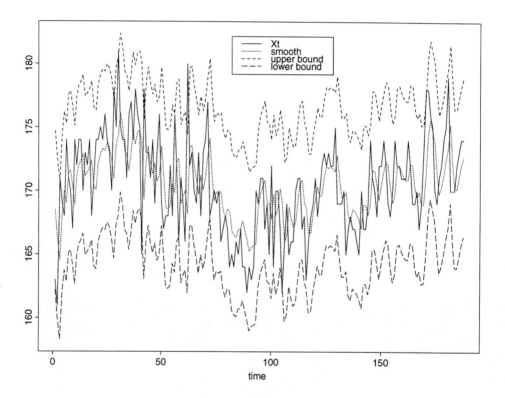

Figure 9.1: Chemical process observations with the optimal smoother

9.5   Summary and Concluding Remarks

Estimating function method is used to construct an optimal smoother for correlated data. As illustrated in Section 9.4, our examples show that the optimal smoother has advantages over the EWMA for a class of linear processes. For a given set of observations one can easily build a model by examining the autocorrelation structure and obtain the corresponding optimal smoother along with the control limits.

Acknowledgement. B. Abraham was partially supported by a grant from the Natural Sciences and Engineering Research Council of Canada.

References

1. Abraham, B., and Kartha, C. P. (1978). Geometric moving charts and Time Series; *ASQC Technical Conference Transactions.*

2. Abraham, B., and Kartha, C. P. (1979). Forecast stability and control charts. *ASQC Technical Conference Transactions.*

3. Abraham, B. and Ledolter, J. (1986). Forecast functions implied by autoregressive integrated moving average models and other related forecast procedures, *International Statistical Review,* **54(1)**, 5–66.

4. Abraham, B. and Thavaneswaran, A. (1991). A Nonlinear time series model and estimation of missing observations, *Annals of the Institute of Statistical Mathematics,* **43**, 493–504.

5. Abraham, B., Thavaneswaran, A. and Peiris , S. (1997). On the prediction for a nonlinear time series models using estimating functions, *Institute of Mathematical Statistics Lecture Notes Series* (to appear).

6. Brockwell, P. J. and Davis, R. A. (1991). *Time Series: Theory and Methods,* Second Edition, New York, Springer Verlag.

7. Godambe, V. P. (1985). The foundations of finite sample estimation in Stochastic Processes, *Biometrika,* **72**, 419–429.

8. Heyde, C. C. (1987). On combining quasi-likelihood estimating functions, *Stochastic Processes and Their Applications,* **25**, 281–287.

9. Hunter, J. S. (1986). The exponentially weighted moving average, *Journal of Quality Technology,* **18**, 203–209.

10. Lowry, C. A. and Montgomery, C. D. (1995). A review of multivariate control charts, *IEEE Transactions,* **27**, 800–810.

11. Montgomery, D. C. and Mastrangelo, C. M. (1991). Some statistical process control methods for autocorrelated data, *Journal of Quality Technology,* **23**, 179–268.

12. Roberts, S. W. (1959). Control chart tests based on geometric moving averages, *Technometrics*, **1**, 239–251.

13. Thavaneswaran, A. and Abraham, B. (1988). Estimation of nonlinear time series models using estimating equations, *Journal of Time Series Analysis*, **9**, 99–108.

14. Thavaneswaran, A. and Thompson, M. E. (1988). A criterion for filtering in semimartingale models, *Stochastic Processes and Their Applications*, **28**, 259–265.

ss *Control and Its Applications in*

es

ae J. Kim

ity, Seoul, Korea

Abstract: The pıı.. ple of Statistical Process Control (SPC) and the construction of an SPC system are introduced in this paper. The goal of SPC activities is explained, and the steps for continuous improvement effort are described. Also the flow of an SPC system and a flow-chart of problem-solving in this system are sketched with figures and explained. Korean experience in industry for SPC is illustrated with a real case study.

Keywords and phrases: Continuous improvement, problem-solving, statistical process control

10.1 Introduction: Definition of SPC

Based on our consulting experience in Korean industries, the meaning of SPC may be explained as follows:

S (statistical): By the help of statistical data and statistical analyzing methods,

P (process): understanding the present process capability and quality specifications,

C (control): control the quality to meet the specifications with minimum variations.

In every process there are inherent variations in quality. SPC intends to minimize the variations with minimum cost. As shown in Figure 10.1, there are four major factors that are most important for successful continuous improvement effort in SPC activities. The four major factors are as follows.

1. **Education and training:** education of statistical thinking, quality control (QC) tools, SPC methods and other scientific management methods such as total productive maintenance (TPM), just-in-time (JIT), value engineering (VE) and industrial engineering (IE).

2. **Total participation and standardization:** 100% participation of all employees, self-motivation atmosphere, suggestion system, and full participation in making and observing standards.

3. **Quality improvement teamwork:** small group activities such as quality circle, quality improvement team, task-force-team and cross-functional team.

4. **Using statistical methods:** proper use of statistical tools such as control charts, process capability index, reliability, correlation and regression methods, design of experiments, etc.

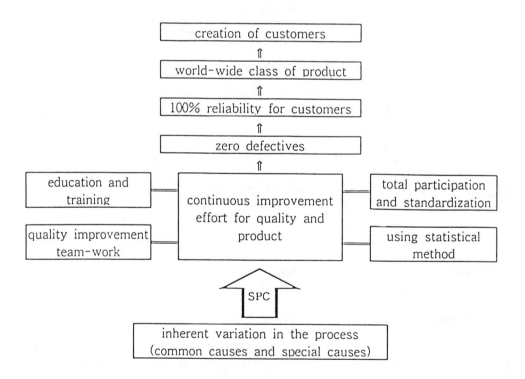

Figure 10.1: The goal and key factors of SPC

10.2   Flow of an SPC System

SPC is for prevention of out-of-specification product, and it is not for detection or inspection. It is much more effective to avoid waste by not producing unusable output in the first place. In the past, the whole manufacturing process mainly depended on production to make the product, on quality control to inspect the final product, and screen out items not meeting specifications, which is wasteful and costly. SPC is a feedback system for prevention. Four elements of SPC system determine the effectiveness of the system. They are as follows:

1. **Process:** The whole combination of suppliers, design quality, input materials, manufacturing methods, manpower, measurement system, machines, environment that work together to produce output.

2. **Information about process performance:** Information about the actual performance of the process can be learned by studying the process output. Quality measurement, evaluation and comparison of the produced output are necessary.

3. **Diagnosis:** Diagnosis on the quality problems, and statistical analysis to find out the causes of quality variations. The causes are divided into common and special causes.

4. **Action on the process:** Actions based on the diagnostic results are required. For common causes, systematic action would be necessary by managers, and for special causes, action by people close to the process can be taken. Some adjustments may be necessary for design quality.

Such an SPC system is explained in Figure 10.2.

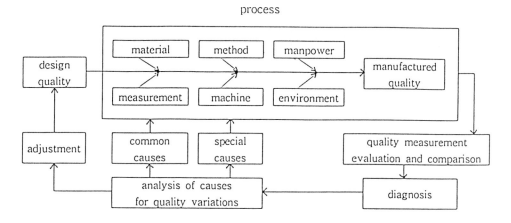

Figure 10.2: Flow of an SPC system

10.3   Team Effort for Quality Improvement in SPC

There are basically two types of teams for quality improvement in SPC. One is the well known QC circle whose members are mainly line workers. Another is the team organized by middle managers such as engineers and researchers. Sometimes a mixed type is desirable depending on the situations encountered. Figure 10.3 shows three types of small group organization for quality problem solving.

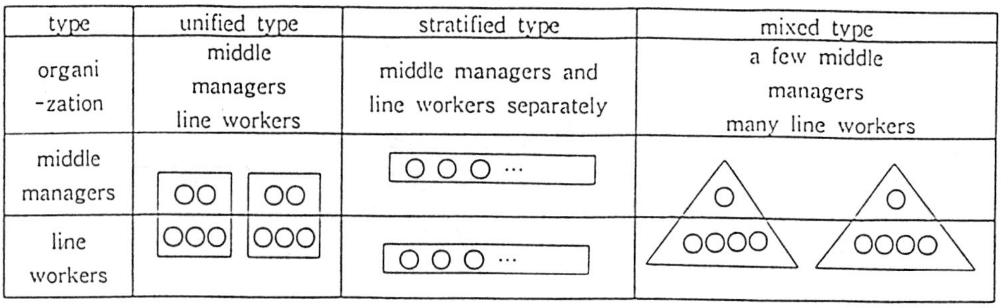

type	unified type	stratified type	mixed type
organi -zation	middle managers line workers	middle managers and line workers separately	a few middle managers many line workers
middle managers	⬜ OO ⬜ OO	⬜ O O O ⋯	△ O △ O
line workers	⬜OOO⬜ ⬜OOO⬜	⬜ O O O ⋯	△OOOO△ △OOOO△

Figure 10.3: Organization of small groups for quality improvement

The authors have been involved in education and consultation for quality programs of many companies for the last several years. From our experience, the following strategy is a good way of planning and implementation for quality improvement by middle managers. It consists of seven steps.

Seven steps of quality improvement program for SPC:

1. Top manager's commitment for quality policy, and organization of quality. improvement team which will achieve quality target.

2. Quality evaluation and selection of projects.

3. Setting up goals for the team to achieve.

4. Findings of trouble sources and determination of possible counter-measures.

5. Implementation of counter-measures and preparation of team reports.

6. Holding a report session in place where all managers participate, and recognition of team efforts for quality improvement results.

7. Evaluation of the gains, standardization of the results and planning for next team efforts.

It is of interest to study the general procedure of quality problem solving. Constant watch of quality variations is necessary in SPC. Such constant watch can be done by statistical tools such as control charts, graphs, and process capability indices. If a quality problem occurs, in general, the following procedure is applied to solve the occurred quality problem. This procedure is explained in Figure 10.4. When a quality problem is detected, the present situation and the causes of trouble are investigated through data collection and analysis. For data analysis, statistical computation using statistical software is needed, since we should ordinarily handle large sets of past data which are related to the trouble sources.

Based on the findings of trouble sources with quality, some counter-measures to solve the problem are considered and implemented. If the problem is solved by these counter-measures, the flow goes to the usual checking process for the mean and variation of quality. If the problem is not solved, some type of experimental designs should be adopted to find either the optimum operating conditions or another way of handling the situation. Taguchi's robust design is a powerful tool for this stage. After such activity, one should check again whether the quality problem has been solved or not. If not, quite different counter-measures such as change of manufacturing engineering or facility replacement should be studied. For more details on quality improvement efforts and other SPC related matters, see Park (1990, 1993, 1996), and Park et al. (1995, 1997).

10.4   Computerization of SPC and Its Difficulties

In order to maintain a good SPC system, we need to have an integrated computerized system for process control, in which several types of statistical computation are necessary. Figure 10.5 shows the rough sketch of the system.

In Figure 10.5, there are 5 systems in which some types of statistical computerization is needed. They are quality design, quality tracking, operating condition monitoring, quality evaluation and quality analysis. If a company wants to have a good system of process control, it must have a good statistical software which is built in its process control facility. Many Korean companies do not have a good integrated computerized SPC system, which may be the case in many other countries as well. The reasons why there are some difficulties to have a good SPC system are as follows.

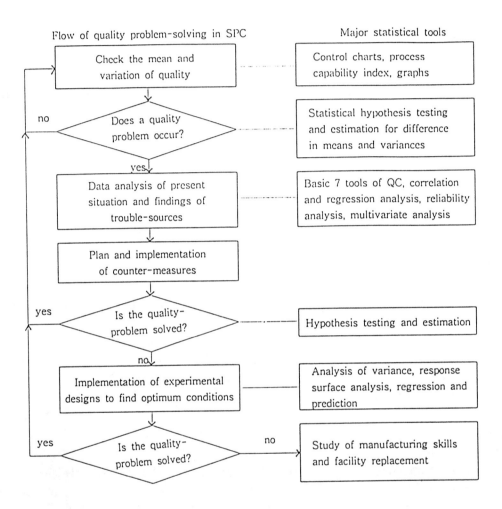

Figure 10.4: Flow-chart of quality problem solving in SPC

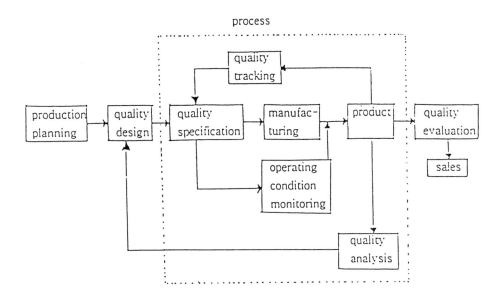

Figure 10.5: SPC integrated system

1. The company does not have a good statistical software written in easy terms nor in the Korean language, which can be conveniently used for engineers and workers in factories.

2. Engineers and workers do not understand statistical tools, so they are afraid of using statistical methods.

3. Managers are not eager to invest in computerization, and to train statistical tools to engineers and workers.

4. SPC related departments in universities such as statistics, industrial engineering and management science do not educate their students well to cope with the statistical computation and quality problems. Therefore, when they face the problems in companies, they are not capable of handling the problems well.

However, recently a number of companies are trying to have a good SPC system which is a positive sign. From the view point that the coming century is the century of quality and productivity, we believe that, to be a competitive company in the international market, the company should have a good SPC system.

10.5   Case Study

This is a case study of an oil refinery company in Korea. One of the authors has been consulting with this company since 1991, and this is a part of one team's report in 1992. The company now has 28 SPC teams, and each team is supposed to solve at least one problem a year.

This company controls the color quality of SW-kerosene which is produced in the kerosene sweetening process U-1700 by the x-Rs control chart. At that time the color quality was not satisfactory, and an SPC team was formed to improve the quality. The quality is expressed by the saybolt number (y) which is a 'larger-the-better' characteristic. It should be at least 20 for customer satisfaction. At the beginning the team considered more than 10 factors, but by a brain-storming discussion they reduced them to 6 factors which might influence the color quality. Among these factors, 4 factors are feeding conditions and 2 factors are operating conditions in the reactor as shown in Figure 10.6.

In order to investigate the relationship between the product color y and the six factors $x_1 - x_6$, data from the previous two months were collected as shown in Table 10.1. The value in each piece of data represents the average of each variable in a day.

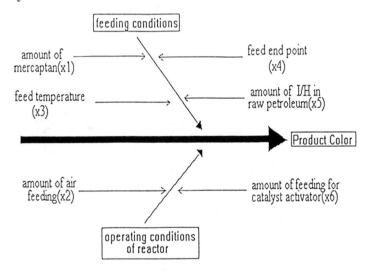

Figure 10.6: Causes-and-effects diagram for color quality

For data analysis, they found the correlation matrix among 7 variables and the means and standard deviations of these variables as in Table 10.2. Note that the average of y is about 21.9 and the standard deviation is 3.2 which is too large for desirable quality control. In this table, * shows 5% significance, ** 1% significance and + 10% significance.

Table 10.1: SW-kerosene color data

data number	x_1 feed mercaptan wt ppm	x_2 air feeding %	x_3 feed temperature ^{0} C	x_4 feed E. P. ^{0} C	x_5 catalyst activator wt ppm	x_6 I/H petroleum Vol %	y color saybolt number
1	111.6	130	46	247	62	100.0	21
2	91.5	135	50	251	70	94.2	21
3	90.8	135	50	245	70	94.1	21
4	80.8	175	43	247	80	16.3	13
5	82.8	145	46	248	80	22.0	20
6	78.6	145	46	255	80	22.0	20
7	93.2	120	43	254	80	22.0	21
8	93.7	120	38	253	80	16.4	23
9	95.9	120	38	253	80	10.9	21
10	97.6	115	42	256	70	10.8	27
11	100.0	115	42	254	70	10.8	26
12	100.0	115	50	260	50	10.0	26
13	100.0	115	51	264	40	0.9	26
14	100.0	119	47	268	40	0.9	27
15	98.7	119	46	259	40	1.0	28
16	100.0	119	44	268	40	1.4	27
17	94.0	135	45	261	50	2.8	23
18	138.7	160	53	268	70	96.9	17
19	157.0	143	56	264	36	95.1	20
20	160.0	150	55	265	36	94.4	18
21	160.0	137	47	261	65	94.4	21
22	119.0	128	59	274	40	86.2	21
23	116.0	139	53	273	52	13.8	19
24	118.6	139	52	265	52	13.8	17
25	117.7	139	50	270	40	13.8	17
26	117.5	150	50	273	40	13.8	21
27	114.3	173	40	259	40	1.9	18
28	133.2	115	58	270	60	9.6	19
29	129.5	107	53	274	40	9.6	23
30	128.4	128	59	262	40	9.6	21
31	121.4	128	57	268	40	11.4	21
32	150.0	110	51	269	40	2.7	23
33	150.8	110	50	276	40	3.0	23
34	150.0	110	49	275	40	3.0	24
35	106.9	110	49	271	40	3.0	26
36	124.3	120	53	271	40	3.0	20
37	139.2	105	52	274	40	2.2	24
38	139.7	119	53	270	40	1.2	22
39	140.4	119	52	266	40	1.2	22
40	140.4	119	48	276	40	1.2	23
41	138.9	119	50	270	40	1.3	26
42	139.5	119	51	271	40	1.8	23
43	105.9	133	52	260	40	4.9	20
44	110.0	133	52	269	40	3.9	23
45	101.6	133	53	273	40	3.9	22

Table 10.2: Correlation matrix, mean and standard deviation

	x_1	x_2	x_3	x_4	x_5	x_6	y
x_1	1.0	-0.17467	0.53243	0.63156	-0.55272	0.15361	-0.07459
x_2	-0.17467	1.0	-0.07690	-0.34252	0.25102	0.39262	-0.74801
x_3	0.53243	-0.07690	1.0	0.55343	-0.56221	0.16607	-0.18758
x_4	0.63156	-0.34252	0.55343	1.0	-0.76345	-0.33973	0.17077
x_5	-0.55272	0.25102	-0.56221	-0.76345	1.0	0.28486	-0.26357
x_6	0.15361	0.39262	0.16607	-0.33973	0.28486	1.0	-0.35967
y	-0.07459	-0.74801	-0.18758	0.17077	-0.26357	-0.35967	1.0

variable	mean	standard deviation
x_1	117.29079	22.80504
x_2	28.26666	16.31419
x_3	49.33333	5.01814
x_4	264.00000	8.81373
x_5	50.95555	15.61558
x_6	23.04657	34.12817
y	21.88889	3.17104

Observe that y is significantly correlated with x_2, x_5 and x_6. However, since some strong correlation exists among the factors, it is difficult to say that x_2, x_5 and x_6 are only the important factors that affect the color quality y. In such situations one way to find a suitable model for y is to use the stepwise regression approach. Table 10.3 shows the results of stepwise regression for the data set of Table 10.1.

Table 10.3: Results of stepwise regression

step	variable entered	regression equation·	coeff. of determination (R^2)
1	x_2	$\hat{y} = 40.53796 - 0.14539x_2$	0.560
2	x_2, x_3	$\hat{y} = 48.68697 - 0.14908x_2 - 0.15580x_3$	0.620
3	x_2, x_3, x_5	$\hat{y} = 55.95178 - 0.13578x_2 - 0.26877x_3 - 0.06648x_5$	0.680
4	$x_2, x_3, x_5,$ x_1	$\hat{y} = 58.68172 - 0.13867x_2 - 0.21791x_3 - 0.08392x_5$ $- 0.3393x_1$	0.725
5	$x_2, x_3, x_5,$ x_1, x_6	$\hat{y} = 565.0461 - 0.15324x_2 - 0.26454x_3 - 0.11058x_5$ $- 0.04517x_1 - 0.02085x_6$	0.755

From the results of stepwise regression in Table 10.3, they decided to adopt the third equation with x_2, x_3 and x_5, since x_1 and x_6 which entered into the equation at the 4th and 5th steps are not easy to control, and the R^2 value of 0.68 is large enough for a regression model. Therefore, the final equation used was

$$\hat{y} = 55.95178 - 0.13578x_2 - 0.26877x_3 - 0.06648x_5.$$

The signs of the coefficients are consistent with those of the correlation coefficients in Table 10.2. Also note that the coefficients of x_2, x_3 and x_5 are not fluctuating much in the five regression equations, which means that the equations are quite stable. For instance, the coefficients of x_2 are -0.14539, -0.14908, -0.13578 and -0.15324.

The above regression equation tells that, in order to increase the saybolt number of color quality, the operating levels of x_2, x_3 and x_5 should be as low as possible. Previously they had used the conditions $x_2 = 140\%$, $x_3 = 50°C$, and $x_5 = 50$ppm. The past data show that x_2, x_3 and x_5 could be as low as 105%, 42°C and 36ppm, respectively. In order to find the optimum operating conditions for x_2, x_3 and x_5, they decided to run an experimental design with a small number of experiments. They chose the orthogonal array experiment, $L_9(3^4)$. The levels of the three factors chosen were as follows.

$$
\begin{aligned}
A(x_2) : A_0 &= 106\%, \ A_1 = 123\%, \ A_2 = 140\% \\
B(x_3) : B_0 &= 42°C, \ B_1 = 46°C, \ B_2 = 50°C \\
C(x_5) : C_0 &= 36\text{ppm}, \ C_1 = 43\text{ppm}, \ C_2 = 50\text{ppm}
\end{aligned}
$$

Table 10.4: Data of $L_9(3^4)$ experiment for color quality

exp. no.	column number 1	2	3	4	data : y_1 (saybolt number)	yield : y_2 (%)
1	0	0	0	0	28	95.5
2	0	1	1	1	25	96.3
3	0	2	2	2	24	96.2
4	1	0	1	2	26	96.6
5	1	1	2	0	23	97.1
6	1	2	0	1	24	96.3
7	2	0	2	1	23	96.5
8	2	1	0	2	22	96.9
9	2	2	1	0	19	96.1
assignment of factors	A	B	C	(e)		

The data obtained are listed in Table 10.4. Since the yield data y_2 is also critical, the yield data were also observed. The other variables such as x_1, x_4, x_6 are fixed at certain levels during the experiments.

The ANOVA table in Table 10.5 indicate that A and B are the only significant factors and similar results hold for the yield data. In order to locate the optimum operating conditions for A and B, they decided to use the response surface contours. By using the following linear transformations,

$$x_1 = [(\text{amount of air feeding}) - 123\%]/17$$
$$x_2 = [(\text{feed temperature}) - 46°C]/4$$

they obtained the following second order polynomial models,

$$\hat{y}_1 = 23.8889 - 2.1667x_1 - 1.6667x_2 - 0.8333x_1^2 + 0.6667x_2^2 \text{ with } R^2 = 0.96$$

$$\hat{y}_2 = 97.0444 + 0.2500x_1 - 0.4167x_1^2 - 0.5667x_2^2 - 0.2750x_1x_2 \text{ with } R^2 = 0.98.$$

The response contours of $\hat{y}_1$ and $\hat{y}_2$ are shown in Figure 10.7.

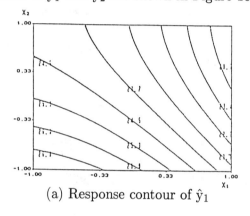

(a) Response contour of $\hat{y}_1$

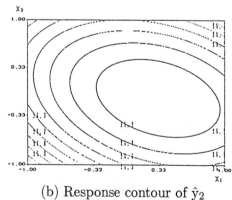

(b) Response contour of $\hat{y}_2$

Figure 10.7: Response contours of $\hat{y}_1$ and $\hat{y}_2$

The team wanted to find the conditions of x_2 and x_3 which maximize the color quality under the restriction that the yield is at least 96.3%. By superimposing one contour on the other, they found that $x_1 = -0.33$ (amount of air feeding $x_2 = 110\%$) and $x_2 = -1$ (feed temperature $x_3 = 42°C$) is the optimum operating condition. Since x_1, x_4, x_5 and x_6 are not significant, they decided to keep the levels of these variables at the current operating conditions. Therefore, the overall standardized conditions for the six variables are;

$$x_1 = 110\text{ppm}, \ x_2 = 110\%, \ x_3 = 42°\text{C}, \ x_4 = 270°\text{C}, \ x_5 = 40\text{ppm}, \ x_6 = 4.0\%.$$

They ran several confirmatory tests at these conditions, and found that the average saybolt number is 27.0 with standard deviation 1.5, and the average yield is 96.5% with standard deviation 0.2, which are quite satisfactory.

10.6 Summary and Concluding Remarks

In this paper we introduced an SPC System and considered steps for Problem Solving and Continuous Improvement in the context of this system. A case study from Korean industry was also presented,to illustrate the usage of this system.

Acknowledgements. The present studies are supported in part by the Basic Science Research Institute Program, Ministry of Education, 1996, Project Number BSRI-96-1415. The authors would like to thank the referees for their valuable comments.

References

1. Park, S. H. and Park, Y. H. (1995). *Statistical Quality Control*, Minyoung-sa, Seoul.

2. Park, S. H., Park, Y. H. and Lee, M. J. (1997). *Statistical Process Control*, Minyoung-sa, Seoul.

3. Park, S. H. (1990). *Applied Design of Experiments*, Youngjimoonwha-sa, Seoul.

4. Park, S. H. (1993). *Quality Engineering*, Minyoung-sa, Seoul.

5. Park, S. H. (1996). *Robust Design and Analysis for Quality Engineering*, Chapman & Hall, London.

11

Multivariate Statistical Process Control of a Mineral Processing Industry

Nihal Yatawara and Jeff Harrison

Curtin University of Technology, Perth, Western Australia

Abstract: Multivariate Statistical Process Control procedures are becoming increasingly popular in process industries due to the need for monitoring a large number of process variables simultaneously. Although extensions to classical univariate control charts such as Shewhart, CUSUM and EWMA to multivariate situations are possible, more recently introduced statistical projection methods such as the Principal Component Analysis (PCA) and Partial Least Squares (PLS) seem to be more suitable for dynamic processes with input-output relationships. These methods not only utilize the product quality data (Y), but also the process variable data (X). This paper gives an overview of the PCA and the PLS methods and their use in monitoring operating performance of a crusher used in a mineral processing plant.

Keywords and phrases: Diagnostic plots, latent variables, partial least squares, predictions, process control

11.1 Introduction

Most mineral processing industries now use computers in their operations and hence large amounts of data are collected routinely. This data may contain a wealth of information that could be used on-line or off-line to monitor the performance of the operations and to improve upon it. The data also provide a measure of the current state of a process, which can be useful in implementing process control procedures. Often, combined engineering and statistical process control techniques are sought in this respect. Another use of data would be to build suitable models giving direct or indirect input-output relationships to better understand the process behaviour.

The classical approach to handling process control through univariate control charting techniques such as the Shewhart, CUSUM, EWMA or multiple linear regression, suffer from several drawbacks. In the presence of complex relationships among process variables as in mineral processing industries, these offer very little in terms of performance monitoring. This is partly because of the difficulties involved in identifying the true state of a process from a display of several variables.

On the other hand the existence of complex correlations among variables poses another serious problem for which univariate control charts render no feasible solution. Most variables in mineral processing industries are correlated as they vary jointly with one another. Usually, the ore properties such as the hardness and the initial particle size affect the feed rate, the pressure exerted by the crushers and the amount that goes through the bypass etc. Missing data and excessive noise can also create further problems, making it difficult to extract accurate information. In engineering process control, plant engineers use phenomenological assumptions to find solutions to these problems.

In this paper the complementary approach based on statistical techniques such as Principal Component Analysis (PCA) and Partial Least Squares (PCS) are discussed. Both these methods are multivariate statistical methods which project information on to low dimensional subspaces containing only the relevant information about the process.

11.2   PCA and PLS Methods

PCA is a standard multivariate technique and is discussed in many textbooks [Mardia et al., (1982), Jackson, (1991)]. Comparatively, the PLS method is more recent, and its applications which originated in the chemometrics area were largely due to S. A. Wold and co-workers [for example see Wold et al. (1984)].

11.2.1   Principal component analysis

Consider a matrix of observations Z. In principal component analysis, the variance of this matrix is explained in terms of new latent variables which are called principal components. The first principal component variable is the linear combination of Z that has the greatest variance. The second principal component variable (PCV) is the linear combination with the next greatest variance among coefficient vectors of unit length that are orthogonal to the first coefficient vector. Proceeding in this manner, one can obtain k possible principal component variables. In mathematical symbols, the calculated PC's can be given by

$$t_1 = p_1'Z \quad \text{subject to } |p_1| = 1,$$
$$t_2 = p_2'Z \quad \text{subject to } |p_2| = 1 \text{ and } p_2'p_1 = 0.$$

etc.

It is easily shown that the principal component loading vectors p are the eigenvectors of the covariance matrix $\sum$ of Z and the corresponding eigenvalues λ_i are the variances of the principal components. Using the loading vectors and the PC's the observation can be written as

$$Z = \sum_{i=1}^{k} t_i p_i' + E \tag{11.1}$$

where k, is the number of Principal Components obtained and E is the residual matrix. In practice the first two or three PC's would explain most of the variability in the data.

In finding the PCV's, one proceeds until the percent variation in the $(k+1)^{\text{th}}$ PC is not significant. Cross-validation is often employed as a suitable procedure for selecting the number of PCV's [MacGregor et al. (1994)].

11.2.2   Method of partial least squares (PLS)

PCA is a method that can be used to select a set of PCV's that express most of the variability among a set of input variables on its own. However, in mineral processing applications, there are input as well as output matrices present and we need to find latent variable models which are capable of not only expressing the variability of input variables but which are most predictive of the output.

Multiple linear regression seems to be the simplest method that can be adopted to fulfill these requirements. However, it does not seem appropriate in the presence of noisy correlated data as in this case. Even the recent extensions of multiple linear regression methods such as ridge or regularization do not seem to answer the question of dimensionality reduction. PLS method seems to be most promising under these circumstances. It creates a set of orthogonal latent vectors from a block of input variables (X block) that maximises the covariance between those vectors and a block of output variables (Y block) (Hoskuldson (1988)). The output variables, which are associated with product quality are included in Y and the others are columns of X.

The mathematical details of PLS have been discussed in detail in a number of articles [Geladi et al. (1986), Hoskuldson (1988), Kresta et al. (1991)]. In summary the PLS algorithm can be described as follows: [see MacGregor (1994)]

1. Set u equal to a column of Y.

2. Obtain w by $w' = u'X/u'u$ (i.e. regress columns of X on u).

3. Normalize w to unit length.

4. Calculate the score vector by $t = Xw/w'w$.

5. Obtain $q' = t'Y/t't$ by regressing columns of Y on t.

6. Calculate a new vector u by $u = Yq/q'q$.

7. Check for convergence, if yes, go to 8 if no go to 2.

8. X loadings: $p = X't/t't$.

9. Regression: $b = u't/t't$.

10. Calculate the residual matrices:

$$
\begin{aligned}
E &= X - tp' \text{ and} \\
F &= Y - btq'
\end{aligned}
$$

11. Repeat steps 1-10 after replacing X and Y by E and F respectively.

One can use NIPALS (non-linear iterative partial least squares) algorithm to perform the above computations. Alternatively the PLS procedure given in the chemometrics tool box in MATLAB can be used.

11.3   PLS Monitoring Methods and Diagnostics

Fundamentally, the PLS monitoring method uses Shewhart charting techniques but plots latent variables against each other rather than a quality measure against time. To be practically useful, the process variability should be described with three or less latent variables to allow simple charting. Also, a data sample when the process is in "control" is needed as a reference data set, to determine control regions. If these data are reasonably accurately measured then the latent variables may be considered to follow a multivariate normal distribution and confidence intervals can be formed. [Kresta et al. (1991)].

As we gather new observations, the corresponding latent variables can be calculated using

$$
\begin{aligned}
t_1 &= Xw_1, \quad t_2 = (X - t_1 p_1 T) w_2 \text{ and} \\
t_3 &= ((X - t_1 p_1) - t_2 p_2) w_3
\end{aligned}
$$

etc. and they can be plotted in the in-control plots. Observations which fall outside the in-control region indicate a departure from "normal" operating mode. Furthermore, squared prediction errors defined by

$$
SPE_y = \sum_{i=1}^{m} \left(Y_i - \hat{Y}_i \right)^2,
$$

where m is the number of Y variables, can be plotted against the latent variables. An observed increase in the value or the SPE_y indicates a change in the X variables, not explained by the reference set. As discussed in Krista et al. (1994), these simple diagnostic capabilities are some of the attractive features of the PLS method that have captured the attention of the practitioners.

11.4 Application of PLS for HPRC Data

11.4.1 The processing plant and data

The mineral processing plant under investigation is in Western Australia. The study here was done mainly to develop an efficient monitoring mechanism for the rolls crusher of the operation. As shown in Figure 11.1a, Ore is conveyed from a storage bin into a surge bin above two cylindrical crusher rollers that are powered by individual electric motors. The Ore is drawn into the crushing gap by the opposing rotation of the rollers. As the Ore is fed through the rollers they are forced apart and an opposing hydraulic pressure is applied to maintain the width between them. Instruments are located to measure the important parameters of the operation (Figure 11.1b). The HPRC lA and lB differ only in that the powered rollers for lB have variable speed drives to allow for variation in total Ore feed through the crushing plant area, and this instrument is represented by the variable VSPD1B.

The other variables are: Feed lX - the (inferred feed rate calculated from conveyor speed (tonnes per hour)), Gap lX - the gap width between the rollers (millimeters), Pres 1X1 and Pres 1X2 - opposing hydraulic pressures for each roller (Bar), PWR1X and PWRlXF - power draw for each roller motor (kilowatts). Note: $X = A$ or B.

There is a conveyor that splits some of the Ore from the stream to the HPRC feed bins and is directed around the HPRC plant and onto conveyor CV53. This is the bypass Ore and is preset to a value depending on the total plant feed target and the current operating conditions of the HPRC and is called BYPASS (tonnes/hr). The screening plant divides the Ore stream into 3 sizes: > 18 mm (cv5); between 3 mm and 18 mm (cv3); and < 3 mm. The < 3 mm material is called "degrit" and is considered as the critical Ore size for downstream plant efficiencies.

The data for the study were taken from 4 days, consisting of 1440 observations taken at 1 minute intervals on each day. Since the measurements are taken from 3 locations a lag correction was introduced, by treating CV53 as the reference point. Day 4 data set appeared to be the best candidate for a reference set and all other days were suitably calibrated (see Figure 11.2). The annotations for each day's data are given in the following table.

Table 11.1: Annotations for data sets

1/8	1A in poor condition due to it being nearly full Y worn. Segment change imminent. 1B operating moderately well.
3/8	1A off line for segment change, 1B operating relatively poorly until later in day, variable ore dayshift
12/9	The ore was relatively soft and the pressures were reduced on 1B to try a new mode of operation
13/9	1A hardly operated during the day because of a segment change but 1B appeared to be operating well

11.4.2   Application of PLS for crusher performance monitoring

The application of PLS to process monitoring requires a reference data set to establish "in-control" boundaries. The data provided came from four different days, with differing operating conditions, none of which showed stability. However by comparison, set four was found to be the most suitable as a reference data set. The analysis was conducted in two parts to demonstrate that a control chart could be produced when either both crushers or only one crusher was operating. These are referred to as dual and single crusher studies respectively. The dual crusher study was not useful for examining the diagnostic capability of the PLS method because crusher 1A was often faulty. Hence the results presented are mainly from the single crusher study with one (degrit) or two (degrit and crushed) response variables.

The F values associated with the PLS algorithm for both studies indicate either 3 or 2 latent variables as significant. (see Figures 11.3 and 11.4 in the Appendix). The t-value plots and the weight plots (see Figures 11.5 and 11.6) indicate that the single response variable model with three latent variables produced a better reference area. Also these plots clearly show two distinct clusters indicating two modes of operation. The SPE charts seem to show that some observations in the reference data are not predicted by the model. Both plots show significnt "drifts" of the observations into the third quadrant. Thus using this model as a reference, the PLS algorithm was run on the data sets 2 and 3 and the corresponding plots (see Figures 11.7 and 11.8) were compared with those of the reference set. A summary of the results for the data set 2 is given below.

The results indicate an increase in PRES1B1 and PRES1B2, when compared with the reference data set. A similar analysis from the data set 3 indicated that the BYPASS was significantly lower than the reference set.

Plot	$t_1 \times t_3$		$WGT_1 \times WGT_3$
Observations	too small clusters, one above and one below the reference area	GAP1B	increase or decrease
		FEED1B	increase or decrease
		VSPD1B	increase or decrease
Plot	$t_1 \times t_2$		$WGT_1 \times WGT_2$
Observations	large group of observations along the diagonal project well into the third quadrant	PRES1B1	PRES1B2 increase
		FEED1B	decrease
		PWR1B	and PWR1BF decrease
		GAP1B	decrease
Plot	$t_2 \times t_3$		$WGT_2 \times WGT_3$
Observations	same group and observations as before projecting in the negative t_2 direction	PRES1B1	and PRES1B2 increase
		FEED1B	decrease
		BYPASS	decrease

11.5 Conclusion

The dual crusher study has shown that the PLS method can be successfully applied to HPRC data, for performance monitoring. However, the diagnostics were much better when the weight charts were plotted from a single crusher analysis. The diagnostic tools provided by PLS are not able to specify a cause for an "out-of-control" signal in the process, but they do indicate those process variables that contribute most weight to a particular shift in the observations on the latent variable planes. This information can assist an experienced plant operator to identify actual operating faults. Clearly the PLS control charting model will be effective only if the reference data set is obtained from a known "in-control" period of operation. A real reference data set might produce a model different to what was developed in this study, where "constructed" reference sets were used. One obvious characteristic seen in the latent variable plots for reference data is the two clusterings. As pointed out by Kresta et al. (1991), this might indicate different modes of operation. Further investigation would also be required into the effects of different Ore types (soft or hard) on the PLS model.

Although, the PLS model developed here considered past process data, the model can be adapted to an on-line monitoring system. An obvious problem associated with the display of the control chart with the PLS seems to be its complexity as compared to univariate Shewhart plotting. However, for all practical purposes it would be sufficient to display only the previous p-observations, where p is determined as the period of interest [see Kresta et al. (1991)].

References

1. Geladi, P. and Kowalski, B. R. (1986). Partial least squares: A tutorial, *Analytical Chemica Acta*, **185**, 1–17.

2. Hoskuldsson, A. (1988). PLS regression methods, *Journal of Chemometrics*, **2**, 211–228.

3. Jackson, J. E. (1991). *User's Guide to Principal Components*, New York: John Wiley & Sons.

4. Kresta, J. V., MacGregor, J. F. and Marlin, T. E. (1991). Multivariate statistical monitoring of process operating performance, *Canadian Journal of Chemical Engineering*, **69**, 35–47.

5. MacGregor, J. F., Jaecle, C., Kiparissides, C. and Koutoudi, M. (1994). Process monitoring and diagnosis by multiblock PLS methods, *AiChE Journal*, **40**, 826–838.

6. Mardia, K. V., Kent, J. T. and Bibby, J. M. (1982). *Multivariate Analysis*, London: Academic Press.

7. Wold, S. A., Rube, A., Wold, H. and Dunn, W. (1984). The collinearity problem in linear regression. The partial least squares (PLS) approach to generalized inverses, *SIAM Journal of Scientific Statistical Computing*, **5**, 735–743.

Appendix - Figures

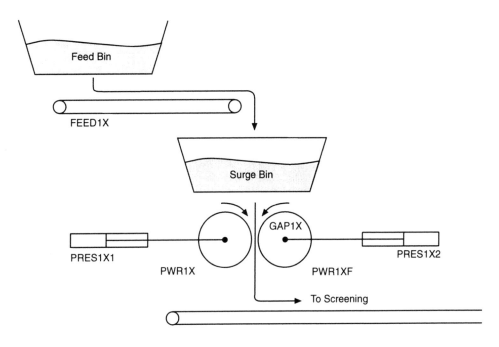

Figure 11.1a: Schematic of the HPRC plant showing process measurement instruments

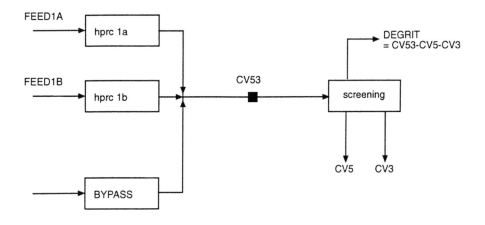

Figure 11.1b: Schematic of the ore crushing process

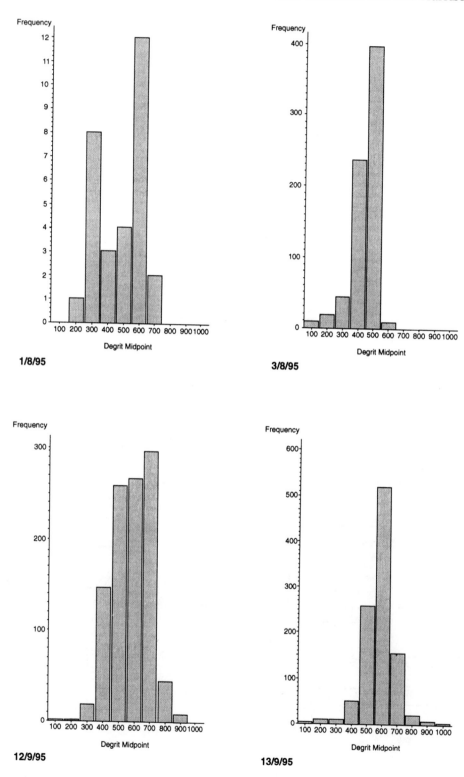

Figure 11.2: Histogram of DEGRIT for crusher 1B each data set

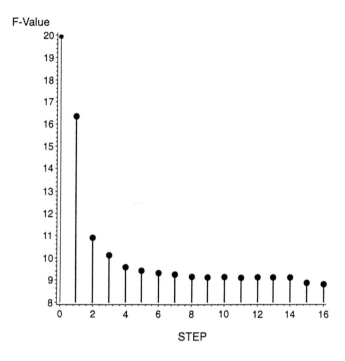

Figure 11.3: F-value by latent variables extracted for dual crushers reference data with 1 response variable and BYPASS as an X variable

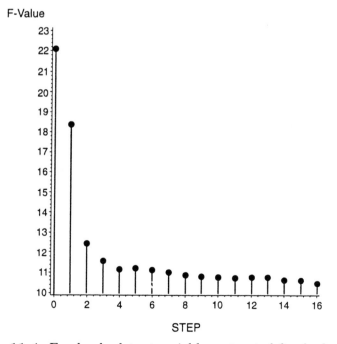

Figure 11.4: F-value by latent variables extracted for dual crushers reference data with 2 response variables including a BYPASS factor

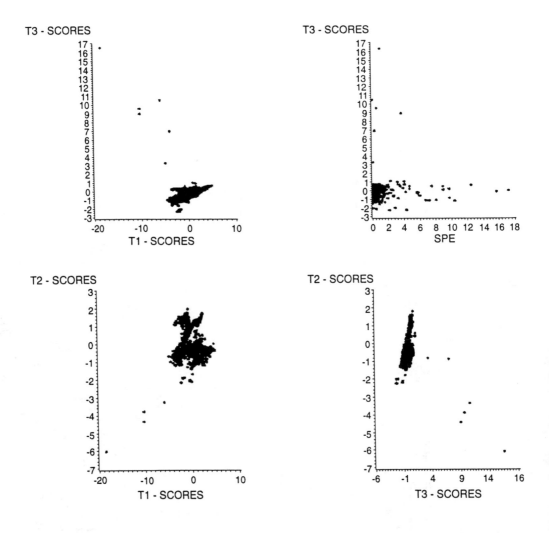

Figure 11.5: Monitoring and SPE charts for reference set - crusher 1B (1 response variable)

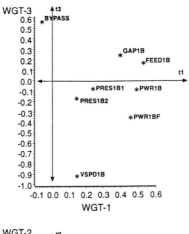

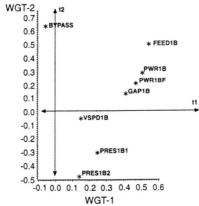

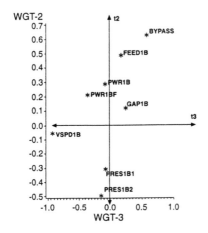

Figure 11.6: Plots of latent variable weights for single crusher 1B (1 response variable)

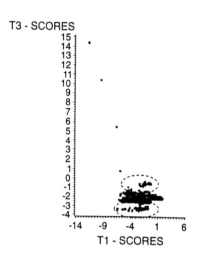
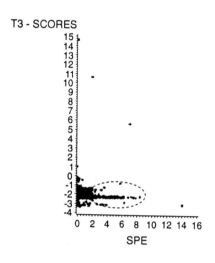
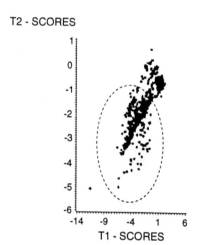
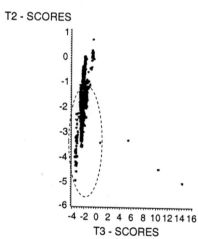

Figure 11.7: Monitoring and SPE charts for 3/8/95 data - crusher 1B (1 response variable)

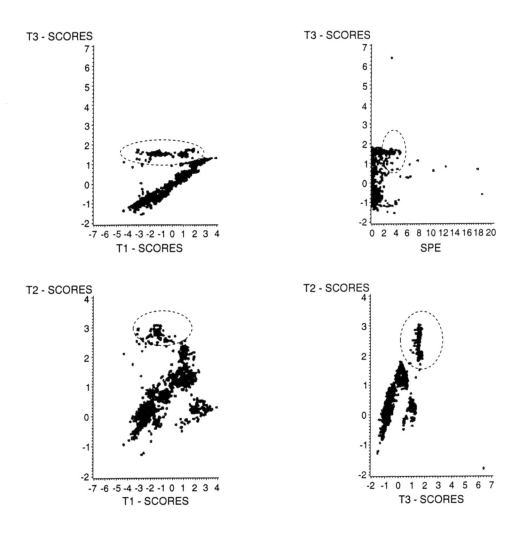

Figure 11.8: Monitoring and SPE charts for 12/9/95 data - crusher 1B (1 response variable)

12

Developing Objective Strategies for Monitoring Multi Input/Single Output Chemical Process

B. K. Pal and V. Roshan Joseph

Indian Statistical Institute, Bangalore, India

Abstract: Processes with multi input and single output are common in chemical industries. Such a process is monitored by taking samples in fixed intervals of time and corrective action is taken whenever the output is found deviating from the specifications. It is required to have an objective criteria for deciding the sampling intervals and the quantum of adjustments to meet the target output. Also if there is a large time lag associated with the change of levels for the output, it is desiarable to get the action in a small interval of time. The methodology is explained through a case study. In the case study the methodology is developed using the mass balancing of the process and the statistical tools like ANOVA, β-correction, regression etc.

Keywords and phrases: Regression, stability concentration, steady state , β-correction, SO_2 gpl

12.1   Introduction

In chemical industries quite often we encounter processes which transform several inputs into a single output. The quality characteristic of the output is required to be maintained at a designed level with certain allowable variations. Obviously this quality is affected by the quality of the inputs as well as the process and environmental conditions. This calls for periodic assessment and adjustment of the adjustable inputs of the process. Experience indicates that often the output cannot be maintained at the desired level [see Pal (1993)]. A typical picture of variations in a chemical process is given in Figure 12.1.

As already mentioned the control procedure followed in most cases comprises of the following

a) Take sample of process output at fixed intervals, say once in an hour and evaluate the output characteristics.

b) Based on the results, adjustments of the input characteristics are made whenever necessary.

output characteristic

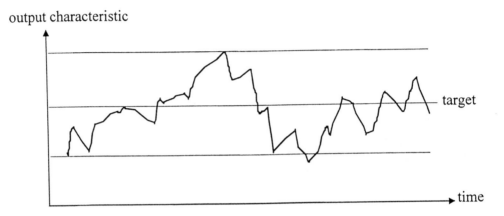

Figure 12.1: A typical graph of a characteristic in a chemical process

Some time is needed for testing the samples. Process adjustments are made only after test results are obtained. If the process is not at a steady state, actions taken may not be appropriate. Further there is also some processing time before the impact of the action is reflected in results. Due to all of the above, some variation from the target is expected. The ideal control procedure should be capable of making the output at target all the time. Practically one should aim at a control procedure which will ensure the characteristic at the target with minimum variation.

Most of the chemical processes are not well engineered for continuos monitoring. While a permanent solution could be modernization of the processes, but this is costly and also requires certain time and investment. This paper attempts to develop an objective strategy for monitoring multi input/single output chemical process using statistical principles without the need for capital investment.

12.2   General Formulation

A multi input/single output chemical process can be shown diagramatically as below

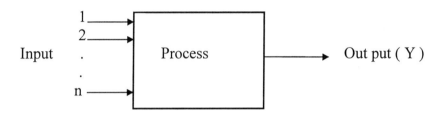

Figure 12.2: Block diagram of a multi input/ single output process

The process input characteristics can be broadly classified under two groups i.e. controllable variables (X) and uncontrollable variables (Z). As the name indicates no control is economically possible on the uncontrollable variables with the present setup of the process. The output (Y) is affected by the controllable and uncontrollable variables. Mathematically we can express Y as a function of X and Z as,

$$Y = f(X, Z). \tag{12.1}$$

One or more of the controllable variables can be adjusted depending on the uncontrollable variables and the rest of the controllable variables so as to get the target output. i.e.,

$$f(X, Z) = T. \tag{12.2}$$

It is not easy to monitor X and Z continuously and to ensure the relationship (12.2) to hold good all the time. Observations were taken at some definite intervals to see the extend of variation in the output. Corrections are effected on the output through the controllable variables if situation demanded. In many chemical processes the deviation of the output from the target cannot be made zero instantaneously due to a time lag associated with the change of levels of the output, which is inherent for a particular process. Suppose Y_0 is the observation of the output at time t_0, then the output at time t, $Y(t)$ can be expressed mathematically as,

$$Y(t) - Y_0 = f(X, Z, t - t_0). \tag{12.3}$$

The above function should be such that as $t \to \infty$, $Y(t) \to f(X, Z)$. i.e., the function which we mentioned before [equation (12.1)] is attained under steady state conditions.

Thus if Y_0 is different from T then adjust X such that $Y(t)$ is at T in a small predetermined interval of time δ, which is practically possible or in other words adjust X such that,

$$f(X, Z, \delta) = T - Y_0. \tag{12.4}$$

There is an unavoidable error in adjusting the process based on the observed output due to the sampling/measurement errors and due to the variations in the output during the time interval of collecting the sample and adjusting the process. Denote the standard deviation of this error as σ. Due to this error, Dr. Genechi Taguchi advises to correct the output by a fraction of the deviation, $\beta(Y - T)$ to minimize the mean square error from the target where,

$$\beta = \begin{cases} 0 & \text{if } (Y - T)^2 \le \sigma^2 \\ 1 - 1/K & \text{otherwise; } K = (\frac{Y-T}{\sigma})^2. \end{cases}$$

Hence adjust X such that,

$$f(X, Z, \delta) = \beta(T - Y_0). \tag{12.5}$$

We will achieve the correction in δ time. After δ time we need to adjust X again so as to satisfy equation (12.1) to maintain the results for a long time.

Thus the control procedure we intend to develop is as follows

i. Objectively decide a sampling interval.

ii. Decide the quantum of action through - correction method.

iii. Develop a relationship between the output and the input either theoretically or empirically.

iv. Adjust the controllable variables to get the action in a small interval of time.

The above methodology can be explained through a case study.

12.3   A Case Study

Sulphur dioxide solution is prepared by mixing sulphur dioxide gas and water in a tank and it is a continous process. The concentration of SO_2 in the solution is to be controlled in 4 ± 1 gpl (gms/litre). The controllable variables of the process are the mass flow rates of SO_2 gas and water. Even though the flow rate can be adjusted to any particular value, it varies over time leading to variations in SO_2 concentration. There are no other variables for this process.

Objective criteria for deciding sampling intervals:

At the time of the study samples were collected in 4 hour intervals to analyze the concentration. To objectively decide the sampling interval we have to understand the smallest interval in which a significant change in the output occurs when the process is left unattended. Samples were taken in 10 minute intervals for about 4 hours without making any deliberate adjustments in the process. The observations are given in Table 12.1.

Table 12.1: Data on SO_2 gpl

S NO.	SO_2 gpl	S NO.	SO_2 gpl	S NO.	SO_2 gpl	S NO.	SO_2 gpl
1	4.22	7	4.35	13	4.42	19	4.22
2	4.35	8	4.41	14	4.42	20	4.28
3	4.29	9	4.42	15	4.42	21	4.28
4	4.35	10	4.48	16	4.48	22	4.35
5	4.35	11	4.48	17	4.35	23	4.42
6	4.35	12	4.42	18	4.22	24	4.35

An Analysis (ANOVA) of variance is performed on SO_2 gpl values and is given in Table 12.2.

Table 12.2: ANOVA for SO_2 gpl

Source	df	SS	MS	F	$F_{.01}$
2hr	1	0.0028	0.0028		
1hr/2hr	2	0.0492	0.0246	7.40*	6.23
1/2hr/1hr/2hr	4	0.0393	0.0098	2.96	4.77
Residual	16	0.0532	0.0033		
Total	23	0.1450			

From the ANOVA we see that the least significant interval is 1 hour. Hence the sampling interval can be chosen between 1/2 hour to 1 hour.

When to adjust and how much to adjust:

Using the error observed during the 1/2 hour, we get,

$$\sigma = \sqrt{0.00333} = 0.06.$$

Now using the β-correction method we can obtain the required action. Table 3 shows the β-correction required for some of the values of SO_2 gpl.

Table 12.3: β-correction			
SO_2 gpl (Y)	$Y - T$	β	$-\beta(Y - T)$
3.0	-1.0	.996	.996
3.8	-0.2	.910	.182
3.9	-0.1	.640	.064
4.0	0	.000	.000
4.1	0.1	.640	-.064
4.2	0.2	.910	-.182
5.0	1.0	.996	-.996

The required correction on the SO_2 gpl can be obtained by adjusting either SO_2 flow or water flow. Usually the water flow is adjusted to 3000 kg/hr or 3600 kg/hr depending on the production requirements and hence the corrections are adjusted through the $S0_2$ flow rate. To know the amount of adjustment required on SO_2 flow to get the β-correction, a relationship between SO_2 gpl and other process parameters is to be established.

Developing the relationship:

Let X_1 and X_2 be the mass flow rates (kg/hr) of SO_2 gas and water. Clearly the concentration of the SO_2 in the solution is expected to be $X_1/(X_1 + X_2)$ (kg/kg) which is the relationship we mentioned as equation (12.1) in Section 12.2.

To obtain the function relating to time (equation 12.3) we can use the mass balance equations of the system. Let m be the mass of the solution in the tank at any time and X_3 be the mass flow rate of SO_2 solution, then we have

$$\frac{dm}{dt} = X_1 + X_2 - X_3 \tag{12.6}$$

for the mass solution and

$$\frac{d(mY)}{dt} = X_1 - X_3Y \tag{12.7}$$

for the mass of SO_2, where Y is the concentration of SO_2 in the solution at any time.

Assume that the mass coming into the tank is going out of the tank or equivalently the mass of the solution in the tank is constant. Then equation (12.7) becomes,

$$m\frac{dY}{dt} = X_1 - (X_1 + X_2)Y.$$

Solving this first order linear differential equation, using the initial condition $Y(t_o) = Y_o$, we obtain,

$$Y(t) - Y_o = (\frac{X_1}{X_1 + X_2} - Y_o)(1 - e^{-(X_1+X_2)(t-t_0)/m}). \tag{12.8}$$

When both flow rates (X_1 and X_2) are kept constants we see from equation (12.8) that as time increases ($t \to \infty$) the SO_2 concentration stabilizes to a value, $Y(t) \to X_1/(X_1 + X_2) = q$ irrespective of the initial concentration as obtained earlier. This behaviour of the process is shown in Figure 12.3.

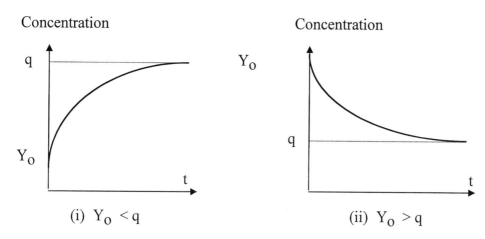

Figure 12.3: Behaviour of the process

Data were collected and is given in Table 12.4. When $X_1 = 16.1$ kg/hr and $X_2 = 3000$ kg/hr, the stability concentration is expected to be 5.33 gpl as per the formula, but the actual stabilized value is around 4.35 gpl. This indicates that the instrument for SO_2 flow rate is not calibrated properly and we suspect that the data recorded were not accurate.

To get a valid relationship a regression equation can be fitted to the data based on equation (12.8) by taking the inspection bias as regression coefficients. Let the actual SO_2 flow $= a\,X_1 + b$, in the usual operating range where a is the scale error and b is the zero error for that instrument. Then equation (12.8) becomes,

$$Y(t) - Y_o = \left(\frac{a\,X_1 + b}{a\,X_1 + b + X_2} - Y_o\right)\left(1 - e^{-(a\,X_1 + b + X_2)(t - t_0)/m}\right). \qquad (12.9)$$

In equation (12.9) a, b and m are unknown constants and can be estimated from the data using nonlinear regression which leads to

$$a \approx 0.816, \quad b \approx 0, \quad m \approx 728 \text{ kgs} \quad \text{with } R^2 = 90\% \text{ and } s = 0.047 \text{ gpl.}$$

Thus the relationship is,

$$Y(t) - Y_0 = \left(\frac{0.816X_1}{0.816X_1 + X_2} - Y_o\right)\left(1 - e^{\frac{-(0.816X_1 + X_2)}{728}(t - t_0)/60}\right) \qquad (12.10)$$

and the stability concentration is,

$$q = \left(\frac{0.816X_1}{0.816X_1 + X_2}\right) 1000 \text{ gpl.} \qquad (12.11)$$

S NO.	SO_2 flow (kg/hr)	water flow (kg/hr)	Y_0 (gpl)	$t_1 - t_0$ (mins)	Y_1 (gpl)
1	15.1	3012	4.1	10	4.03
2	15.1	2850	4.03	10	4.09
3	13.1	3000	4.09	10	3.84
4	13.1	2988	3.84	9	3.71
5	13.1	3012	3.71	10	3.65
6	13.0	2850	3.65	10	3.46
7	13.1	3030	3.46	10	3.46
8	13.1	2994	3.46	10	3.52
9	17	2910	3.52	10	4.16
10	17.1	3108	4.16	10	4.35
11	17.1	2892	4.35	11	4.54
12	17.1	2970	4.54	9	4.61
13	17.1	2880	4.61	11	4.61
14	17.1	3030	4.22	11	4.61
15	16.0	3042	4.35	10	4.35
16	16.1	3060	4.29	10	4.29
17	16.1	2958	4.35	10	4.35
18	16.1	2976	4.35	10	4.35
19	16.1	3006	4.35	10	4.35
20	16.1	3018	4.35	10	4.35
21	16.1	3018	4.35	10	4.35
22	16.1	3000	4.41	10	4.42
23	16.3	2994	4.42	10	4.48
24	16.3	2982	4.48	10	4.48
25	16.3	3036	4.48	12	4.42
26	16.3	2982	4.42	8	4.42
27	16.4	2976	4.42	10	4.42
28	16.3	3000	4.42	10	4.42
29	16.3	3006	4.42	12	4.48
30	16.3	3012	4.48	10	4.35

Table 12.4: Flow data

Adjustment procedure:

The procedure is to adjust the SO_2 gas flow rate depending on the water flow rate to obtain the target SO_2 gpl. i.e., adjust X_1 such that,

$$\frac{816X_1}{0.816X_1 + X_2} = 4$$

$$X_1 = 0.00492X_2 \text{ kg/hr.}$$

Thus when water flow = 3000 kg/hr, SO_2 flow should be at 14.8 kg/hr. Under this adjustment procedure the time required to get the β-correction can be solved from equation (12.10) and is shown in Figure 12.4.

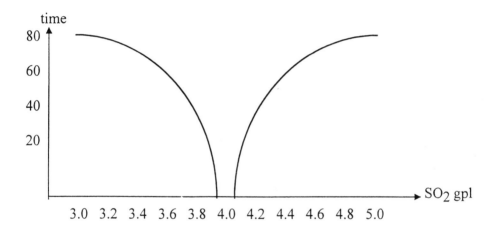

Figure 12.4: Time to get -correction with $X_1 = 14.8$ kg/hr and $X_2 = 3000$ kg/hr

We see that when SO_2 gpl is less than 3.8 or greater than 4.2, it takes more than half an hour to get the β-correction. This time can be reduced by increasing the SO_2 flow rate above 14.8 kg/hr if SO_2 gpl is less than 4 gpl or by decreasing the SO_2 flow rate below 14.8 kg/hr if SO_2 gpl is greater than 4 gpl. Suppose we are interested in getting the β-correction say in 10 minutes. Now the required SO_2 flow rate for this 10 minutes can be solved from equation (12.10) for different possible values of water flow.

$$\left(\frac{0.816X_1}{0.816X_1 + X_2} - Y_o\right)\left(1 - e^{-\frac{(0.816X_1 + X_2)}{728}10/60}\right) = \beta(0.004 - Y_0). \quad (12.12)$$

After 10 minutes the SO_2 flow has to be readjusted to the flow to get stability concentration at 4 gpl (14.8 kg/hr when X_2=3000 kg/hr). The behaviour of the process during the adjustments is shown in Figure 12.5.

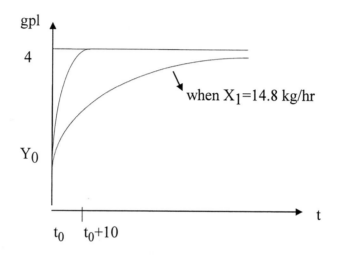

Figure 12.5: Behaviour of process during adjustments ($Y_0 < 4$ gpl and X_2=3000 kg/hr)

To make the method operator friendly the SO_2 flow is solved from equation (12.12) and a ready reckoner chart is prepared as given in Table 12.5. Thus the control procedure is

i. Observe SO_2 gpl every hour.

ii. If the SO_2 gpl is less than 3.9 or greater than 4.1 adjust the SO_2 flow as shown in the chart for 10 minutes.

iii. After 10 minutes readjust the SO_2 flow to the value corresponding to 4.0 gpl.

Consecutive SO_2 gpl values taken before and after implementation of the chart is shown in Figure 12.6. It can be seen that the variations in SO_2 gpl is very small compared to the previous time.

SO$_2$ gpl	SO$_2$ flow in Kg/hr	
	water flow = 3000 kg/hr	water flow = 3600 kg/hr
3.0	18.5	21.2
3.1	18.1	20.7
3.2	17.7	20.4
3.3	17.4	20.0
3.4	17.0	19.7
3.5	16.6	19.3
3.6	16.2	19.0
3.7	15.8	18.6
3.8	15.4	18.2
3.9	14.8	17.7
4.0	14.8	17.7
4.1	14.8	17.7
4.2	14.1	17.1
4.3	13.7	16.7
4.4	13.3	16.3
4.5	12.9	16.0
4.6	12.5	15.6
4.7	12.1	15.3
4.8	11.7	14.9
4.9	11.4	14.6
5.0	11.0	14.2

Table 12.5: Ready reckoner chart for the control of SO$_2$ in 4±1 gpl

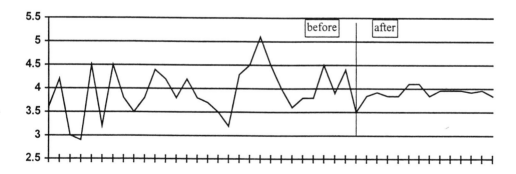

Figure 12.6: SO$_2$ gpl values before and after implementation of the ready reckoner

12.4   Conclusions

The control procedure discussed above can be applied to any process which have the following properties.

i. Multi input and single output.

ii. Status of all input characteristics are estimable from time to time, though some of them may not be controllable.

iii. Testing time is small.

In any chemical industry, there are many processes which come under this category. The methodology adopted in the paper can be suitably deployed to control the output characteristic at the desired level without too much capital investment. There are two main constraints. Firstly this method requires estimates of all the factors affecting the output. Some times this is not feasible. In such situations attempts are to be made to reduce the effect of nonestimable factors on the output using Taguchi's robust process design concept. Secondly if the time needed for testing is too large as compared to the sampling interval of process control, this method is not effectively applicable. New strategies have to be developed for such cases.

Reference

1. Pal, B. K. (1993). *Taguchi methods in Indian industries–An experience, A collection of Quality engineering applications-Indian industries*, ISI, Bangalore, India.

13

Properties of the C_{pm} Index for Seemingly Unstable Production Processes

Olle Carlsson

University of Örebro/Uppsala, Sweden

Abstract: In this paper the use of the C_{pm} index is extended to cover situations where the process mean, for example, due to shift-to-shift effects, varies on or around the target value. Approximate lower confidence intervals for the C_{pm} index are derived for cases involving fixed and random factors and a simple relation between C_{pm} and the minimum probability of conforming items is given. Finally, a sensitivity analysis is performed to examine the importance of the presence of a random process mean on the index and its lower confidence level.

Keywords and phrases: Chi-square approximation, conforming items, process capability, target value, unstable production process

13.1 Introduction

Capability indices have been widely used during the last ten years. Some of the reasons for this popularity are: most manufacturing companies have become aware of the importance of variability; the indices are easy to compute and a single-number summary considerably simplifies the supervision of a production process or of forming an opinion about the homogeneity of delivered lots; and indices provide appropriate summary information about alternative processes or suppliers, performance problems etc. In many cases, however, capability indices have been uncritically used. Some authors, for example Gunter (1989) have severely criticised this practice.

The choice between different capability indices can always be discussed from several points of view including that of the customer. If the customer wants deliveries from a homogeneous production and is unconcerned about the process mean (level) as long as it stays within the specification limits, the index C_p

should be preferred. If, in addition, the customer also wants the deliveries to be close to the target value, C_{pm} is recommended. Boyles (1991) and Chan et al. (1988) point out that the index C_{pk} becomes arbitrarily large when the variability decreases, and that a large value of C_{pk} does not in itself say anything about the distance between the process level and the target value. Some practitioners simultaneously use C_p and C_{pk} to overcome this weakness, but then the gain in simplicity is lost. If any pair of statistics should be used, it seems obvious to use the pair of sufficient statistics $(\bar{X}, S)$, where $\bar{X}$ is the arithmetic mean and S is the standard deviation. On the other hand, C_{pk} has the great advantage that it can be rewritten and interpreted as a one- or two-sided statistical tolerance interval.

At first, capability indices were intuitively constructed measures, but during recent years a number of theoretical studies have appeared concerning expected values, variances, distributions, confidence intervals etc., see for instance Kane (1986), Marucci and Beazley (1988), Chan et al. (1988), Spiring (1991), Boyles (1991), Kushler and Hurley (1992), Rodriguez (1992) and Vännman (1995). An extensive list of research within this field can be found in Kotz and Johnson (1993).

The underlying assumptions for these studies are that the quality characteristic in general is assumed to be independently, identically and normally distributed and that the process is stable (in statistical control). Shewhart (1931) and Deming (1986) say that if a process is predictable then it is stable, but Deming (1986, p. 340) also requires *"Quality characteristics...remain nearly constant hour after hour, day after day"*. He concludes that capability is definable if and only if

1. the quality characteristic can be properly described by some probability distribution, and

2. the process mean and the process variance both are independent of time.

Shewhart (1931, p. 6) is less restrictive. *"A phenomenon will be said to be controlled when, through the use of past experience, we can predict, at least within limits, how the phenomenon may be expected to vary in the future."* This corresponds to condition 1 above and implies that if it is possible to find a suitable statistical model describing the quality characteristic, then capability is definable and can be interpreted.

Shewhart's weaker definition allows capability studies on a larger set of production processes. In this paper the study is restricted to the common situation where an industrial process may well be centred at a process level on or close to the target value but can not be steadily kept at that level. Such occasional deviations from the process level can depend on more or less distinguishable causes e.g. shift-to-shift, machine-to-machine differences or variations in input raw material etc. Ford Motor Company (1990) takes account to the presence of subgroup variation by adding a term denoted X_{add} in the denominator of the actual index. X_{add} is defined as $X_{add} = \bar{X}_{max} + \bar{X}_{min} - 2\hat{\sigma}/\sqrt{n}$, where $\bar{X}_{max}$ and $\bar{X}_{min}$ are the largest and the smallest group value, respectively, and $\hat{\sigma}$ is the estimated within group standard deviation.

In this paper it is assumed that the distribution of the quality characteristic can be properly modelled by a one-way analysis of variance model, where the response variable is the measured difference between the quality characteristic and the target value, the constant term is the difference between the intended process level and the target value, the factor is the occasional deviations of the mean process levels from intended process level and finally, the error term is the common causes. This approach allows, compared to Ford Motor Company's (1990) approach, testing the presence and estimating the size of subgroup variation. Further, approximate lower confidence intervals for the C_{pm} index can be derived and a simple relation between C_{pm} and the minimum probability of conforming items can be given. Finally, sensitivity analysis of the impact of the presence of occasional deviations on the index and its lower confidence level can be made.

A one-way analysis of variance approach would have been more obvious and the corresponding estimates have a clear interpretation; but there is a psychological aspect. At least in Sweden, upper management in general is not familiar with analysis of variance while capability indices are well known. Thus, when the impact of a (hypothetical) variance reduction can be quantified in terms of a higher index and increased yield, the management may change their somewhat restrictive attitude to support further studies.

13.2 Different Models

Chan et al. (1988) defines the index C_{pm} as $C_{pm} = (U - L)/6\sqrt{M\hat{S}E}$, where $M\hat{S}E = n^{-1}\sum_{i=1}^{n}(X_i - T)^2$; U and L are the upper and lower specification limits, respectively; T denotes the target value, σ^2 is the variance and μ is the expected value (process level) of the actual quality characteristic. Now, suppose that a shipment consists of t lots and that there are r observations of the quality characteristic from each lot. For distributional reasons it is necessary to separate the study into two cases, a fixed factor case and a random factor case. An example of the first case is machine-to-machine effects while an example of the second case is variations in input raw material.

13.2.1 Random effects model

Assume that the quality characteristics are generated from the model,

$$X_{ij} = \mu + \alpha_i + \varepsilon_{ij}, \tag{13.1}$$

where $\alpha_i \sim N(0,\omega^2)$, ω^2 is the variance of the shifts, $\varepsilon_{ij} \sim N(0,\sigma^2)$ and σ^2 is the variance generated by the common causes, $i = 1,2,\ldots,t$, $j = 1,2,\ldots,r$. Now, let C_{pm} be defined as $C_{pm} = (U-L)/6\sqrt{\sigma^2 + \omega^2 + (\mu-T)^2}$ and let $\hat{C}_{pm} = (U-L)/6\hat{\tau}$ be an estimator of C_{pm}, where $\hat{\tau}^2$ denotes the estimated mean square error, $\hat{\tau}^2 = (tr)^{-1} \sum_{i=1}^{t} \sum_{j=1}^{r} (X_{ij} - T)^2$. It is easy to show that $E(\hat{\tau}^2) = \tau^2 = \sigma^2 + \omega^2 + (\mu-T)^2$. Further, it is well known that $tr\hat{\tau}^2/(\sigma^2 + \omega^2) \sim \chi^2(tr, tr\theta^2)$, i.e., $tr\hat{\tau}^2/(\sigma^2 + \omega^2)$ follows a non-central chi-square distribution with tr degrees of freedom (d.f.), and non-centrality parameter $tr\theta^2$ where, $\theta = |\mu-T|/(\sigma^2 + \omega^2)^{1/2}$. A non-central chi-square distribution can be approximated by a constant times an ordinary chi-square distribution, Patnaik (1949), by equating means and variances, i.e., $\chi^2(tr, tr\theta^2)$ is approximately distributed as $c \cdot \chi^2(v)$. The equations for the means and variances are

$$cv = tr(1+\theta^2) \text{ and } 2c^2 v = 2tr(1+2\theta^2), \tag{13.2}$$

and the number of degrees of freedom becomes

$$v = tr(1+\theta^2)^2/(1+2\theta^2). \tag{13.3}$$

The square of the ratio between the capability index and its estimator can be written as

$$C_{pm}^2/\hat{C}_{pm}^2 = \hat{\tau}^2/\tau^2 = (tr)^{-1} \sum_{i=1}^{t} \sum_{j=1}^{r} (X_{ij} - T)^2/[\sigma^2 + \omega^2 + (\mu - T^2]. \tag{13.4}$$

Insertion of the first equation in (13.2) into (13.4) implies after some simplification that

$$C_{pm}^2/\hat{C}_{pm}^2 \text{ is approximately } \chi^2(v)/v. \tag{13.5}$$

The number of d.f., v, is not known, but can be estimated in different ways. Boyles (1991), for the case $\mu \neq T$ and $\omega = 0$, uses the estimator $\hat{\theta}_B = |\bar{X}_{..} - T|/\hat{\sigma}$ for the non-centrality parameter, where $\bar{X}_{..} = \sum_{i=1}^{r} \sum_{j=1}^{t} X_{ij}/tr$ is the shipment mean and $\hat{\sigma} = [n^{-1} \sum_{l=1}^{n} (X_i - \bar{X})^2]^{1/2}$ is the standard deviation. The number of degrees of freedom becomes

$$\hat{v}_B = tr[1 + (\hat{\theta}_B)^2]^2/[1 + 2(\hat{\theta}_B)^2] \tag{13.6}$$

However, for the general case i.e. $\mu \neq T$ and $\omega > 0$, θ^2 and subsequently v, can be estimated by a common approach used in analysis of variance. The total sum of squares can be partioned as $SS_{TOT} = SS_{ERROR} + SS_{LOTS} + SS_{TARGET}$ where the corresponding sums of squares are $\sum_{i=1}^{t} \sum_{j=1}^{r} (X_{ij} - T)^2$, $\sum_{i=1}^{t} \sum_{j=1}^{r} (X_{ij} - \bar{X}_{i.})^2$, $r\sum_{i=1}^{t} (\bar{X}_{i.} - \bar{X}_{..})^2$ and $tr(\bar{X}_{..} - T)^2$, respectively, where $\bar{X}_{i.} = \sum_{j=1}^{r} X_{ij}/r$ is the lot mean. The corresponding ANOVA table with the expected mean squares is:

Source	SS	d.f.	MSE	E(MSE)
Target	SS_T	1	MS_T	$\sigma^2 + r\omega^2 + tr(\mu - T)^2$
Lots	SS_L	$t-1$	MS_L	$\sigma^2 + r\omega^2$
Error	SS_E	$t(r-1)$	MS_E	σ^2

Hence, the estimator of the non-centrality parameter, $\hat{\theta}^2$, becomes

$$\hat{\theta}^2 = (MS_T - MS_L)/\{t[(r-1)MS_E + MS_L]\} \tag{13.7}$$

In this case it is not possible to derive the expected value of $\hat{\theta}^2$. Properties of the proposed estimator can be compared with Boyles' estimator for the special case when $\mu \neq T$, $\omega = 0$ and $n = tr$, where the estimator (13.7) simplifies to $\hat{\theta}_0^2 = (MS_T - MS_E)/n \cdot MS_E$. The square of Boyles' estimator can be written as $\hat{\theta}_B^2 = [n/(n-1)](\hat{\theta}_0^2 + n^{-1})$, which relates the two estimators. The mean squares MS_E and MS_T are distributed as $\sigma^2\chi^2(n-1)$ and $\sigma^2\chi^2(1, n\theta^2)$, respectively. Then, $MS_T/MS_E \sim F'(1, n-1, n\theta^2)$ where $F'(\cdot, \cdot, \cdot)$ denotes the singly non-central F-distribution, Stuart and Ord (1991). The expected value of the estimators are $E(\hat{\theta}_B^2) = [1/(n-3)](n\theta^2+1)$ and $E(\hat{\theta}_0^2) = [1/(n-3)][(n-1)\theta^2 + 2/n]$, respectively. Both estimators are positively biased and $\hat{\theta}_B^2$ has a larger bias than $\hat{\theta}_0^2$. Finally, $[(n-3)/(n-1)][(MS_T - MS_E)/n\,MS_E]$ is an unbiased estimator of θ^2 and the estimated number of d.f. is found by insertion in (13.3).

13.2.2 Fixed effects model

Assume that the quality characteristic, as in Section 13.2.1, can be described by the model

$$X_{ij} = \mu + \alpha_i + \varepsilon_{ij} \tag{13.8}$$

where $\varepsilon_{ijk} \sim N(0, \sigma^2)$, but α_i are constants subject to the restriction $\sum_{l=1}^{t} \alpha_i = 0$, which implies that $X_{ij} \sim N(\mu + \alpha_i, \sigma^2)$, $i = 1, 2, \ldots, t$, $j = 1, 2, \ldots, r$. As above, let $\hat{\tau}^2 = (tr)^{-1}\sum_{i=1}^{t}\sum_{j=1}^{r}(X_{ij} - T)^2$ denote the estimated mean square error, and let $\hat{C}_{pm} = (U - L)/6\hat{\tau}$ be an estimator of C_{pm}. It is easy to show that the expected value of $\hat{\tau}^2$ is equal to $E(\hat{\tau}^2) = \sigma^2 + (t^{-1})\sum_{i=1}^{t}\alpha_i^2 + (\mu - T)^2$. As above, $tr\hat{\tau}^2/\sigma^2 \sim \chi^2(tr, tr\theta^2)$, but here the non-centrality parameter becomes $\theta = \left[t^{-1}\sum_{i=1}^{t}\alpha_i^2 + (\mu - T)^2\right]^{1/2}/\sigma$. Paitnak's (1949) approximation can also be used for this case, and the number of d.f. becomes the same as in (13.3) but of course with a different θ. The ratio corresponding to (13.4) can be written as

$$C_{pm}^2/\hat{C}_{pm}^2 = \hat{\tau}^2/\tau^2 = (tr)^{-1}\sum_{i=1}^{t}\sum_{j=1}^{r}(X_{ij}-T)^2 / \left[\sigma^2 + t^{-1}\sum_{i=1}^{t}\alpha_i^2 + (\mu-T)^2\right] \tag{13.9}$$

and

$$C_{pm}^2/\hat{C}_{pm}^2 \text{ is approximately } \chi^2(v)/v. \tag{13.10}$$

The total sum of squares can be partioned in the same way as before, but the ANOVA table with the expected mean squares now becomes:

Source	SS	d.f.	MSE	$E(MSE)$
Target	SS_T	1	MS_T	$\sigma^2 + tr(\mu - T)^2$
Lots	SS_L	$t - 1$	MS_L	$\sigma^2 + r(t-1)^{-1}\sum_{i=1}^{t}\alpha_i^2$
Error	SS_E	$t(r-1)$	MS_E	σ^2

This leads to the non-centrality parameter estimator

$$\hat{\theta}^2 = (tr)^{-1}[(t-1)MS_L/MS_E + MS_T/MS_E] - r^{-1} \qquad (13.11)$$

Both MS_T/MS_E and MS_L/MS_E follow non-central F distributions, viz., $F'(1, t(r-1))$, $tr(\mu-T)^2/\sigma^2)$ and $F'(t-1, t(r-1), r\sum_{i=1}^{t}\alpha_i^2/\sigma^2)$, respectively. Hence the expected value of $\hat{\theta}^2$ is

$$E(\hat{\theta}^2) = [t(r-1)\theta^2 + 2r^{-1}]/[t(r-1) - 2] \qquad (13.12)$$

The bias is positive and equal to $2(\theta^2 + r^{-1})/[t(r-1) - 2]$. Hence, an unbiased estimator of θ^2 is given by

$$\hat{\theta}^2_{UNB} = [1 - 2/t(r-1)]\hat{\theta}^2 - 2/tr(r-1) \qquad (13.13)$$

13.3   Confidence Intervals

$100(1 - \alpha)\%$ confidence intervals for C_{pm} for both kinds of distributions can be constructed in the same way. Let $\chi^2_{1-\alpha}(v)$ and z_α denote the upper-tail percentile points of the chi- square and normal distributions respectively. An approximate $100(1 - \alpha)\%$ lower confidence limit based on (13.5) becomes

$$\hat{C}_{pm} \cdot [\chi^2_{1-\alpha}/(2\hat{v})^{1/2}]. \qquad (13.14)$$

For large samples $\sqrt{2\chi^2(v)}$ is approximately normally distributed, $N(\sqrt{2v-1}, 1)$ and a corresponding confidence limit is given by

$$\hat{C}_{pm}[1 - z_\alpha/(2\hat{v})^{1/2}]. \qquad (13.15)$$

The chi-square and normal percentiles can be found in statistical tables, and Boyles (1991) gives tables for the coefficient of $\hat{C}_{pm}$, corresponding to the interval (13.15) as a function of $\hat{v} = 3$ (1)100, and for 90, 95 and 99 percent confidence levels. For a restricted parameter set, Boyles (1991) also shows by Monte Carlo simulation that the percent coverage rate for the lower confidence limits deviates from the nominal confidence level with at most .9% for sample sizes 20, 50, 100.

The estimated number of d.f. is found by insertion in (13.3), and it can be seen that $\hat{v} = tr(1 + \hat{\theta}^2)^2/(1 + 2\hat{\theta}^2) = tr[1 + (\hat{\theta}^4/(1 + 2\hat{\theta}^2))] \geq tr$. It may appear surprising that $\hat{v}$ is larger than tr, but it must be kept in mind that the new distribution is no longer a chi-square distribution but a two-parameter gamma distribution, hence $\hat{v}$ can no longer be seen as the number of d.f. in a traditional sense. However, $\hat{v}$ is only slightly larger than tr, e.g. for $\hat{\theta}^2 < .20$ we have $\hat{v}/tr < 1.0015$, and for $\hat{\theta}^2 < .50$, $\hat{v}/tr < 1.042$.

Now assume that C_{pm} is given. It can be assumed, without any loss of generality that the square root term in the denominator of C_{pm} is equal to one and that the target value is equal to zero. For both models the denominator can be written as $\sigma_G^2 + \mu_G^2$, where the index G stands for general. For the random effects model $\sigma_G^2 = \sigma^2 + \omega^2$ and $\mu_G^2 = \mu^2$, while for the fixed effects model $\sigma_G^2 = \sigma^2$ and $\mu_G^2 = (t^{-1})\sum_{i=1}^{t} \alpha_i^2 + \mu^2$. The length of the lower confidence interval, $C(\mu_G, \sigma_G)$ (which should be taken as a point estimate of the lower confidence limit) depends on the non-centrality parameter and it is not symmetric in the location and variability parameters. Let the relative length, $R(\mu_G, \sigma_G)$, be defined as the ratio $1 - C(\mu_G, \sigma_G)/C_{pm}$, which for large samples becomes $R(\mu_G, \sigma_G) = z_\alpha (2v_G)^{-1/2}$ where, $v_G = tr(1 + \theta_G^2)^2/(1 + 2\theta_G^2)$ and $\theta_G^2 = \mu_G^2/(1 - \mu_G^2)$, $0 \leq \mu_G^2 < 1$. The relative length decreases with μ_G^2, because $R(\mu_G, \sigma_G)$ decreases with v_G which increases with θ_G and, finally, θ_G increases with μ_G. This implies that if a choice can be made between μ_G and σ_G along the circle $\mu_G^2 + \sigma_G^2 = 1$, a decrease in σ_G is to be preferred because it shortens the confidence interval. In other words, a process which is stable on the wrong level has more consistent capability indices than a process which has varying process levels around the target value. Finally, for given μ and σ, the non-centrality parameter for the fixed effects model is larger than that of the random effects model. This implies that the fixed effects model generates shorter confidence intervals than the random effects model.

13.4   Probability of Conforming Items

Suppose that U and L are symmetrical around $T = 0$, such that $U = -L = d > 1$ and that $\mu_G^2 + \sigma_G^2 = 1$. The probability of a conforming item, $P(L \leq X \leq U)$, can be written

$$
\begin{aligned}
P(L \leq X \leq U) &= \Phi[(U - \mu_G)/\sigma_G] - \Phi[(L - \mu_G)/\sigma_G] \\
&= \Phi[(d - \mu_G)/\sigma_G] + \Phi[(d + \mu_G)/\sigma_G]^{-1} \quad (13.16)
\end{aligned}
$$

where Φ denotes the distribution function of a normal distribution.

The first derivative with respect to μ_G is

$$
\begin{aligned}
(\partial/\partial\mu_G)P(L \leq X \leq U) &= [(1 + \mu_G d)/\sigma_G^3]\phi[(d + \mu_G)/\sigma_G] \\
&\quad - [(1 - \mu_G d)/\sigma_G^3]\phi[(d - \mu_G)/\sigma_G] \quad (13.17)
\end{aligned}
$$

where ϕ = first derivative of Φ.

The first derivative is equal to zero for $\mu_G = 0$ and positive for $\mu_G \geq 1/d$. The second and the third derivatives are equal to zero for $\mu_G = 0$, but Taylor series expansion shows that the fourth derivative is positive for $d > \sqrt{3}$ or equivalently for $C_{pm} > 1/\sqrt{3}$. This, and the fact that the function $P(L \leq X \leq U)$ is symmetric around $\mu_G = 0$, implies that $P(L \leq X \leq U)$ has a minimum for $\mu_G = 0$ when $C_{pm} > 1/\sqrt{3}$. Then, $min[P(L \leq X \leq U)] = 2\Phi(3C_{pm}) - 1$ for $C_{pm} > 1/\sqrt{3}$. For smaller C_{pm} see Ruczinski (1995).

Maximum Probability of Non-Conforming Items for Different C_{pm}			
C_{pm}	*ppm*	C_{pm}	*ppm*
1.00	2699.796	1.50	6.795
1.10	966.848	1.60	1.587
1.20	318.217	1.70	.340
1.30	96.193	1.80	.067
1.33	66.073	1.90	.012
1.40	26.691	2.00	.002

It can be seen from the table that the probability of non-conforming items decreases rapidly and that $C_{pm} = 1$ (1.33) corresponds to the $\pm 3\sigma (\pm 4\sigma)$ limits. Further, Motorola's (1991) 6σ limits corresponds to $C_{pm} = 2$ and and their worst case limits to $C_{pm} = 1.5$. Motorola's approach corresponds to the C_{pm} index studied by Boyles (1991) and corresponds to $\omega = 0$ in this study.

13.5   Summary and Discussion

In general, the denominator of the C_{pm} index consists of two terms: the process variation due to common causes, σ^2, and the term due to the permanent shift, $(\mu - T)^2$. However, a production process often can not be kept at a predeterminded level on, or off, the target value. In order to cover this situation the index is generalised by including a second kind of variability describing occasional shifts in the process mean due assignable causes, which can be statistically modelled.

   Simple approximate confidence intervals for both fixed and random effects models were derived; they can be implemented by using statistical tables. It was shown for both cases that the relative length of the interval was more sensitive to changes in the occasional factors, , than for changes in the permanent shift, $(\mu - T)^2$. Further, it was shown that the confidence intervals, given the same noise, σ^2, and $(\mu - T)^2$ were shorter for the fixed factor model than the corresponding interval for the random effects model. The reason is that shifts of the process level, α_i, are within a given range in the fixed effects model, while there is a small probability for very large shifts in the random effects model. It was also shown, for given C_{pm}, that a production process which is stable at the wrong level has more consistent capability indices than a process which has occasionally varying process levels around the target value. Finally, a simple lower limit of the probability of a conforming item was given when $C_{pm} > 1/\sqrt{3}$. The minimum probability of a conforming item occurs when the process level is on the target value, $\mu = T$. In other words: For given C_{pm}, the larger $|\mu - T|$, the larger $P(L \leq X \leq U)$.

References

1. Boyles, R. A. (1991). The Taguchi capability index, *Journal of Quality Technology*, **23**, 17–26, 162–175.

2. Chan, L. K., Cheng, S. W. and Spiring, F. A. (1988). A new measure of process capability C_{pm}, *Journal of Quality Technology*, **20**, 162–175.

3. Deming, W. E. (1986). Out of the crisis, *Massachusetts Institute of Technology*, Cambridge, Mass: Center for Advanced Engineering Study.

4. Ford Motor Company (1990). *Process Capability*, Detroit.

5. Gunter, B. (1989). The use and abuse of C_{pk}: parts1-4, *Quality Progress*, Jan. 1989, 72-73, March 1989, 108–109, May 1989, 79–80, July 1989, 86–87.2.

6. Kane, V. E. (1986). Process capability indices, *Journal of Quality Technology*, **18**, 41–52.

7. Kotz, S. and Johnson, N. L. (1993). *Process capability indices*, Chapman & Hall, London.

8. Kushler, R. and Hurley, P. (1992). Confidence bounds for capability indices, *Journal of Quality Technology*, **24**, 188–195.

9. Marucci, M. D. and Beazley, C. E. (1988). Capability indices: process performance measures, *ASQC Quality Conference Transactions*, 516–523.

10. Motorola (1991). Six-sigma quality-TQC, the American style, Hinshitsu, *Journal of JSQC*, July, Japan.

11. Paitnak, P. B. (1949). The noncentral χ^2 and F-distributions and their applications, *Biometrica*, **36**, 202–232.

12. Rodriguez, R. N. (1992). Recent developments in process capability analysis, *Journal of Quality Technology*, **24**, 176–185.

13. Ruczinski, I. (1995). *The degree of includence and other process capability indices. Thesis.* University of Würzburg, Germany.

14. Shewhart, W. A. (1931). Economic control of quality of manufactured product, Van Nostrand, New York.

15. Spiring, F. A. (1991). Assessing process capability in the presence of systematic assignable causes, *Journal of Quality Technology*, **23**, 125–134.

16. Stuart, A. and Ord, J. K. (1991). *Kendall's Advanced Theory of Statistics*, **2**, 5th ed., Edward Arnold, London.

17. Vännman, K. (1995). A unified approach to capability indices, *Statistica Sinica*, **5**, 805–822.

14

Process Capability Indices for Contaminated Bivariate Normal Populations

Subrahmaniam Kocherlakota and Kathleen Kocherlakota

University of Manitoba, Winnipeg, Manitoba, Canada

Abstract: In this paper we examine the robustness of bivariate generalizations (CP_1, CP_2) of the process capability index when sampling is from a bivariate contaminated normal distribution. The effect of the contaminated model is studied in terms of expected values, variances and correlation coefficient of $\left(\widehat{CP}_1, \widehat{CP}_2\right)$. The behavior of the test of hypothesis suggested in an earlier paper [Kocherlakota and Kocherlakota (1991)] is also considered.

Keywords and phrases: Bivariate normal distribution, bivariate χ^2 distribution, contaminated normal population, process capability index, robustness, tests of hypotheses

14.1 Introduction

An excellent discussion of the process capability indices and related problems is given in Kane (1986) and Kotz and Johnson (1993). As pointed out by these authors, the assumption of normality is questionable in some situations; hence, the behavior of the indices when the normality assumption is no longer valid is of considerable interest. In this paper a study is made of these effects when the population being sampled is a bivariate contaminated normal distribution. Kocherlakota and Kocherlakota (1991) have examined the joint distribution of the process capability indices

$$\widehat{CP}_i = \frac{U_i - L_i}{6s_i} = CP_i \frac{\sqrt{v}}{\sqrt{W_i}}, \quad i = 1, 2$$

based on the pairs of characters (X_1, X_2) having the bivariate normal distribution $(\mu_1, \mu_2, \sigma_1, \sigma_2, \rho)$. In this definition of the indices $\widehat{CP}_1, \widehat{CP}_2, U_i,$ and L_i are the upper and lower specification limits for the ith character, s_i^2 is the unbiased estimator of σ_i^2 and ν is the degrees of freedom of s_i^2. Further

$$W_i = \nu s_i^2 / \sigma_i^2, \quad CP_i = (U_i - L_i / 6\sigma_i.$$

For this problem Kocherlakota and Kocherlakota (1991) have shown that the joint distribution of $\left(\widehat{CP}_1, \widehat{CP}_2\right)$ has the probability density function (pdf)

$$g\left(t_1, t_2\right) = \sum_{r=0}^{\infty} nb\left(r; \frac{\nu}{2}, \rho^2\right) \prod_{j=1}^{2} \left\{ \left(\frac{2\theta_j^2}{t_j^3}\right) f_{\nu+2r}\left[\frac{\theta_j}{t_j}\right]^2 \right\}, \qquad (14.1)$$

where

$$t_i = \widehat{CP}_i, \theta_i = CP_i \sqrt{\nu} / \sqrt{(1 - \rho^2)},$$

$$nb(j; \alpha, p) = \frac{\Gamma(\alpha + j)}{\Gamma(\alpha)j!} p^j (1 - p)^\alpha, \qquad j = 0, 1, 2, \cdots$$

and

$$f_n(x) = \frac{e^{-\frac{x}{2}} x^{\frac{n}{2}-1}}{2^{\frac{n}{2}} \Gamma\left(\frac{n}{2}\right)}, \qquad x \geq 0.$$

The marginal distributions of $\widehat{CP}_1$ and $\widehat{CP}_2$ are discussed in Chou and Owen (1989). From (14.1) it can be shown that

$$E\left(t_1 t_2\right) = E_N\left(\widehat{CP}_1\right) E_N\left(\widehat{CP}_2\right) \, {}_2F_1\left[\frac{1}{2}, \frac{1}{2}; \frac{\nu}{2}; \rho^2\right], \qquad (14.2)$$

where

$$E_N\left(\widehat{CP}_i\right) = \sqrt{\frac{\nu}{2}} \frac{\Gamma\left(\frac{\nu-1}{2}\right)}{\Gamma\left(\frac{\nu}{2}\right)} \, CP_i, \quad i = 1, 2$$

as shown by Chou and Owen (1989). In (14.2) the notation ${}_2F_1\left[\frac{1}{2}, \frac{1}{2}; \frac{\nu}{2}; \rho^2\right]$ is a generalized hypergeometric function defined by

$$ {}_2F_1\left(\alpha, \beta; \gamma; z\right) = \sum_{r=0}^{\infty} \frac{\Gamma(\alpha + r)}{\Gamma(\alpha)} \frac{\Gamma(\beta + r)}{\Gamma(\beta)} \frac{\Gamma(\gamma)}{\Gamma(\gamma + r)} \frac{z^r}{r!}.$$

It can be seen from the above that

$$Cov\left(\widehat{CP}_1, \widehat{CP}_2\right) = CP_1 CP_2 C_{11}^2 (\kappa - 1) \qquad (14.3)$$

where

$$\kappa = {}_2F_1\left[\frac{1}{2}, \frac{1}{2}; \frac{\nu}{2}; \rho^2\right], \quad C_{11} = \sqrt{\frac{\nu}{2}} \frac{\Gamma\left(\frac{\nu-1}{2}\right)}{\Gamma\left(\frac{\nu}{2}\right)}.$$

Hence, the correlation between $\widehat{CP}_1$ and $\widehat{CP}_2$ is

$$\bar{\omega} = \left\{ \frac{\kappa - 1}{\nu/(\nu - 2) - C_{11}^2} \right\} C_{11}^2.$$

14.2 Contaminated Bivariate Normal Distribution

The contamination model was introduced by Tukey in 1960. In this model the observations from one normal population are 'contaminated' by a small proportion of observations from another normal population. Hence, the observations arise from a mixture of two normal distributions. In this section we will examine the distribution of $\left(\widehat{CP}_1, \widehat{CP}_2\right)$ when (X_1, X_2) has the pdf

$$f(\boldsymbol{x}) = p\phi\left(\boldsymbol{x}; \boldsymbol{0}, \Sigma\right) + (1 - p)\phi\left(\boldsymbol{x}; \boldsymbol{\mu}, \Sigma\right), \tag{14.4}$$

where $\phi\left(\boldsymbol{x}; \boldsymbol{\mu}, \Sigma\right)$ represents the bivariate normal pdf with mean $\boldsymbol{\mu}$ and variance matrix Σ. In (14.4) $\boldsymbol{x}' = (x_1, x_2)$, $\boldsymbol{\mu}' = (\mu_1, \mu_2)$ and

$$\Sigma = \left[\begin{array}{cc} \sigma_1^2 & \rho\sigma_1\sigma_2 \\ \rho\sigma_1\sigma_2 & \sigma_2^2 \end{array}\right].$$

As $\left(\widehat{CP}_1, \widehat{CP}_2\right)$ are functions of $\left(\frac{S_1^2}{\sigma_1^2}, \frac{S_2^2}{\sigma_2^2}\right)$ with $S_1^2 = \sum\limits_{i=1}^{n}(x_{1i} - \bar{x}_1)^2$ and $S_2^2 = \sum\limits_{i=1}^{n}(x_{2i} - \bar{x}_2)^2$, we have to determine their joint distribution.

14.2.1 The distribution of $(S_1^2/\sigma_1^2, S_2^2/\sigma_2^2)$

The joint moment generating function (mgf) of $(S_1^2/\sigma_1^2, S_2^2/\sigma_2^2)$ can be shown to be

$$\begin{aligned}
\psi\left(\theta_{11}, \theta_{22}\right) = & \sum_{r=0}^{n} b(r; n, p)\left\{(1 - 2\theta_{11})(1 - 2\theta_{22}) - 4\rho^2\theta_{11}\theta_{22}\right\}^{-\frac{v}{2}} \\
& \exp\left[\frac{r(n-r)}{n\Delta}\left\{\left(\frac{\mu_1}{\sigma_1}\right)^2\theta_{11}(1 - 2\theta_{22}) + 4\frac{\mu_1}{\sigma_1}\frac{\mu_2}{\sigma_2}\theta_{11}\theta_{22}\rho\right.\right. \\
& \left.\left. + \left(\frac{\mu_2}{\sigma_2}\right)^2\theta_{22}(1 - 2\theta_{11})\right\}\right],
\end{aligned} \tag{14.5}$$

where $\Delta = (1 - 2\theta_{11})(1 - 2\theta_{22}) - 4\rho^2\theta_{11}\theta_{22}$. Thus, for each r, the joint distribution of $(S_1^2/\sigma_1^2, S_2^2/\sigma_2^2)$ is that of noncentral bivariate chi-square random variables with $v = n - 1$ degrees of freedom, coefficient of correlation ρ and noncentrality parameters

$$\lambda_{1r} = \frac{r(n-r)}{n}\left(\frac{\mu_1}{\sigma_1}\right)^2, \quad \lambda_{2r} = \frac{r(n-r)}{n}\left(\frac{\mu_2}{\sigma_2}\right)^2.$$

Two cases can be seen to arise from (14.5).

Case I: $\sigma_{12} = 0$: In this case

$$\psi(\theta_{11}, \theta_{22}) = \sum_{r=0}^{n} b(r; n, p) \{(1 - 2\theta_{11})(1 - 2\theta_{22})\}^{-\frac{\nu}{2}}$$

$$\exp\left[\lambda_{1r}\frac{\theta_{11}}{1 - 2\theta_{11}} + \lambda_{2r}\frac{\theta_{22}}{1 - 2\theta_{22}}\right], \qquad (14.6)$$

which shows that, for each r, the random variables $(S_1^2/\sigma_1^2, S_2^2/\sigma_2^2)$ are independent noncentral univariate chi-square variables on ν degrees of freedom and noncentralities $\lambda_{1r}, \lambda_{2r}$, respectively.

Case II: $\sigma_{12} \neq 0$ or $\rho \neq 0$: As stated above, for each r, the joint distribution is that of noncentral bivariate chi-square variables. Although this distribution is quite complicated, we can use the results of Steyn and Roux (1972) to approximate it by a central bivariate chi-square.

Steyn and Roux (1972) examine the problem of approximating the noncentral Wishart distribution with ν degrees of freedom, variance matrix $\sum$ and noncentrality matrix $\boldsymbol{\mu}\sum^{-1}\boldsymbol{\mu}'$. They suggest approximating this distribution by a central Wishart with the same degrees of freedom ν and variance matrix $\sum + \frac{1}{\nu}\boldsymbol{\mu}\boldsymbol{\mu}'$.

Applying their procedure to the present problem, we see the variance matrix becomes

$$\sum + \begin{pmatrix} \mu_1\sqrt{\frac{r(n-r)}{n\nu}} \\ \mu_2\sqrt{\frac{r(n-r)}{n\nu}} \end{pmatrix} \begin{pmatrix} \mu_1\sqrt{\frac{r(n-r)}{n\nu}}, & \mu_2\sqrt{\frac{r(n-r)}{n\nu}} \end{pmatrix}$$

$$= \begin{pmatrix} \sigma_1^2 + \frac{r(n-r)}{n\nu}\mu_1^2 & \rho\sigma_1\sigma_2 + \frac{r(n-r)}{n\nu}\mu_1\mu_2 \\ & \sigma_2^2 + \frac{r(n-r)}{n\nu}\mu_2^2 \end{pmatrix}; \qquad (14.7)$$

that is

$$\frac{S_1^2}{\sigma_1^2 + \frac{r(n-r)}{n\nu}\mu_1^2} = \frac{S_1^2/\sigma_1^2}{1 + \frac{\lambda_{1r}}{\nu}} \quad \text{and} \quad \frac{S_2^2}{\sigma_2^2 + \frac{r(n-r)}{n\nu}\mu_2^2} = \frac{S_2^2/\sigma_2^2}{1 + \frac{\lambda_{2r}}{\nu}}$$

have the central bivariate chi-square distribution on ν degrees of freedom and correlation coefficient

$$\rho^* = \frac{\rho + \sqrt{\frac{\lambda_{1r}\lambda_{2r}}{\nu}}}{\left\{\left(1 + \frac{\lambda_{1r}}{\nu}\right)\left(1 + \frac{\lambda_{2r}}{\nu}\right)\right\}^{\frac{1}{2}}}. \qquad (14.8)$$

14.2.2 PDF of $(S_1^2/\sigma_1^2, S_2^2/\sigma_2^2)$

Case I: $\sigma_{12} = 0$: Writing $X = S_1^2/\sigma_1^2, Y = S_2^2/\sigma_2^2$, from (14.6) the pdf of (X, Y) is readily seen to be

$$f(x, y) = \sum_{r=0}^{n} b(r; n, p) f_{1r}(x; \nu, \lambda_{1r}) f_{2r}(y; \nu, \lambda_{2r}), \tag{14.9}$$

where, for $i = 1, 2$,

$$f_{ir}(z; \nu, \lambda_{ir}) = \sum_{s=0}^{\infty} P_s\left(\frac{\lambda_{ir}}{2}\right) \frac{e^{-\frac{z}{2}} z^{\frac{\nu}{2}+s-1}}{2^{\frac{\nu}{2}+s}\Gamma\left(\frac{\nu}{2}+s\right)} \tag{14.10}$$

with $P_s(\mu)$ indicating the Poisson probability at s with parameter μ. From (14.9) we obtain

$$E(X) = \sum_{r=0}^{n} b(r; n, p)(\nu + \lambda_{1r}) = \nu + \sum_{r=0}^{n} b(r; n, p)\lambda_{1r}$$

$$\mathrm{Var}(X) = \sum_{r=0}^{n} b(r; n, p)(2\nu + 2\lambda_{1r}) = 2v + 2\sum_{r=0}^{n} b(r; n, p)\lambda_{1r}$$

with similar expressions for the variable Y.

From the joint pdf of (X, Y) we have

$$E(XY) = \sum_{r=0}^{n} b(r; n, p)(\nu + \lambda_{1r})(\nu + \lambda_{2r})$$

$$= \nu^2 + \sum_{r=0}^{n} b(r; n, p)(\lambda_{1r} + \lambda_{2r}) + \sum_{r=0}^{n} b(r; n, p)\lambda_{1r}\lambda_{2r};$$

hence,

$$\mathrm{Cov}(X, Y) = \sum_{r=0}^{n} b(r; n, p)\lambda_{1r}\lambda_{2r} - \left\{ \sum_{r=0}^{n} b(r; n, p)\lambda_{1r} \sum_{r=0}^{n} b(r; n, p)\lambda_{2r} \right\}. \tag{14.11}$$

Case II: $\sigma_{12} \neq 0$ or $\rho \neq 0$: As seen above, for each r, the joint distribution of (X, Y) is noncentral bivariate chi-square. We shall approximate this by the bivariate central chi-square distribution. Thus, writing in the general case ν_r^*, ρ_r^* for the degrees of freedom and coefficient of correlation for the transformed variables $\frac{X}{c_{1r}}, \frac{Y}{c_{2r}}$ with c_{1r} and c_{2r} as constants, for each r, the joint pdf of (X, Y) can be approximated by

$$h_r(x, y) = \frac{1}{(1 - \rho_r^{*2})^2} \sum_{s=0}^{\infty} nb\left(s; \frac{\nu_r^*}{2}, \rho_r^{*2}\right) \tag{14.12}$$

$$g_{\nu_r^*+2s}\left(\frac{x}{c_{1r}(1 - \rho_r^{*2})}\right) g_{\nu_r^*+2s}\left(\frac{y}{c_{2r}(1 - \rho_r^{*2})}\right) \frac{1}{c_{1r}c_{2r}}.$$

Therefore, the joint pdf of X and Y is approximately $f(x, y) = \sum_{r=0}^{n} b(r; n, p) h_r(x, y)$.

14.3   Behavior of $\widehat{CP}_1, \widehat{CP}_2$ Under Contamination

In this section we will discuss how the distribution function and moments of $\widehat{CP}_1$ and $\widehat{CP}_2$ are affected by the contamination model given in equation (14.4).

14.3.1   Distribution function of $\widehat{CP}_1, \widehat{CP}_2$

Writing $U = \widehat{CP}_1, V = \widehat{CP}_2$, we have $U = CP_1\sqrt{\nu}/\sqrt{X}, V = CP_2\sqrt{\nu}/\sqrt{Y}$. Hence, the joint pdf of (U, V) is obtained by the transformation $X = \nu CP_1^2/U^2$, $Y = \nu CP_2^2/V^2$ in the joint distribution of (X, Y).

The distribution function of (U, V) is

$$G(u, v) = P\{U \le u, V \le v\} \qquad (14.13)$$

$$= P\left\{\frac{CP_1\sqrt{\nu}}{\sqrt{X}} \le u, \frac{CP_2\sqrt{\nu}}{\sqrt{Y}} \le v\right\}$$

$$= P\{X \ge x, Y \ge y\},$$

where $x = \nu CP_1^2/u^2$, $y = \nu CP_2^2/v^2$

Case I: $\sigma_{12} = 0$: In this case

$$P\{X \ge x, Y \ge y\} = \sum_{r=0}^{n} b(r; n, p) \int_x^\infty f_{1r}(t_1; \nu, \lambda_{1r})\, dt_1 \int_y^\infty f_{2r}(t_2; \nu, \lambda_{2r})\, dt_2$$

$$= \sum_{r=0}^{n} b(r; n, p)\left[1 - F_{1r}(x)\right]\left[1 - F_{2r}(y)\right], \qquad (14.14)$$

where $f_{ir}(z; \nu, \lambda_{ir})$ is given in (14.10) and $F_{ir}(\cdot)$ is its distribution function.

Case II: $\sigma_{12} \ne 0$: We will consider the approximation introduced earlier for the noncentral bivariate chi-square distribution. For each r, the joint distribution of (X, Y) has the pdf given in (14.12). From this the distribution function of (U, V) is

$$P(U \le u, V \le v) = P(X \ge x, Y \ge y)$$

$$= \sum_{r=0}^{n} b(r; n, p) \frac{1}{(1 - \rho_r^{*2})^2} \sum_{s=0}^{\infty} nb\left(s; \frac{\nu_r^*}{2}, \rho_r^{*2}\right) \frac{1}{c_{1r}c_{2r}}\left[1 - G_{r,s}^{(1)}(x)\right]\left[1 - G_{r,s}^{(2)}(y)\right],$$

where

$$1 - G_{r,s}^{(1)}(x) = \int_x^\infty \frac{\exp\left(-\frac{t}{2c_{1r}(1-\rho_r^{*2})}\right)}{\Gamma\left(\frac{\nu_r^*}{2} + s\right)} \left\{\frac{t}{c_{1r}(1 - \rho_r^{*2})}\right\}^{\frac{\nu_r^*}{2}+s-1} \frac{1}{2^{\frac{\nu_r^*}{2}+s}}\, dt$$

$$= c_{1r}\left(1 - \rho_r^{*2}\right) \int_{x_1}^\infty \frac{\exp\left(-\frac{t}{2}\right) t^{\frac{\nu_r^*}{2}+s-1}}{2^{\frac{\nu_r^*}{2}+s}\Gamma\left(\frac{\nu_r^*}{2} + s\right)}\, dt$$

with $x_1 = \frac{x}{c_{1r}(1-\rho_r^{*2})} G_{r,s}^{(2)}(y)$ can be represented by a similar expression. Hence

$$P\{U \le u, \; V \le \nu\} = \sum_{r=0}^{n} b(r; n, p) \sum_{s=0}^{\infty} nb\left(s; \frac{\nu_r^*}{2}, \rho_r^{*2}\right)$$

$$\left[1 - H_{r,s}\left(\frac{x}{c_{1r}(1-\rho_r^{*2})}\right)\right]\left[1 - H_{r,s}\left(\frac{y}{c_{2r}(1-\rho_r^{*2})}\right)\right], \tag{14.15}$$

where

$$H_{r,s}(z) = \int_0^z \frac{\exp\left(-\frac{t}{2}\right) t^{\frac{\nu_r^*}{2}+s-1}}{2^{\frac{\nu_r^*}{2}+s}\Gamma\left(\frac{\nu_r^*}{2}+s\right)} \, dt.$$

14.3.2   Moments of $\widehat{CP}_1, \widehat{CP}_2$

Using the results of Section 14.2, the moments of U and V can be found from the basic relations

$$U = \frac{CP_1\sqrt{\nu}}{\sqrt{X}}, \qquad V = \frac{CP_2\sqrt{\nu}}{\sqrt{Y}}.$$

Case I: $\sigma_{12} = 0$: For each r

$$E\left[\frac{1}{\sqrt{X}}\right] = \sum_{s=0}^{\infty} P_s\left(\frac{\lambda_{1r}}{2}\right) \frac{1}{\sqrt{2}} \frac{\Gamma\left(\frac{\nu-1}{2}+s\right)}{\Gamma\left(\frac{\nu}{2}+s\right)}$$

and hence

$$E[U] = CP_1 \frac{\sqrt{\nu}}{\sqrt{2}} \sum_{r=0}^{n} b(r; n, p) \sum_{s=0}^{\infty} P_s\left(\frac{\lambda_{1r}}{2}\right) \frac{\Gamma\left(\frac{\nu-1}{2}+s\right)}{\Gamma\left(\frac{\nu}{2}+s\right)}.$$

Since the inner summation can be written as

$$\exp\left(-\frac{\lambda_{1r}}{2}\right) \frac{\Gamma\left(\frac{\nu-1}{2}\right)}{\Gamma\left(\frac{\nu}{2}\right)} \sum_{s=0}^{\infty} \frac{\Gamma\left(\frac{\nu-1}{2}+s\right)}{\Gamma\left(\frac{\nu-1}{2}\right)} \frac{\Gamma\left(\frac{\nu}{2}\right)}{\Gamma\left(\frac{\nu}{2}+s\right)} \left(\frac{\lambda_{1r}}{2}\right)^s \frac{1}{s!}$$

$$= \frac{\Gamma\left(\frac{\nu-1}{2}\right)}{\Gamma\left(\frac{\nu}{2}\right)} \exp\left(-\frac{\lambda_{1r}}{2}\right) {}_1F_1\left[\frac{\nu-1}{2}; \frac{\nu}{2}; \frac{\lambda_{1r}}{2}\right];$$

therefore,

$$E(U) = E_N\left[\widehat{CP}_1\right] \sum_{r=0}^{n} b(r; n, p) \exp\left(-\frac{\lambda_{1r}}{2}\right) {}_1F_1\left[\frac{\nu-1}{2}; \frac{\nu}{2}; \frac{\lambda_{1r}}{2}\right], \tag{14.16}$$

where $E_N\left(\widehat{CP}_1\right)$ represents the expected value of $\widehat{CP}_1$ in sampling from a normal distribution and is equal to $CP_1\sqrt{\frac{\nu}{2}} \frac{\Gamma\left(\frac{\nu-1}{2}\right)}{\Gamma\left(\frac{\nu}{2}\right)}$. Similarly

$$E\left[U^2\right] = CP_1^2 \frac{\nu}{2} \sum_{r=0}^{n} b(r; n, p) \sum_{s=0}^{\infty} P_s\left(\frac{\lambda_{1r}}{2}\right) \frac{1}{\left(\frac{\nu}{2}+s-1\right)}. \tag{14.17}$$

The variance of U can be obtained from (14.16) and (14.17). The expectation of V and V^2 are similar in nature except that λ_{1r} is replaced by λ_{2r} and CP_1 by CP_2.

To determine the $\text{Cov}(U, V)$ we will need the joint moment $E(UV)$:

$$E(UV) = \nu CP_1 CP_2 E\left(\frac{1}{\sqrt{X}} \frac{1}{\sqrt{Y}}\right).$$

Since for each r, X and Y are independent noncentral χ^2 variables on ν degrees of freedom and noncentrality parameters λ_{1r} and λ_{2r}, respectively, we have

$$E[UV] = CP_1 CP_2 \frac{\nu}{2} \sum_{r=0}^{n} b(r; n, p) \sum_{s=0}^{\infty} P_s \left(\frac{\lambda_{1r}}{2}\right) \frac{\Gamma\left(\frac{\nu-1}{2} + s\right)}{\Gamma\left(\frac{\nu}{2} + s\right)}$$

$$\sum_{s=0}^{\infty} P_s \left(\frac{\lambda_{2r}}{2}\right) \frac{\Gamma\left(\frac{\nu-1}{2} + s\right)}{\Gamma\left(\frac{\nu}{2} + s\right)}. \tag{14.18}$$

Again using (14.16) and (14.17), the Cov (U, V) can be obtained.

The effects of contamination will be illustrated in Tables 14.1 and 14.2. The contamination is measured through the parameters p and a. For these calculations the form of nonnormality is assumed to be the same in the two marginal distributions; that is

$$\lambda_{1r} = \lambda_{2r} = \frac{r(n-r)}{n} a^2 \text{ with } a = \frac{\mu_1}{\sigma_1} = \frac{\mu_2}{\sigma_2}.$$

When a is small, there is little effect of the contamination. If a is large and p increases from 0 to 0.5, the effect of the nonnormality is accentuated.

From Table 14.1 we see that

(i) in the normal case there is always a positive bias for $\widehat{CP}_1$ and $\widehat{CP}_2$.

(ii) as the values of a and p increase (leading to a higher degree of nonnormality), the values of the bias and variance of $\widehat{CP}_1$ and $\widehat{CP}_2$ tend to decrease.

(iii) for larger values of a there is a more marked decrease in the bias and variance as p increases.

Table 14.1: Expectation and variance of $\widehat{CP}_i$ $(i = 1, 2)$ for Case I

	$E\left(\widehat{CP}_1\right) = E\left(\widehat{CP}_2\right)$				$\mathrm{Var}\left(\widehat{CP}_1\right) = \mathrm{Var}\left(\widehat{CP}_2\right)$			
	$CP_1 = CP_2$				$CP_1 = CP_2$			
$n = 10$	1.0	1.3	1.5	1.7	1.0	1.3	1.5	1.7
Normal	1.094	1.422	1.641	1.860	.0884	.1493	.1988	.2553
a,p								
.25, .1	1.091	1.418	1.637	1.855	.0878	.1485	.1977	.2539
.25, .5	1.086	1.412	1.629	1.846	.0870	.1470	.1957	.2514
1.00, .1	1.049	1.364	1.574	1.784	.0822	.1389	.1849	.2375
1.00, .5	0.976	1.269	1.464	1.659	.0683	.1155	.1538	.1975
2.00, .1	0.951	1.236	1.426	1.617	.0768	.1297	.1727	.2218
2.00, .5	0.759	0.987	1.139	1.291	.0336	.0568	.0757	.0972
$n = 30$								
Normal	1.027	1.335	1.540	1.746	.0197	.0333	.0443	.0569
a,p								
.25, .1	1.024	1.331	1.536	1.741	.0196	.0331	.0441	.0567
.25, .5	1.019	1.324	1.528	1.732	.0194	.0328	.0437	.0561
1.00, .1	0.984	1.279	1.475	1.673	.0184	.0311	.0414	.0531
1.00, .5	0.918	1.193	1.376	1.560	.0152	.0256	.0341	.0438
2.00, .1	0.884	1.150	1.327	1.504	.0173	.0292	.0389	.0499
2.00, .5	0.722	0.938	1.082	1.226	.0074	.0125	.0166	.0213

Table 14.2: Correlation coefficient of $\widehat{CP}_1, \widehat{CP}_2$ for Case I

	$n = 10$	$n = 30$
Normal	.0000	.0000
a,p		
.25, .1	$.7786 \times 10^{-4}$	$.1002 \; 10^{-3}$
.25, .5	$.1810 \times 10^{-4}$	$.7323 \; 10^{-5}$
1.00, .1	.01719	.02180
1.00, .5	.003793	.001371
2.00, .1	.1700	.2045
2.00, .5	.04061	.01190

In Table 14.2 we see that the correlation coefficient of $\widehat{CP}_1$ and $\widehat{CP}_2$

(i) is independent of the population values of CP_1 and CP_2.

(ii) is usually larger than that in the normal case.

(iii) is close to zero for small values of the parameter a.

(iv) approaches the value under normality if a is large and p increases.

Case II: $\sigma_{12} \neq 0$: Using the central approximation discussed in Section 14.2

$$E[U] = CP_1 \frac{\sqrt{\nu}}{\sqrt{2}} \sum_{r=0}^{n} b(r; n, p) \frac{1}{\sqrt{c_{1r}}} \frac{\Gamma\left(\frac{\nu_r^*}{2} - 1\right)}{\Gamma\left(\frac{\nu_r^*}{2}\right)} \tag{14.19}$$

and

$$E[U^2] = CP_1^2 \nu \sum_{r=0}^{n} b(r; n, p) \frac{1}{c_{1r}} \frac{1}{(\nu_r^* - 2)} \tag{14.20}$$

with similar results for $E(V)$ and $E(V^2)$.

As in Case I, $E(UV) = CP_1 CP_2 \nu E\left(\frac{1}{\sqrt{X}} \frac{1}{\sqrt{Y}}\right)$. For each r, $\left(\frac{X}{c_{1r}}, \frac{Y}{c_{2r}}\right)$ have the central bivariate chi-square distribution. This implies that for each r and s, $\frac{X}{c_{1r}}$ and $\frac{Y}{c_{2r}}$ are independent yielding, for each r

$$E\left[\frac{1}{\sqrt{X}} \frac{1}{\sqrt{Y}}\right] = \sum_{s=0}^{\infty} nb\left(s; \frac{\nu_r^*}{2}, \rho_r^{*2}\right) \frac{1}{(1 - \rho_r^{*2})^2} \frac{1}{c_{1r}} \frac{1}{c_{2r}}$$

$$\int_0^\infty \int_0^\infty \frac{1}{\sqrt{x}} \frac{1}{\sqrt{y}} g_{\nu_y^*+2s}\left\{\frac{x}{c_{1r}(1 - \rho_r^{*2})}\right\} g_{\nu_r^*+2s}\left\{\frac{y}{c_{2r}(1 - \rho_r^{*2})}\right\} dx dy, \tag{14.21}$$

where

$$g_{\nu_r^*+2s}(z) = \frac{e^{-\frac{z}{2}} z^{\frac{\nu_r^*}{2}+s-1}}{2^{\frac{\nu_r^*}{2}+s} \Gamma\left(\frac{\nu_r^*}{2} + s\right)}.$$

Hence, for each r

$$E\left[\frac{1}{\sqrt{X}} \frac{1}{\sqrt{Y}}\right] = \frac{1}{2} \frac{1}{(1 - \rho_r^{*2})^2} \frac{1}{\sqrt{c_{1r} c_{2r}}} \sum_{s=0}^{\infty} nb\left(s; \frac{\nu_r^*}{2}, \rho_r^{*2}\right) \frac{\Gamma^2\left(\frac{\nu_r^*-1}{2} + s\right)}{\Gamma^2\left(\frac{\nu_r^*}{2} + s\right)}$$

which yields

$$E[UV] = CP_1 CP_2 \frac{\nu}{2} \sum_{r=0}^{n} b(r; n, p) \frac{1}{(1 - \rho_r^{*2})^2} \frac{1}{\sqrt{c_{1r} c_{2r}}}$$

$$\times \sum_{s=0}^{\infty} nb\left(s; \frac{\nu_r^*}{2}, \rho_r^{*2}\right) \frac{\Gamma^2\left(\frac{\nu_r^*-1}{2} + s\right)}{\Gamma^2\left(\frac{\nu_r^*}{2} + 2\right)}. \tag{14.22}$$

Substituting for $nb\left(s; \frac{\nu_r^*}{2}, \rho_r^{*2}\right)$, the inner summation in (14.22) becomes

$$\sum_{s=0}^{\infty} \frac{\Gamma\left(\frac{\nu_r^*-1}{2} + s\right)}{\Gamma\left(\frac{\nu_r^*}{2} + s\right)} \frac{\Gamma\left(\frac{\nu_r^*-1}{2} + s\right)}{\Gamma\left(\frac{\nu_r^*}{2} + s\right)} \frac{\Gamma\left(\frac{\nu_r^*}{2} + s\right)}{\Gamma\left(\frac{\nu_r^*}{2}\right)} \frac{(\rho_r^{*2})}{s!} \left(1 - \rho_r^{*2}\right)^{\frac{\nu_r^*}{2}-1}$$

$$= \frac{\Gamma^2\left(\frac{\nu_r^*-1}{2}\right)}{\Gamma^2\left(\frac{\nu_r^*}{2}\right)} {}_2F_1\left(\frac{1}{2}, \frac{1}{2}; \frac{\nu_r^*}{2}; \rho_r^{*2}\right) \tag{14.23}$$

Table 14.3: Expectation and variance of $\widehat{CP}_i(i = 1, 2)$ for Case II

$n = 10$	$E\left(\widehat{CP}_1\right) = E\left(\widehat{CP}_2\right)$ $CP_1 = CP_2$				$E\left(\widehat{CP}_1\right) = E\left(\widehat{CP}_2\right)$ $CP_1 = CP_2$			
	1.0	1.3	1.5	1.7	1.0	1.3	1.5	1.7
Normal	1.094	1.423	1.641	1.860	.0884	.1493	.1988	.2553
a,p								
.25, .1	1.091	1.419	1.637	1.855	.0879	.1485	.1977	.2539
.25, .5	1.086	1.412	1.629	1.846	.0870	.1470	.1958	.2515
1.00, .1	1.053	1.372	1.583	1.794	0832	.1407	.1873	.2406
1.00, .5	0.999	1.299	1.499	1.699	.0738	.1248	.1661	.2134
2.00, .1	0.997	1.296	1.495	1.694	.0792	.1339	.1783	.2290
2.00, .5	0.895	1.163	1.342	1.521	.0594	.1004	.1336	.1716
$n = 30$								
Normal	1.027	1.335	1.540	1.746	.0197	.0333	.0443	.0569
a,p								
.25, .1	1.024	1.331	1.536	1.741	.0196	.0331	.0441	.0567
.25, .5	1.019	1.325	1.529	1.732	.0194	.0328	.0437	.0561
1.00, .1	0.988	1.284	1.482	1.680	.0185	.0313	.0417	.0536
1.00, .5	0.937	1.219	1.406	1.594	.0164	.0278	.0370	.0475
2.00, .1	0.920	1.196	1.380	1.564	.0173	.0292	.0389	.0500
2.00, .5	0.838	1.090	1.258	1.426	.0132	.0222	.0296	.0380

as obtained in Kocherlakota and Kocherlakota (1991). Therefore,

$$E[UV] = CP_1 CP_2 \sum_{r=0}^{n} b(r; n, p) \frac{1}{\sqrt{c_{1r} c_{2r}}} \frac{\nu}{\nu_r^*} c_{11r}^2 \kappa_r, \qquad (14.24)$$

where

$$c_{11r} = \sqrt{\frac{\nu_r^*}{2}} \frac{\Gamma\left(\frac{\nu_r^* - 1}{2}\right)}{\Gamma\left(\frac{\nu_r^*}{2}\right)}$$

and

$$\kappa_r = {}_2F_1\left(\frac{1}{2}, \frac{1}{2}, \frac{\nu_r^*}{2}; \rho_r^{*2}\right).$$

From these results the covariance of (U, V) can be obtained as usual.

The effects of contamination in Case II will be illustrated in Tables 3 and 4. As for Case I the form of nonnormality is assumed to be the same in the two marginal distributions; that is

$$\lambda_{1r} = \lambda_{2r} = \frac{r(n-r)}{n} a^2 \text{ with } a = \frac{\mu_1}{\sigma_1} = \frac{\mu_2}{\sigma_2}.$$

The results in Table 14.3 are similar to those obtained when $\sigma_{12} = 0$; that is, the values of the expectation and variance tend to decrease as the nonnormality increases and/or the sample size increases. The effect on the variance is again more marked than that on the expectation.

Table 14.4: Correlation coefficient of $\widehat{CP}_1, \widehat{CP}_2$ for Case II

	$n = 10$			$n = 30$		
ρ_{xy}	0.1	0.5	0.9	0.1	0.5	0.9
Normal	.0075	.1987	.7521	.0082	.2350	.7966
a,p						
.25, .1	.0084	.2011	.7532	.0103	.2377	.7976
.25, .5	.0098	.2053	.7553	.0120	.2424	.7995
1.00, .1	.0353	.2417	.7709	0436	.2866	.8147
1.00, .5	.0616	.2944	.7960	.0733	.3416	.8351
2.00, .1	.1456	.3518	.8111	.1788	.4234	.8569
2.00, .5	.2446	.4832	.8660	.2860	.5421	.8947

In Table 14.4 when ρ_{xy} is small, the estimators, $\widehat{CP}_1$ and $\widehat{CP}_2$, tend to be uncorrelated. However, when ρ_{xy} increases, the correlation between $\widehat{CP}_1$ and $\widehat{CP}_2$ becomes non-zero and tends to be quite large when $\rho_{xy} = .9$. The correlation is larger when n increases from 10 to 30.

14.4   Tests of Hypotheses

Kocherlakota and Kocherlakota (1991) consider the test of the hypothesis

$$H_0 : CP_1 \leq c_0 \ \text{ or } \ CP_2 \leq c_0 \ \text{ (process is } not \text{ capable)}$$

against

$$H_1 : CP_1 > c_0 \ \text{ and } \ CP_2 > c_0 \ \text{ (process is capable)},$$

with c_0 specified so that the desired level of process capability is attained. Note that H_0 is equivalent to 'at least one of CP_1 or $CP_2 \leq c_0$' with the alternative 'both CP_1 and $CP_2 > c_0$'. The critical region appropriate to this test is $\{\widehat{CP}_1 > c, \widehat{CP}_2 > c\}$, where c is determined so that the probability of rejection of H_0 when H_0 is true is α. Using the bivariate χ^2 distribution, they have given the following table of c for $\alpha = 0.05$ when $n = 10, 30$ and $\rho = 0.1(0.1)0.9$ with $c_0 = 1.3$.

For the test suggested here, the power when $CP_1 = CP_2 = CP$ is given by $P\{\widehat{CP}_1 > c, \widehat{CP}_2 > c | CP_1 = CP_2 = CP\}$. In this paper we confine our study to the case when $c_0 = 1.3$. This facilitates the use of the values of c given in Table 14.5.

Table 14.5: Values of c, determined from the bivariate χ^2 distribution such that $P\left\{\widehat{CP}_1 > c, \widehat{CP}_2 > c | c_0 = 1.3\right\} = .05$

ν	0.1	0.2	0.3	0.4	ρ_{xy} 0.5	0.6	0.7	0.8	0.9
9	1.6460	1.6528	1.6644	1.6815	1.7050	1.736	1.7782	1.8354	1.9196
29	1.4608	1.4644	1.4705	1.4793	1.4910	1.5061	1.5254	1.5503	1.5845

Details of the determination of c can be found in Kocherlakota and Kocherlakota (1991). Values of the power of the test for a variety of values of CP and for various levels of nonnormality are presented in Table 14.6.

Table 14.6: Power function for Case II using cut-off points based on the bivariate χ^2 distribution

		$n = 10$ CP			$n = 30$ CP		
	ρ_{xy}	1.3	1.5	1.7	1.3	1.5	1.7
Normal	.1	.0500	.1712	.3812	.0500	.3789	.8166
	.5	.0500	.1611	.3506	.0500	.3457	.7728
	.9	.0500	.1338	.2756	.0500		
a, p .25, .1	.1	.0487	.1678	.3758	.0475	.3690	.8092
	.5	.0489	.1584	.3460	.0478	.3370	.7651
	.9	.0491	.1317	.2681			
.25, .5	.1	.0466	.1620	.3663	.0434	.3520	.7962
	.5	.0471	.1536	.3379	.0442	.3222	.7513
	.9	.0475	.1280	.2620			
1.00, .1	.1	.0371	.1339	.3159	.0252	.2557	.6995
	.5	.0388	.1300	.2952	.0275	.2388	.6543
	.9	.0399	.1098	.2296			
1.00, .5	.1	.0221	.0221	.2296	.0083	.1305	.5158
	.5	.0255	.0910	.2218	.0107	.1296	.4805
	.9	.0274	.0789	.1737			

Two particular points of interest in Table 14.6 are: (i) For $CP = 1.3$, we have the value of the size of the test; that is, the observed level of significance. (ii) When a and p are small, the power is very close to that attained when sampling from a bivariate normal distribution. (iii) As n increases the power increases substantially.

Acknowledgements. The authors wish to thank the Natural Sciences and Engineering Research Council of Canada for its support of this research through grants made to each of them.

References

1. Chou, Youn-min and Owen, D. B. (1989). On the distribution of the estimated process capability indices, *Communication in Statistics - Theory and Methods*, **18**, 4549–4560.

2. Kane, V. E. (1986). Process capability indices, *Journal of Quality Technology*, **18**, 41–52.

3. Kocherlakota, S. and Kocherlakota K. (1991). Process capability index: bivariate normal distribution, *Communication in Statistics - Theory and Methods*, **20**, 2529–2547.

4. Kocherlakota, S., Kocherlakota, K. and Kirmani, S. N. U. A. (1992). Process capability indices under nonnormality, *International Journal of Mathematical and Statistical Sciences*, **1**, 175–210.

5. Kotz, S. and Johnson, N. L. (1993). *Process Capability Indices*, London: Chapman & Hall.

6. Steyn, H. S. and Roux, J. J. J. (1972). Approximations for the noncentral Wishart distribution, *South African Statistical Journal*, **6**, 165–173.

7. Tukey, J. W. (1960). A survey of sampling from contaminated distributions, In *Contributions to Probability and Statistics—Essays in Honor of Harold Hotelling* (Eds., Olkin et al.), Stanford, CA: Stanford University Press.

detection. The probability of successful detection,

$$PSD\left(\tau, d\right) = \Pr\left(t_A - \tau \leq d \, | t_A \geq \tau\right)$$

was suggested by Frisén (1992) as a measure of the performance. The PSD is better for the CUSUM than for the Shewhart method for $d = 3$. The shape of the curve for CUSUM with the present parameters is very similar to the constant curve for Shewhart. The Shiryaev-Roberts and the LR methods have a worse probability of detection of a change which happens early but better for late changes.

15.3.3   Predicted value

The predictive value $PV\left(t\right) = \Pr\left(\tau \leq t_A \, | t_A = t\right)$ has been used as a criterion of evaluation by Frisén (1992), Frisén and Åkermo (1993) and Frisén and Cassel (1994). The price for the high probability of detection of a change in the beginning of the surveillance for the CUSUM and the Shewhart method is that the early alarms are not reliable.

The predicted value of an alarm reflects the trust you should have in an alarm. The Shiryaev-Roberts method has a relatively constant predicted value. This means that the same kind of action is appropriate both for early and late alarms.

15.4   Summary and Concluding Remarks

Control methods based on likelihood ratios are known to have several optimality properties. When the methods are used in practice, knowledge about several characteristics of the method is important for the judgement of which action is appropriate at an alarm. At large, the properties differ between the LR methods on one hand and the Shewhart and the CUSUM methods on the other. In comparison with this, the choice of the intensity parameter of the LR method has very little influence on the performance. The predicted value of an alarm is relatively constant for the Shiryaev-Roberts method so that the same kind of action is appropriate both for early and late alarms.

Acknowledgements. This work was supported by the Swedish Council for Research in the Humanities and Social Sciences.

References

1. Frisén, M. (1992). Evaluations of methods for statistical surveillance, *Statistics in Medicine*, **11**, 1489–1502.

2. Frisén, M. (1994a). Statistical Surveillance of Business Cycles, *Research report*, 1994:1, Department of Statistics, Göteborg University.

3. Frisén, M. (1994b). A classified bibliography on statistical surveillance, *Research report*, 1994, Department of Statistics, Göteborg University.

4. Frisén, M. (1996). Characterization of methods for surveillance by optimality, *Unpublished manuscript*, Department of Statistics, Göteborg University.

5. Frisén, M. and Cassel, C. (1994). Visual evaluations of statistical surveillance, *Research report*, 1994:3, Department of Statistics, Göteborg University.

6. Frisén, M. and de Maré, J. (1991). Optimal surveillance, *Biometrika*, **78**, 271–280.

7. Frisén, M. and Åkermo, G. (1993). Comparison between two methods of surveillance: exponentially weighted moving average vs CUSUM, *Research report*, 1993:1, Department of Statistics, Göteborg University.

8. Frisén, M. and Wessman, P. (1996). Evaluations of likelihood ratio methods for surveillance, *Research report*, 1993:1, Department of Statistics, Göteborg University.

9. Girshick, M. A. and Rubin, H. (1952). A Bayes approach to a quality control model, *Annals of Mathematical Statistics*, **23**, 114–125.

10. Maré, J. de (1980). Optimal prediction of catastrophes with application to Gaussian processes, *Annals of Probability*, **8**, 841–850.

11. Mevorach, Y. and Pollak, M. (1991). A small sample size comparison of the CUSUM and Shiryaev-Roberts approaches to change point detection, *American Journal of Mathematical and Management Sciences*, **11**, 277–298.

12. Page, E. S. (1954). Continuous inspection schemes, *Biometrika*, **41**, 100–114.

13. Roberts, S. W. (1966). A comparison of some control chart procedures. *Technometrics*, **8**, 411–430.

14. Shewhart, W. A. (1931). *Economic control of quality of manufactured product*. Van Nostrand, New York.

15. Shiryaev, A. N. (1963). On optimum methods in quickest detection problems, *Theory Probab. Appl.*, **8**, 22–46.

16. Srivastava, M. S. and Wu, Y. (1993). Comparison of EWMA, CUSUM and Shiryaev-Roberts procedures for detecting a shift in the mean, *Annals of Statistics*, **21**, 645–670.

17. Wetherill, G. B. and Brown, D. W. (1990). *Statistical process control*, London: Chapman and Hall.

18. Wessman, P. (1996). Some principles for surveillance adopted for multivariate processes with a common change point, *Research report*, 1996:2, Department of Statistics, Göteborg University.

19. Zacks, S. (1983). Survey of classical and Bayesian approaches to the change-point problem: Fixed sample and sequential procedures of testing and estimation, In *Recent Advances in Statistics*, 245–269.

16

Properties of the Taguchi Capability Index for Markov Dependent Quality Characteristics

Erik Wallgren

University of Örebro/Uppsala, Örebro, Sweden

Abstract: Most process capability indices have been constructed under the assumption that quality characteristics are measurements of independent, identically distributed normal variables. Confidence intervals for the Taguchi process capability index, commonly denoted C_{pm}, is derived in this paper when consecutive measurements are observations of dependent variables according to a Markov process in discrete time.

Keywords and phrases: Autocorrelation, capability index, coverage rates, lower confidence limits, simulation

16.1 Introduction

Process Capability Indices (PCI) are used in manufacturing processes for the double purpose of monitoring a production process and, according to contract agreements, supplying customers with a condensed measure of a lot's quality. The PCIs have become widely used for several reasons. They are easy to compute (but perhaps not so easy to interpret), and they provide a single number summary of a process, which is attractive for producers as well as customers.

The variability of a quality characteristic is in general caused by many different factors. Shewhart (1931) makes a distinction between different causes of variation, and he identifies two components, one steady component which appears to be inherent in the process and another one, more or less distinguishable, due to chance.

The process is said to be stable, or to be 'in a steady state', or 'in statistical control' if the process is free of intermittent components. Shewhart (1931) formulated this as: *'A phenomenon will be said to be controlled when,*

*through the use of past experience, we can predict, at least within limits, how
the phenomenon may be expected to vary in the future.'* To be able to predict
the future the process must behave in such a way that it is possible to assign
a probability model to the quality characteristic, and a process with this property is then said to be stable. Deming (1986) strengthens this further and also
requires that the quality characteristics, at least with respect to their means
and standard deviations, must stay constant over time. Deming also discusses
the advantages of a stable process and says, *'it gives an identity to the process,
it has a measurable, communicable capability.'*

In this paper we study the implications, with regard to distribution, calculation and interpretation, of different estimators of the Taguchi index [Chan et
al. (1988)] when the sample consists of serially correlated random variables.

16.2   Earlier Results

The history of PCI is now about twenty years long. It emanates from the
field of statistical process control where several PCIs, and their advantages
and disadvantages are discussed. A comprehensive account is given in the
book by Kotz and Johnson (1993) which covers the following three indices with
modifications

$$C_p = \frac{\text{USL - LSL}}{6\sigma} \tag{16.1}$$

$$C_{pk} = \frac{\min(\text{USL} - \mu, \ \mu - \text{LSL})}{3\sigma} \tag{16.2}$$

$$C_{pm} = \frac{\text{USL - LSL}}{6\sqrt{\text{MSE}}} \tag{16.3}$$

T, USL and LSL are the target and the upper and the lower specification
limits for a quality characteristic of a production process; $\text{MSE} = \sigma^2 + (\mu - T)^2$
where μ and σ are the mean and standard deviation for the actual characteristic.

Vännman (1995) makes a unified approach to capability indices, by introducing the index $C_p(u, v)$, defined as

$$C_p(u, v) = \frac{d - u|\mu - M|}{3\sqrt{\sigma^2 + v(\mu - T)^2}} \tag{16.4}$$

where $d = (\text{USL-LSL})/2$, $M = (\text{USL} + \text{LSL})/2$ and u and v are non-negative
constants. For different values of u and v, the earlier capability indices become
special cases of $C_p(u, v)$:

$$C_p(0, 0) = C_p, \ C_p(0, 1) = C_{pm} \text{ and } C_p(1, 1) = C_{pk}$$

Criticisms have been raised regarding PCI. Gunter (1989) says: *'The greatest abuse of C_{pk} that I have seen is that it becomes a mindless effort that managers confuse with real statistical process control efforts.'* But with knowledge of a specific PCI and its interpretation, and with an awareness of its weaknesses, the use of a PCI can contribute to the communication between a customer and a producer.

So far the PCIs studied have been based on the assumption that measurements of a process can be treated as observations of independent, identically distributed (IID) random variables, most commonly normally distributed with the parameters μ and σ. However, this assumption is very restrictive, and during recent years efforts have been made to broaden these assumptions in different directions.

Franklin and Wasserman (1992) use bootstrap methods to study the effect of non-normality of the quality characteristic. They estimate coverage rates for bootstrap confidence limits for the indices C_p, C_{pk} and C_{pm}. For skewed distributed quality characteristics, they show that the coverage rate is closer to the nominal confidence level than if the confidence limit is estimated as described in the literature, e.g. Kotz and Johnson (1993).

Carlsson (1995) enlarges the area of application by showing that PCIs also can be interpreted in the common industrial situation with an instability in the form of repeatedly occurring shifts in a production, i.e. one series of observations, varying around one level of process mean, is followed by a series varying around another process mean.

16.3 The Taguchi Index for Autocorrelated Variables

Another common industrial situation is where consecutive measurements from a process are serially correlated, i.e. the condition of *independent* random variables is not fulfilled. A typical example of such a process is the bursting strength of craft liner where consecutive measurements tend be to serially correlated. Such random variables can be modeled in different ways. In this paper, however, the study is restricted to an autoregressive model of order 1.

16.4 The Model

Suppose that n consecutive observations, X_t, $t = 1, 2, \ldots, n$ are taken from a production process with the upper and the lower specification levels USL and

LSL respectively, and assume that the observations follow a first-order autoregressive model, AR(1), with identically, independently and normally distributed errors with variance σ^2, $\varepsilon_t \sim IIDN(0,\sigma^2), t = 1,\ldots,n$. Further assume that the process is, at least, second order stationary. That is,

$$X_t - \mu = \rho(X_{t-1} - \mu) + \varepsilon_t \qquad (16.5)$$

where μ is the actual process mean , ρ the 1st order autocorrelation. Then

$$V(X_t) = \sigma^2 \cdot \frac{1}{1-\rho^2} \qquad (16.6)$$

and the covariances

$$\mathrm{cov}(X_t, X_{t+h}) = \rho^{|h|} \frac{\sigma^2}{1-\rho^2}, \quad h = \pm 1, \pm 2, \ldots \qquad (16.7)$$

By using the variance (1.6) the C_{pm} index for a quality characteristic generated by an AR(1)-process can be defined as

$$C_{pmr} = \frac{\text{USL - LSL}}{6\sqrt{\mathrm{MSE}_r}} \qquad (16.8)$$

where $\mathrm{MSE}_r = \sigma_r^2 + (\mu - T)^2$ and

$$\sigma_r^2 = \frac{\sigma^2}{(1-\rho^2)} \qquad (16.9)$$

The extra subscript 'r' is introduced to indicate the presence of autocorrelation.

16.5   Estimation of C_{pmr}

Both the C_{pm} and the C_{pmr} indices have $\tau = (\mathrm{MSE})^{1/2}$ in the denominator. Following Taguchi (1985), who suggested $\hat{\tau} = \{\frac{1}{n}\sum_{i=1}^n (X_i - T)^2\}^{1/2}$ as a natural estimator for τ, an estimator for the C_{pmr} index is then $\hat{C}_{pmr} = $ (USL - LSL)$/(6\hat{\tau})$. The mean and variance of $\hat{\tau}^2$ can be found, after simple but tedious calculations. The expected value becomes

$$E(\hat{\tau}^2) = \sigma_r^2(1 + \theta^2) \qquad (16.10)$$

where

$$\theta^2 = \left(\frac{\mu - T}{\sigma_r}\right)^2 \qquad (16.11)$$

and the variance of $\hat{\tau}^2$, for large samples, is approximately

$$V(\hat{\tau}^2) = \frac{2}{n}(\sigma_r^2)^2 \cdot \left[\frac{1+\rho^2}{1-\rho^2} + 2\theta^2\frac{1+\rho}{1-\rho}\right] \qquad (16.12)$$

The distribution of $\hat{\tau}^2$ depends on whether the process is on target or not, and if the process is autocorrelated or not. The distribution of $n\hat{\tau}^2/\sigma_r^2$ is summarized in the following table:

	$\rho = 0$	$\rho \neq 0$
$\mu = T$	central χ^2 distribution	not known (but resembles a central χ^2 distribution)
$\mu \neq T$	non-central χ^2 distribution	not known (but resembles a non-central χ^2 distribution)

It can be found by studying the moment generating function that, when $\rho \neq 0$, $\hat{\tau}^2$ is not non-central χ^2-distributed, not even asymptotically. But due to the resemblance it may be reasonable to believe that it can be approximated to a scaled central χ^2-distribution by equating the first two moments, Patnaik (1949). With this assumption, and following Boyles (1991), an approximate $100(1-\alpha)\%$ lower confidence limit for C_{pmr} is

$$\hat{C}_{pmr} \cdot (\chi_\alpha^2(\hat{V})/\hat{V})^{1/2} \qquad (16.13)$$

where $\chi_\alpha^2(\hat{v})$ is the lower $\alpha\%$ percentile for a central χ^2-variable with $\hat{v}$ d.f.,

$$\hat{v} = n(1+\hat{\theta}^2)^2/(\frac{1+\hat{\rho}^2}{1-\hat{\rho}^2} + 2\hat{\theta}^2\frac{1+\hat{\rho}}{1-\hat{\rho}}), \qquad (16.14)$$

$$\hat{\theta}^2 = (\hat{\mu}-T)^2/(\frac{\hat{\sigma}^2}{1-\hat{\rho}^2}) \qquad (16.15)$$

and $\hat{\mu}$, $\hat{\sigma}$ and $\hat{\rho}$ are ML-estimators.

For large sample sizes, the chi-square distribution can be approximated by a normal distribution, and a lower confidence limit is given by

$$\hat{C}_{pmr} \cdot (1 - z_{1-\alpha}/(2\hat{v})^{1/2}) \qquad (16.16)$$

where z_α is the lower $\alpha\%$ percentile of a standard normal distribution.

An upper confidence limit as well as a symmetric two-sided confidence interval for C_{pmr} can be calculated in a similar way.

16.6   A Simulation Study

A large simulation was done to study the empirical percentage coverage rate (EPCR) for the lower confidence limit for C_{pmr}. The objective was to compare the case where the autocorrelation in the observations was *ignored* in the estimation, with the case where the autocorrelation in the observations was *included* in the estimation.

Two different processes, A and B, also used by Boyles (1991), were considered. In both processes the lower and the upper specification limits are 35 and 65 respectively and the process target was 50. For process A, the process mean μ=50 and standard deviation σ=5 which implies that $C_{pm} = 1$. For process B μ=57.5 and σ=2.5 with C_{pm}=.63

The results of the simulations are summarized in the following four figures.

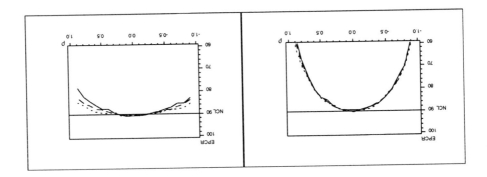

Figure 16.1: EPCR for autocorrelation (-.9, .9). Process A. Sample sizes 20 (—), 50 (- - -) and 100 (···). NCL=90%. Left panel: autocorrelation ignored. Right panel: autocorrelation included

In Figure 16.1 process A is used at NCL = 90%. The left panel shows that the empirical coverage rate is only close to the nominal confidence level for correlations closed to 0, but in the right panel, the empirical coverage rate is close to the nominal confidence level for correlations $\rho < |.6|$.

In Figure 16.2 the same pattern can be found but the empirical coverage rate is now close to the nominal confidence level for a wider range of correlations. Here the NCL = 99%.

In Figure 16.3 process B is used and the NCL is 90%. The left panel shows that the empirical coverage rate is close to the NCL for correlations only close to 0 or -.8. In the right panel, where autocorrelation is included, the empirical coverage rate is close to the NCL for correlations $\rho < |.6|$, especially for large sample sizes.

In Figure 16.4 the same pattern as in Figure 16.3 can be found. Note that, in the left panel, the short interval between the 99% NCL and the maximal 100% coverage rate makes the result for negative correlations look deceivingly good. In the right panel it can be seen that the coverage rate is close to the NCL for almost all correlations.

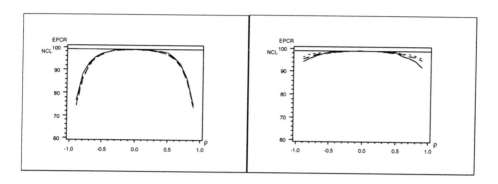

Figure 16.2: EPCR for autocorrelation (-.9, .9). Process A. Sample sizes 20 (—), 50 (- - -) and 100 (···). NCL=99%. Left panel: autocorrelation ignored. Right panel: autocorrelation included

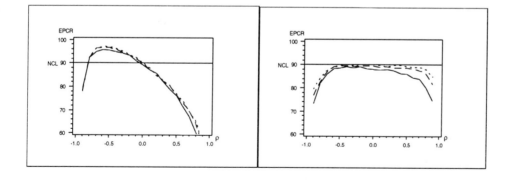

Figure 16.3: EPCR for autocorrelation (-.9, .9). Process B. Sample sizes 20 (—), 50 (- - -) and 100 (···). NCL=90%. Left panel: autocorrelation ignored. Right panel: autocorrelation included

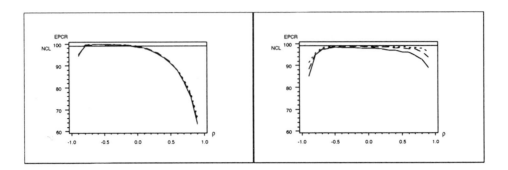

Figure 16.4: EPCR for autocorrelation (-.9, .9) Process B. Sample sizes 20 (—), 50 (- - -) and 100 ($\cdots$). NCL=99%. Left panel: autocorrelation ignored. Right panel: autocorrelation included

16.7 Summary and Concluding Remarks

When the Taguchi index for a quality characteristic is estimated from consecutive measurements that are autocorrelated, the coverage rate for the lower confidence limit for the index is too low. Including an estimate of the autocorrelation in the confidence limits gives coverage rates very close to nominal levels, especially for $\rho < |.7|$ and for large sample sizes.

References

1. Boyles, R. A. (1991). The Taguchi capability index, *Journal of Quality Technology*, **23**, 18–26.

2. Carlsson, O. (1995). On the use of the Cpm index for detecting instability of a production process, *Economic Quality Control*, **10**, 39–50.

3. Chan, L. K., Cheng, S. W. and Spiring, F. A. (1988). A new measure of process capability C_{pm}, *Journal of Quality Technology*, **20**, 160–175.

4. Deming, W. E. (1986). *Out of the crisis*, Cambridge University Press, Cambridge.

5. Franklin, L. A. and Wasserman, G. S. (1992). A note on the conservative nature of the tables of lower confidence limits for C_{pk} with a suggested correction, *Communications in Statistics—Simulation and Computation*, **23**, 1165–1169.

6. Gunter, B. (1989). The use and abuse of C_{pk}: parts 1-4, *Quality Progress*, 1989.

7. Kotz, S. and Johnson, N. L. (1993). *Process capability indices*, Chapman & Hall, London.

8. Patnaik, P. B. (1949). The non-central χ^2 and F-distributions and their applications, *Biometrika*, **36**, 202–232.

9. Shewhart, W. A. (1931). *Economic control of quality of manufactured product*, Van Nostrand, New York.

10. Taguchi, G. (1985). A tutorial on quality control and assurance- the Taguchi methods, *1985 ASA Annual Meeting*, Las Vegas, Nevada.

11. Vännman, K. (1995). A unified approach to capability indices, *Statistica Sinica*, **5**, (**2**), 805–822.

PART III
DESIGN AND ANALYSIS OF EXPERIMENTS

17

Fast Model Search for Designed Experiments with Complex Aliasing

Hugh A. Chipman

University of Waterloo, Waterloo, Ontario, Canada

Abstract: In screening experiments, run size considerations can necessitate the use of designs with complex aliasing patterns. Such designs provide an opportunity to examine interactions and other higher order terms as possible predictors, as Hamada and Wu (1992) propose. The large number of model terms mean that many models may describe the data well. Bayesian stochastic search methods that incorporate preferences for certain model structures (Chipman, Hamada, and Wu, 1997) are one effective method for identifying a variety of models. This paper introduces priors that allow the posterior to be simplified, speeding up the search and eliminating Monte Carlo error in the evaluation of model probabilities. Default choices of prior parameters are proposed. Several new plots and summaries are illustrated using simulated data in a Plackett-Burman 12-run layout.

Keywords and phrases: Automatic prior selection, Gibbs sampler, interactions, partial aliasing, Plackett-Burman designs

17.1 Introduction

When screening many variables, designs with economical run sizes are desirable. Quite often these designs will not be orthogonal if terms other than main effects are considered. A common example is the 12-run Plackett-Burman (1946) (PB) design (see Table 17.1), in which each main effect is correlated (partially aliased) with the 45 two-factor interactions not involving that effect.

Table 17.1: Screening experiment with Plackett-Burman 12-run design and response data

A	B	C	D	E	F	G	H	I	J	K	Y_1	Y_2
+	+	−	+	+	+	−	−	−	+	−	1.058	-1.358
+	−	+	+	+	−	−	−	+	−	+	1.004	1.228
−	+	+	+	−	−	−	+	−	+	+	-5.200	2.291
+	+	+	−	−	−	+	−	+	+	−	5.320	9.432
+	+	−	−	−	+	−	+	+	−	+	1.022	-5.719
+	−	−	−	+	−	+	+	−	+	+	-2.471	-2.417
−	−	−	+	−	+	+	−	+	+	+	2.809	-2.494
−	−	+	−	+	+	−	+	+	+	−	-1.272	2.674
−	+	−	+	+	−	+	+	+	−	−	-0.955	-5.943
+	−	+	+	−	+	+	+	−	−	−	0.644	1.596
−	+	+	−	+	+	+	−	−	−	+	-5.025	6.682
−	−	−	−	−	−	−	−	−	−	−	3.060	-5.973

Hamada and Wu (1992) viewed designs with complex aliasing as an opportunity to identify promising interactions as well as main effects. With a modified stepwise algorithm, they showed that promising models could be identified when only a few effects were large (the assumption of *effect sparsity*), and attention was focused on interactions between large main effects (the assumption of *effect heredity*). While their stepwise algorithm finds promising models, it is not always able to identify all promising models. Chipman, Hamada, and Wu (1997) (CHW) gave a more thorough model search algorithm. This approach used a Bayesian model, and the Stochastic Search Variable Selection (SSVS) algorithm of George and McCulloch (1993) (henceforth GM93). This methodology was able to search the model space more completely while incorporating effect heredity.

This paper considers a number of enhancements to CHW. First, it uses a "fast" version of SSVS proposed by George and McCulloch (1997) (henceforth GM97). This formulation allows analytical evaluation of relative probabilities on models, rather than relying upon Monte Carlo integration, as in CHW. The ability to analytically evaluate relative probabilities also helps identify stopping conditions for the algorithm. Second, the priors used are discussed in more detail, and automatic choices for all prior parameters are given, simplifying the use of this approach for non-Bayesians who wish to identify promising models. Third, a number of graphics and summaries are introduced to identify promising models found by the procedure. These include a C_p - like plot that divides up model probability according to model size, a time series plot of the probability of models visited, and reweighting of the posterior to assess prior influence.

The paper is organized as follows. Section 17.2 outlines the priors used, and in Section 17.3 the Gibbs sampler is described for this situation. An example in Section 17.4 illustrates automatic choices of prior parameter. Section 17.5

gives an example in which multiple models fit the data well. The impact of different priors on models is examined, and several new plots and summaries are introduced to examine the promising models identified. Section 17.6 discusses related issues.

17.2 Priors

For the regression model, $Y = X\beta + \sigma\epsilon$, with ϵ a standard normal random variable, the parameters of interest are β, σ. Some of the k columns of X (such as interactions) may be formed from the original variables.

In variable selection, some of the k elements of β may be negligible. This is captured in GM93 by augmenting the model with an unobserved k-vector γ of 0's and 1's. A 0 corresponds to a negligible coefficient, and a 1 corresponds to an important coefficient. The prior is specified as

$$\Pr(\gamma, \sigma, \beta) = \Pr(\beta|\sigma, \gamma) \Pr(\gamma) \Pr(\sigma)$$

with
$$\beta_i \sim \text{i.i.d. } N(0, \sigma^2 \tau_i^2 (1 - \gamma_i) + \sigma^2 \tau_i^2 c_i^2 \gamma_i), \qquad i = 1, \dots, k$$

By choosing $c_i \gg 1$, the prior variance of β_i is $\sigma^2 \tau_i^2$ when $\gamma_i = 0$ and c_i^2 times larger when $\gamma_i = 1$. The hyperparameters c_i, τ_i are chosen to indicate magnitudes of small and large effects. Section 17.4 suggests how these may be selected automatically.

As in GM93,
$$\sigma^2 \sim \text{IG}(\nu/2, \nu\lambda/2),$$

where IG denotes an inverted gamma distribution. This is equivalent to $\nu\lambda/\sigma^2 \sim \chi_\nu^2$.

This prior formulation differs from that of GM93 and CHW, in that the variance of β_i depends on σ. With this prior, the posterior for (β, σ, γ) can be integrated over σ, β. As discussed in Section 17.3, this facilitates model search. The posterior for γ is given by (see GM97):

$$\Pr(\gamma|Y) \propto |\tilde{X}'\tilde{X}|^{-1/2} |D_\gamma|^{-1} (\lambda\nu + S_\gamma^2)^{(n+\nu)/2} \Pr(\gamma) = g(\gamma), \qquad (17.1)$$

where D_γ is diagonal with ith element $\tau_i(1 - \gamma_i) + c_i\tau_i\gamma_i$,

$$S_\gamma^2 = \tilde{Y}'\tilde{Y} - \tilde{Y}'\tilde{X}(\tilde{X}'\tilde{X})^{-1}\tilde{X}'\tilde{Y},$$

$$\tilde{Y} = \begin{bmatrix} Y \\ 0 \end{bmatrix} \quad \text{and} \quad \tilde{X} = \begin{bmatrix} X \\ D_\gamma^{-1} \end{bmatrix}.$$

The remaining component of the prior is for γ, which puts probability on the space of all possible models. When interactions and other related predictors are present, it is not practical to assume that all elements of γ are independent. Instead, a dependence structure for related predictors [Chipman (1996)] is used. This prior consists of a product of k probabilities:

$$\Pr(\gamma) = \prod_{i=1}^{k} \Pr(\gamma_i | \text{Parents}(\gamma_i)) \tag{17.2}$$

The probability that a given term is active or inactive depends on its "parent" terms, typically taken to be those terms of the next lowest order from which the given term may be formed. For example, main effects A and B would have no parents, and an interaction AB would have two parents (A and B). The corresponding elements of (17.2) would be $\Pr(\gamma_A), \Pr(\gamma_B)$ and $\Pr(\gamma_{AB}|\gamma_A, \gamma_B)$. The prior is specified by choosing marginal probabilities that a main effect is active, and the conditional probability that an interaction is active:

$$P(\gamma_{AB} = 1 | \gamma_A, \gamma_B) = \begin{cases} p_{00} & \text{if } (\gamma_A, \gamma_B) = (0,0) \\ p_{01} & \text{if } (\gamma_A, \gamma_B) = (0,1) \\ p_{10} & \text{if } (\gamma_A, \gamma_B) = (1,0) \\ p_{11} & \text{if } (\gamma_A, \gamma_B) = (1,1) \end{cases} . \tag{17.3}$$

Choosing $p_{00} = p_{10} = p_{01} = 0, p_{11} > 0$ allows an interaction to be active only if both corresponding main effects are active (referred to as *strong heredity*). Choosing $p_{00} = 0, p_{01}, p_{10}, p_{11} > 0$ allows an interaction to be active if one or more of its parents are active (*weak heredity*). Models obeying strong heredity are usually easier to interpret, while weak heredity may help the stochastic search explore the model space. Typical values of $(p_{00}, p_{10}, p_{01}, p_{11})$ might be (0, 0, 0, 0.25) for strong heredity, and (0, 0.10, 0.10, 0.25) for weak heredity. Values less than 0.5 represent the belief that only a few effects are likely to be active [see Box and Meyer (1986) for a similar argument].

17.3 Efficient Stochastic Search

The large number of models makes it impractical to exhaustively evaluate the posterior probability for each model. Instead the Gibbs sampler is used to make draws from the posterior distribution. In the context of model selection, this algorithm may be thought of as a stochastic search.

For this problem, the Gibbs sampler (see Smith and Roberts (1993) and Gelfand and Smith (1990)) may be summarized as follows. The algorithm makes use of "full conditional" distributions, which specify the conditional distribution of one parameter given all others (and the data). The algorithm starts with initial values of all parameters, and then repeatedly draws each parameter conditional on all the others and the data:

0. Start with $\gamma^0 = (\gamma_1^0, \gamma_2^0, \ldots, \gamma_k^0)$

1. Draw γ_1^1 from $p(\gamma_1 | \gamma_2^0, \ldots, \gamma_k^0, Y)$.

2. Draw γ_2^1 from $p(\gamma_2 | \gamma_1^1, \gamma_3^0, \ldots, \gamma_k^0, Y)$

$\vdots$

k. Draw γ_k^1 from $p(\gamma_k | \gamma_1^1, \ldots, \gamma_{k-1}^1, Y)$.

Each draw is from a Bernoulli distribution. Steps 1 to k are repeated a large number of times, each time conditioning on the most recently drawn values of the other elements of γ. The sequence $\gamma^0, \gamma^1, \gamma^2, \ldots$ converges to the posterior for γ.

A similar scheme was used in CHW, but with the parameters β and σ included in the Gibbs sampler rather than being integrated out. Elimination of β and σ from the Gibbs sampler speeds up the algorithm [see GM97 and Liu, Wong, and Kong (1995)]. By reducing the dimensionality of the parameter space, the Gibbs sampler is able to move around faster.

An additional advantage is the ability to analytically evaluate the posterior probability of a model up to a normalizing constant, using (17.1). The posterior probability of a model γ' is then

$$\Pr(\gamma' | Y) = g(\gamma') \bigg/ \sum_{i=1}^{2^k} g(\gamma_i). \tag{17.4}$$

The normalizing constant in the denominator entails evaluation of the posterior probability for all models, which is prohibitive. For any given set of models, relative probabilities can be evaluated by replacing the denominator of (17.4) with a sum of g over all models visited so far. Probabilities estimated in this way will be too large, but if most of the "good" models have been visited, this provides a way to remove sometimes sizeable Monte Carlo error. CHW experienced this error because their posterior probability estimates were based on frequencies.

17.4 Automatic Selection of Prior Parameters

In this section, an example illustrates automatic choices for prior parameters $c_i, \tau_i, \nu, \lambda$, and the model prior for γ. The choices proposed here are easier to implement than those described in CHW.

The data are simulated from a simple model considered by Hamada and Wu (1992): $Y = A + 2AB + 2AC + \sigma\epsilon$ with $\sigma = 0.25$ and ϵ a standard normal.

Eleven predictors are arranged in a 12-run PB design, shown in Table 17.1 along with the response Y_1.

In choosing the prior for σ^2, ν and λ are chosen so the upper tail is some large value and the middle is near the anticipated residual variance. The prior expected value of σ^2 is

$$E(\sigma^2) = \frac{\lambda\nu}{\nu - 2} \qquad \text{for } \nu > 2,$$

suggesting that λ be chosen near the expected residual variance. In the absence of expert knowledge, some fraction of the unconditional variance of the response could be used to estimate λ. CHW propose

$$\boxed{\lambda = \text{Var}(Y)/25.}$$

In this example, $\text{Var}(Y) = 10$ so $\lambda = 0.40$ is selected.

The parameter ν acts as degrees of freedom, with larger values corresponding to a distribution that is tighter about λ. A sufficiently diffuse prior may be selected by choosing ν so that the upper tail (say the 99th percentile) is roughly equal to the unconditional variance. Table 17.2 gives various quantiles for an inverse gamma with $\lambda = 1$. Choosing $\nu = 5$ would place the 99th percentile of the prior at 9.02λ, for example. Smaller degrees of freedom are possible (for example values of $\nu = 1.5$ were used in CHW), although they can lead to unreasonably long tails, because $\text{Var}(\sigma^2)$ is not defined for $\nu \le 4$.

Table 17.2: Quantiles of an Inverse Gamma distribution with $\lambda = 1$

ν	mean	0.01	0.1	0.5	0.9	0.99
1	–	0.15	0.37	2.2	63.33	6365
2	–	0.22	0.43	1.44	9.49	99.50
3	3	0.26	0.48	1.27	5.13	26.13
4	2	0.30	0.51	1.19	3.76	13.46
5	1.67	0.33	0.54	1.15	3.10	9.02
6	1.5	0.36	0.56	1.12	2.72	6.88
7	1.4	0.38	0.58	1.10	2.47	5.65
8	1.33	0.40	0.60	1.09	2.29	4.86
9	1.29	0.42	0.61	1.08	2.16	4.31
10	1.25	0.43	0.63	1.07	2.06	3.91

In general, one would choose

$$\boxed{\nu = 5 \ \text{ or from Table 17.2.}}$$

The prior parameters c_i and τ_i may be chosen as

$$\boxed{c_i = 10, \tau_i = \frac{1}{3 \times \text{range}(X_i)}.} \qquad (17.5)$$

Box and Meyer (1986) suggest $c = 10$, separating large and small coefficients by an order of magnitude. Choice of τ_i is motivated by the fact that a small coefficient has standard deviation $\sigma\tau_i$, and will lie within $0 \pm 3\sigma\tau_i = 0 \pm \sigma/\text{range}(X_i)$ with very high probability. Even a large change in X_i (say of magnitude comparable to range(X_i)) will result in a change in Y of no more than σ, which is presumably small.

The variable selection procedure can be quite sensitive to the choice of τ_i (see CHW and GM93). Equation (17.5) captures the relative magnitudes of the τ_i for different variables, but the overall magnitude may need tuning. Box and Meyer (1993) and CHW propose methods for tuning based on runs of the search algorithm. A faster alternative based on predictive distributions is proposed here. For any given model γ, the expected value of Y for a given X may be calculated. The magnitude of τ_i will determine the degree of shrinkage for coefficient β_i, in a manner similar to ridge regression. A simple way to assess the value of τ_i is to see how the predictions vary for a range of values $r\tau_i$ for a single given model. A good τ_i value would be the smallest value not shrinking predictions too much.

The posterior mean for β is given by

$$\hat{\beta}_\gamma = (X'X + D_\gamma^{-2})^{-1}X'Y$$

where D_γ is diagonal with elements $\tau_i(1 - \gamma_i) + \tau_i c_i \gamma_i$. See GM97 for details.

Figure 17.1 plots predicted values for the original 12 design points for $r \in (1/10, 10)$. The model used (A, B, C, AB, AC) was identified by stepwise regression and a subsequent "hereditization". The "1" value on the horizontal axis is the default choice (17.5) for τ. In this case, the default seems quite reasonable, as any smaller multiples would shrink too much.

The prior on γ may be selected by considering the prior expected number of active terms. If we have

$$k = \text{number of main effects}$$
$$p = \text{Pr(main effect active)}$$
$$p_{01} = p_{10} = \text{Pr(interaction active|one of its parents is active)}$$
$$p_{11} = \text{Pr(interaction active|both its parents are active)},$$

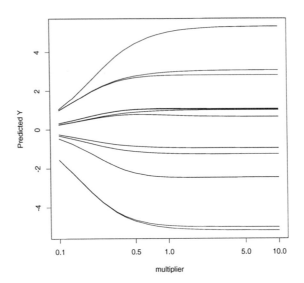

Figure 17.1: Predicted response under original design for various multiples of parameter τ

then under weak heredity

$$\text{E}(\# \text{ effects}) = \text{E}_f[\text{E}(\# \text{ effects}|f \text{ main effects active})]$$

$$= \text{E}_f(f + \binom{f}{2}p_{11} + f(k-f)p_{01}) \qquad (17.6)$$

The first element is the number of main effects. The second is the expected number of active interactions with both parents active. There are $\binom{f}{2} = f(f-1)/2$ such terms and each one has a probability of p_{11} of being active. The third term is the expected number of active interactions with exactly one parent active. These are interactions between one of f active parents and $k-f$ inactive parents. Consequently there are $f(k-f)$ allowable terms, and each has probability p_{01} of being active. Since f is Binomial(k, p), $\text{E}(f) = kp$ and $\text{E}(f^2) = kp(1-p+kp)$. Simplification of (17.6) gives:

$$\text{E}(\# \text{ effects}) = kp + p_{11}\text{E}(f(f-1)/2) + p_{01}\text{E}(fk - f^2)$$

$$= kp + p_{11}k(k-1)p^2/2 + p_{01}k(k-1)p(1-p) \quad (17.7)$$

In this example strong heredity is used so 2.75 active main effects and 0.86 active interactions are expected.

1000 draws from the posterior on γ are generated using the Gibbs sampler and the integrated posterior. A total of 333 different models were visited. The most probable model is A, B, C, AB, AC (probability = 0.55), followed by A,

B, C, AB, AC, BC (0.06) and eight other models, each of which contains A, B, C, AB, AC, and one other main effect (roughly 0.03 probability each). It is quite clear that A, B, C, AB, AC are active. Although main effects B and C are not important, they are included by the strong heredity prior.

17.5 A More Difficult Problem

The design for this example remains the same, and the new response Y_2 is given in Table 17.1. Data are generated from $Y = 2A + 4C + 2BC + 2CD + \sigma\epsilon$ with $\sigma = 0.5$ and ϵ a standard normal.

Prior parameters $\nu = 5, \lambda = \text{Var}(Y)/25, c_i = 10, \tau_i = 1/3\text{range}(X_i)$ were chosen as in the previous section. A number of priors on the model space are explored:

1. Strong heredity with $(p, p_{11}, p_{01}) = (.25, .25, 0)$, yielding 3.61 effects expected to be active.

2. Weak heredity with $(p, p_{11}, p_{01}) = (.25, .25, .10)$, yielding 5.67 effects expected to be active.

3. Independence prior with $(p, p_{11}, p_{01}) = (0.25, 0.1, 0.1)$ and $p_{00} = 0.10$. This yields $2.75 + 55 \times (0.10) = 8.25$ effects expected to be active.

1000 iterations of the Gibbs sampler were used in each of the three cases. The best models of each size for each of the three priors are displayed in Tables 17.3 and 17.4. Since a variety of priors on model size seem plausible, this dependence can be reduced by reporting results conditional on model size, rather than just giving the most probable models (as in CHW).

The main effect for C is active, but models containing a variety of other terms explain the data well. Relaxing the prior (from strong to weak or weak to independence) produces models with better fit for a given number of effects. These models may be more difficult to interpret, especially in the independence case where numerous two way interactions are included without any corresponding main effects.

The true model (A, C, BC, CD) is the most likely of its size in the weak heredity case. The true model has no probability in the strong heredity case. The closest model under strong heredity would be A, B, C, D, BC, CD, which was not visited in the first 1000 steps. The size of this model (almost twice the number of active terms as expected) and the many six term models that fit well may explain why it was not visited.

Table 17.3: Models identified under strong and weak heredity priors. The most probable models of each size are given

Strong Heredity

Model	Prob	R^2
C	0.183	0.712
C J	0.077	0.794
C H	0.059	0.782
C G	0.045	0.77
C H J	0.027	0.864
C D C:D	0.024	0.84
C G J	0.019	0.852
B C H B:H	0.033	0.923
C G J C:G	0.019	0.928
C H J C:H	0.012	0.914
B C H B:C B:H	0.017	0.958
C D G C:D D:G	0.01	0.948
B C H J B:H	0.006	0.957
B C H I B:C B:H	0.003	0.989
B C H B:C B:H C:H	0.002	0.961
C D G C:D C:G D:G	0.002	0.961

Weak Heredity

Model	Prob	R^2
C	0.221	0.712
C C:D	0.065	0.812
C J	0.04	0.794
C B:C	0.031	0.782
C H B:H	0.028	0.887
C E E:I	0.021	0.917
C I E:I	0.021	0.917
A C B:C C:D	0.02	0.992
C D C:D D:G	0.012	0.927
B C B:C B:H	0.01	0.923
A C D B:C C:D	0.002	0.994
C G H B:H G:H	0.002	0.953
B C B:C B:H C:F	0.002	0.961

Table 17.4: Models identified under independence prior. The most probable models of each size are given

Independence

Model	Prob	R^2
C H	0.068	0.782
C E	0.016	0.712
C J:K	0.013	0.746
C J E:I	0.115	0.935
C B:H E:J	0.101	0.923
C H E:I	0.09	0.929
C H J E:I	0.03	0.953
C H E:I I:K	0.025	0.965
C G J E:I	0.022	0.947
B C J A:F D:F	0.007	0.993
C G H A:B D:G	0.005	0.991
A B C D:G I:J	0.004	0.955
C D J E:K F:I G:H	0.002	0.995
C E H B:E B:H E:J	0.001	0.991
C D E B:H C:F E:J	0.001	0.992

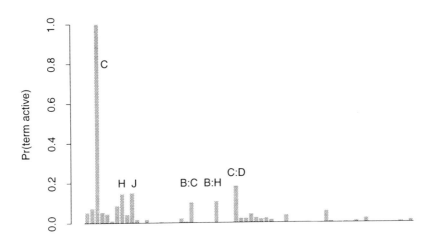

Figure 17.2: Marginal probability that an effect is large, weak heredity case

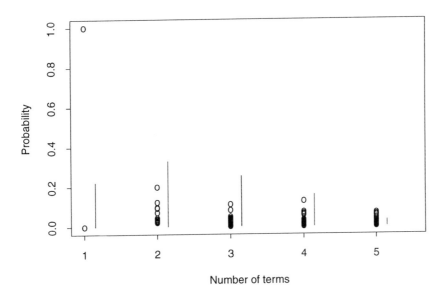

Figure 17.3: The most probable models ordered by size. The vertical bars give total probability for all models of this size and the dots give probabilities for specific models conditional on that model size

Figure 17.2 gives marginal probabilities for the weak heredity case. The main effect for C has the largest probability of being active, and other possibly active effects include H, J, BC, BH, CD. The fact that none of the latter effects has a large probability indicates considerable uncertainty about which effects in addition to C are active. Figure 17.3 is a C_p-like plot, giving the conditional probability of models, conditional on model size (plotted as circles). The marginal probability associated with all models of a certain size is represented with a vertical line. This plot indicates that the most likely model sizes are 1, 2, 3 or 4 terms. Among the one term models, the model with C only dominates, but in other cases a number of models are close.

The different models identified in Tables 17.3 and 17.4 indicate the strong influence of the prior. To illustrate this effect, another weak heredity prior is used with $p = p_{01} = p_{11} = 0.5$. This prior is uniform on all weak heredity models. Rather than re-run the Gibbs sampler, probabilities are recalculated (Table 17.5) for models found with the original prior. The large change in the posterior probabilities indicates that the prior penalizes larger models heavily.

Table 17.5: Posterior probabilities on models, weak heredity case. Second "unweighted" probability column corresponds to a weak heredity prior with all nonzero probabilities equal to 0.5

Model	Prob	Unweighted Prob	R^2
C	0.221	0	0.712
C C:D	0.065	0.001	0.812
C J	0.04	0.001	0.794
C B:C	0.031	0.001	0.782
C H	0.03	0.001	0.782
C H B:H	0.028	0.005	0.887
C G	0.023	0	0.77
C E E:I	0.021	0.004	0.917
C I E:I	0.021	0.004	0.917
A C B:C C:D	0.02	0.031	0.992

One practical issue is how long to run the chain. When probable models are no longer being discovered, the chain can be stopped. Figure 17.4 presents these relative probabilities for an extended run of 5000 iterations, under strong heredity priors. Each vertical line is the probability of a model the first time it is visited. After the first 1000 iterations, new models are still being visited, but none with appreciable probability, suggesting that 1000 runs is sufficient. The solid line gives the cumulative probability, and indicates that about 75% of the probability thus far identified was discovered in the first 1000 steps.

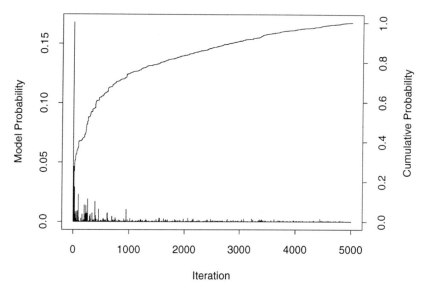

Figure 17.4: Time series plot of relative probabilities of models on the first time each is visited. The solid line represents the cumulative probability of all models visited so far

In many problems the algorithm is quite fast and computing time inexpensive enough that longer runs are practical. In situations where as many models as possible are desired, long runs would be preferred. Each of the 1000 iteration runs here took 50 seconds on a 200 Mhz Pentium Pro system. The relatively small number of iterations used here (1000) was chosen to illustrate that considerable information could be quickly extracted from the data.

17.6   Discussion

This paper illustrates that choice of prior parameters for variable selection is relatively straightforward. Considerations such as the implied shrinkage of predictions, anticipated number of active terms, and anticipated residual error can be used to provide semi-automatic choices of prior parameters.

Interpretation of output is also an issue, especially when many models may describe the data well. The posterior can sometimes favor models with a certain number of active terms. To explore a wider range of model sizes, graphics and tables that give the best models conditional on size are useful.

Stochastic model search using an integrated posterior offers several advantages over Markov chain Monte Carlo methods in which all parameters are drawn. Convergence will be faster because fewer parameters are sampled at each iteration. The relative log posterior probability of a model is available at any point in the simulation, aiding the decision of when to stop the chain.

One disadvantage of this approach is that it will not generalize as easily to other models (generalized linear models, for example). In such cases it may be difficult to integrate the posterior over parameters such as β. Gibbs sampling methods which draw all parameters may be more promising in such situations.

References

1. Box, G. E. P. and Meyer, R. D. (1986). An analysis for unreplicated fractional factorials, *Technometrics*, **28**, 11–18.

2. Box, G. E. P. and Meyer, R. D. (1993). Finding the active factors in fractionated screening experiments, *Journal of Quality Technology*, **25**, 94–104.

3. Chipman, H. (1996). Bayesian variable selection with related predictors, *Canadian Journal of Statistics*, **24**, 17–36.

4. Chipman, H., Hamada, M. and Wu, C. F. J., (1997). A Bayesian variable selection approach for analyzing designed experiments with complex aliasing, *Technometrics*, **39**, 372–381.

5. Gelfand, A. E. and Smith, A. F. M. (1990). Sampling-based approaches to calculating marginal densities, *Journal of the American Statistical Association*, **85**, 398–409.

6. George, E. I. and McCulloch. R. E. (1993). Variable selection via Gibbs sampling, *Journal of the American Statistical Association*, **88**, 881–889.

7. George, E. I. and McCulloch. R. E. (1997). Approaches to variable selection, *Statistica Sinica*, **7**, 339–373.

8. Hamada, M. and Wu, C. F. J. (1992). Analysis of designed experiments with complex aliasing, *Journal of Quality Technology*, **24**, 130–137.

9. Liu, J. S., Wong, W. H. and Kong, A. (1995). Covariance structure and convergence rate of the Gibbs sampler with various scans, *Journal of the Royal Statistical Society, Series B*, **57**, 157–169.

10. Plackett, R. L. and Burman, J. P. (1946). The design of optimum multifactorial experiments, *Biometrika*, **33**, 305–325.

11. Smith, A. F. M. and Roberts, G. O. (1993). Bayesian computation via the Gibbs sampler and related Markov chain Monte Carlo methods, *Journal of the Royal Statistical Society, Series B*, **55**, 3–23.

18

Optimal 12 Run Designs

K. Vijayan and K. R. Shah
University of Western Australia, Perth, Australia
University of Waterloo, Waterloo, Ontario, Canada

Abstract: In this paper we consider twelve run designs with four factors each at 2 levels. We present a specific design and show that it is optimal when one wishes to estimate all the main effects and a subset of two factor interactions. The optimality criteria used are D-optimality and a form of E-optimality

Keywords and phrases: D-optimality, E-optimality, fractional factorial designs, hidden projection, interaction, Plackett-Burman designs

18.1   Introduction

Plackett and Burman (1946) gave optimal designs for the estimation of main effects where the interactions are assumed to be absent or negligible. Recently, Lin and Draper (1992) and independently Box and Bisgaard (1993) pointed out some interesting projection properties of these designs. Lin and Draper showed that the 12 run Plackett and Burman design when projected on any four factors, needs only one additional run to make it into a 2^{4-1} design with resolution four.

More recently, Wang and Wu (1995) pointed out that even when the projection does not lead to a full factorial design or to a fractional factorial design with a high resolution, one may be able to estimate certain interactions without needing any additional runs. They showed that in a 12 run Plackett and Burman design with four factors all the two factor interactions can be estimated. They called this property of the Plackett and Burman designs the hidden projection property.

In industrial experiments the cost of running an experiment can often be very high and hence it is important to use a highly efficient design using only a limited number of runs. In this paper, we are concerned with optimal 12 run

designs with four factors each at two levels. We put forward a design which has
D-optimality and a form of E-optimality for the estimation of the main effects
and for some of the two factor interactions assuming the absence of higher
order interactions. More specifically, it is shown that the proposed design is
D-optimal if we wish to estimate at least one interaction and is E-optimal if
we wish to estimate at least two interactions.

In Section 18.2 we introduce the notation and establish some preliminary
results on the structure of the designs. A candidate for the optimal design is
presented in Section 18.3. Optimality for the estimation of 4, 5 or 6 interactions
is established in Sections 18.4 and 18.5. In Section 18.6 we prove the optimality
for the estimation of 1, 2 or 3 interactions. As noted earlier, our design is
not E-optimal when one wishes to estimate all the main effects and only one
interaction (assuming the remaining effects to be negligible).

18.2   Notation and Preliminaries

We consider designs with 12 runs involving four factors each at two levels de-
noted by ± 1. The levels of factor i will be denoted by a 12×1 vector $\boldsymbol{x}_i$ each
element of which is ± 1. Thus, the *design matrix* is given by the 12×4 matrix

$$D = [\boldsymbol{x}_1 | \boldsymbol{x}_2 | \boldsymbol{x}_3 | \boldsymbol{x}_4]. \tag{18.1}$$

In the model for the experiment if we include all the interaction effects we
have 16 parameters and since we have only 12 observations, these cannot be
all estimated. We shall assume that the interactions involving three or four
factors are absent. Thus, we have at most 11 parameters. Effects which are
present in the model will be called *admissible* whereas the others will be called
inadmissible.

In our notation $\boldsymbol{xy}$ (to be called the product of $\boldsymbol{x}$ and $\boldsymbol{y}$) will indicate a
vector with elements obtained by taking component-wise product of elements
of $\boldsymbol{x}$ and $\boldsymbol{y}$. Similarly, we define $\boldsymbol{xyz}$ etc. The 12×1 vector with each element
unity will be denoted by $\boldsymbol{1}$.

Let p denote the number of admissible effects. We shall assume that the
main effects of the four factors (and the general mean) are always admissible
and hence p takes values between 5 and 11. Our observational equations are

$$E(\boldsymbol{y}) = \boldsymbol{X\beta} \tag{18.2}$$

where $\boldsymbol{y}$ is a 12×1 vector of observations, $\boldsymbol{X}$ is a $12 \times p$ matrix with columns given
by $\boldsymbol{1}$, the four vectors of factor levels and the $(p-5)$ vectors which correspond
to the products of the $\boldsymbol{x}_i$'s corresponding to the two-factor interactions to be
included in the model. Finally, $\boldsymbol{\beta}$ is a $p \times 1$ vector of all the parameters for
admissible effects.

The matrix $K = X^t X$ will be called the *information matrix* for the design. Under the usual assumptions for the linear models, the variance-covariance matrix for the estimates of β is given by $K^{-1}\sigma^2$ where σ^2 is the error variance for a single observation.

We shall consider two optimality criteria in this paper. The first criterion refers to the maximum variance over all the elements of the estimate of β (other than the general mean). Thus, we shall call a design E^*-optimal if it minimizes this variance. The other criterion which we shall consider is the D-optimality which requires maximization of det K (where det A denotes the determinant of the matrix A).

We shall call an effect odd (even) if the corresponding vector has odd (even) number of +1's. This will apply to all the 16 effects including the inadmissible ones. It is easy to see that the product of two effects of the same parity (both odd or both even) is even whereas the product of two effects of unlike parity is odd. A design has all even effects or it has 8 odd effects and 8 even effects.

If x and y have the same property, $x^t y = 0$ mod 4 whereas if they have different parity $x^t y = 2$ mod 4. In what follows, two effects x and y will be said to be orthogonal (not orthogonal) if $x^t y = 0$ $(\neq 0)$. Further, x and y are said to be connected (not connected) if $x^t y = 4$ mod 8 (0 mod 8). A design will be said to have property Δ if there exists three effects x_1, x_2 and x_3 having the same parity such that no two of them are orthogonal i.e. $x_i^t x_j \neq 0$ for $i \neq j$. The above concepts turn out to be very useful in studying the structure of the twelve run designs. We shall now prove the following Lemmas which will be used later.

Lemma 18.2.1 *If x, y, z and w are four even effects such that no two of them are connected then 1 is connected with $xyzw$.*

PROOF. We note that $x^t\{(x + y)(x + z)(x + w)\}$ must be 0 mod 8. Further, $x = x^3$ and $x^2 = x^4 = 1$. Using $x^t y = x^t z = x^t w = y^t z = y^t w = z^t w = 0$ mod 8 it follows that $1^t xyzw = x^t yzw = 4$ mod 8. ∎

We now consider the case where all the 16 effects are even.

Lemma 18.2.2 *Let $S(x)$ denote the set of all effects which are connected with effect x. If the design does not have property Δ, $S(x)$ contains 5 effects. Further, these effects are pairwise orthogonal.*

PROOF It is clear that the number of elements in $S(x)$ is the same for all x. Thus, without loss of generality, we can take $x = 1$.

Let $x_1, x_2, ...x_k$ denote a maximal set of effects in $S(1)$ so that no effect in this set is generated by the others in this set. We shall prove that $k = 4$.

Suppose $k < 4$. Since $|S(x)|$ is the same for all x, it follows that any effect in the set G generated by $x_1, x_2, ...x_k$ is not connected with any effect not in G. Further, since the design does not have property Δ, $x_1, x_2, ...x_k$ are pairwise orthogonal. Let $x_{k+1}, x_{k+2}, ... x_4$ denote the set of effects such that x_1, x_2, x_3, x_4 is a complete set of generators. It is easy to see that these can always be chosen to be pairwise not connected. Since x_1, x_2, x_3, x_4 are pairwise not connected, by Lemma 2.1, $1^t(x_1x_2x_3x_4) = 4$ mod 8 which means that 1 is connected with $x_1x_2x_3x_4$. This contradicts the assumption that $k < 4$. Thus, $k = 4$.

Since x_1, x_2, x_3, x_4 all belong to $S(1)$, these are pairwise orthogonal. It is easy to show that $1^t(xyz) = 0$ mod 8 for i, j, k distinct. Thus, only effects connected with 1 are x_1, x_2, x_3, x_4 and $x_1x_2x_3x_4$. Again, since the design does not have property Δ, these five effects are pairwise orthogonal. Thus, every effect is connected with five effects and is not connected to ten effects. ■

We shall now establish a result which gives the structure of the design when all the effects are even.

Lemma 18.2.3 *If all the effects are even and if the design does not have property Δ, the sixteen effects can be partitioned into sets G_1, G_2, G_3, G_4 containing four effects each such that (i) effects in each set are pairwise orthogonal, (ii) each effect in $G_1(G_2)$ is connected with 3 effects in $G_2(G_1)$ but is connected with only one effect in G_3 or G_4 and (iii) each effect in $G_3(G_4)$ is connected with 3 effects in $G_4(G_3)$ but is connected with only one effect in G_1 or G_2.*

PROOF. It is easy to see that we can find effects a and b which are mutually orthogonal and are also orthogonal to 1. Let G_1 be the subgroup $(1, a, b, ab)$. Let G_2^* denote a coset (e, ea, eb, eab). Since $1^t(1+a)(1+b)(1+e) = 0$ mod 8, it follows that 1 is connected with odd number of effects in G_2^*. Thus, if we write G_2, G_3, G_4 as cosets of G_1 1 is connected with odd number of effects in each of G_2, G_3 and G_4. Since 1 is connected to exactly five effects it must be connected with three effects in one of these. We can choose G_2 to be the coset with which 1 has three connections. Further, effects in each G_i can be seen to be mutually orthogonal. This completes the proof.   ■

When the design has both odd effects and even effects and does not have property Δ we can still form a subgroup $G_1 = (1, a, b, ab)$ containing mutually orthogonal even effects. We can also form a coset G_2 containing the other even effects. However, each effect in $G_1(G_2)$ is connected to p effects in $G_2(G_1)$ where p is either 1 or 3.

It would be useful to make some remarks on the arguments used in the proofs given in Sections 18.4, 18.5 and 18.6. In regression analysis, it is well known that if an effect is estimated using only a subset of the effects in the actual model, the variance is less than the variance of the estimate using all the effects in the model. It is also well known that the determinant of the information matrix is not greater than the product of the determinants of the diagonal sub-matrices. The first result is used for E^*-optimality whereas the second result is used for D-optimality.

18.3 Optimal Design

In the succeeding sections we shall prove the optimality of the design described in this section. This optimality holds for $p > 6$. When $p = 6$ our design is D-optimal but another D-optimal design is also available. Further, our design is not E^*-optimal for $p = 6$. When $p = 5$, i.e. when there are no interactions, a Plackett-Burman design is known to be optimal.

Let d^* be the design for which the design matrix $\boldsymbol{D}_{d*}$ is given by

$$
\boldsymbol{D}_{d*}^{t} =
\begin{bmatrix}
1 & 1 & 1 & 1 & 1 & 1 & -1 & -1 & -1 & -1 & -1 & -1 \\
1 & 1 & 1 & -1 & -1 & -1 & 1 & 1 & 1 & -1 & -1 & -1 \\
1 & 1 & -1 & -1 & -1 & 1 & -1 & -1 & 1 & 1 & 1 & -1 \\
1 & 1 & 1 & 1 & -1 & -1 & 1 & -1 & -1 & 1 & 1 & 1
\end{bmatrix}
$$

For $p = 11$, if we write the effects in the order $(F_1, F_2 F_3, F_1 F_4)$, $(F_2, F_1 F_3, F_2 F_4)$, $(F_3, F_1 F_2, F_3 F_4)$, (F_4, μ) where μ denotes the general mean, it is easy to verify that the information matrix $\boldsymbol{K}$ is given by

$$
\boldsymbol{K} =
\begin{pmatrix}
A & 0 & 0 & 0 \\
0 & A & 0 & 0 \\
0 & 0 & A & 0 \\
0 & 0 & 0 & B
\end{pmatrix}
$$

where $\boldsymbol{A} = \begin{pmatrix} 12 & 4 & 4 \\ 4 & 12 & 4 \\ 4 & 4 & 12 \end{pmatrix}$ and $\boldsymbol{B} = \begin{pmatrix} 12 & 4 \\ 4 & 12 \end{pmatrix}$.

It is easy to verify that the maximum variance for the estimate of any effect is $1/10\sigma^2$. Further, $\det \boldsymbol{K} = 2^{31} \cdot 5^3$.

To compare our design with the Plackett-Burman design given in Wang and Wu (1995) we compute the D-efficiency for the overall design and the D_s-efficiencies for the individual effects. These are given in the following table. Corresponding values for the Plackett-Burman design as given by Wang and Wu are given in the parentheses. For the sake of completeness we give here the expressions for the D-efficiency and the D_s-efficiency in terms of our notation.

$$
\begin{aligned}
D\text{-efficiency} &= (\det \boldsymbol{K}/12)^{1/k} \quad \text{where, } k = p - 1. \\
D_s\text{-efficiency} &= 1/12\mathrm{Var}(\text{estimate}).
\end{aligned}
$$

Table 18.1: Estimation efficiencies for the overall design
and for individual effects

Case	Overall	1	2	3	4	12	13	14	23	24	34
1	.95 (.95)	1 (1)	1 (1)	.89 (.88)	.89 (.88)	.89 (.78)	-	-	-	-	-
2	.94 (.92)	1 (1)	.89 (.87)	.89 (.87)	.89 (.75)	.89 (.76)	.89 (.76)	-	-	-	-
3	.93 (.89)	1 (.85)	1 (.85)	.83 (.85)	.89 (.85)	.83 (.63)	-	-	-	-	.83 (.63)
4	.93 (.89)	.89 (1)	.89 (.74)	.89 (.74)	.89 (.74)	.89 (.74)	.89 (.74)	.89 (.74)	-	-	-
5	.93 (.89)	.89 (.87)	.89 (.87)	.89 (.87)	.89 (.62)	.89 (.74)	.89 (.74)	-	.89 (.74)	-	-
6	.93 (.87)	1 (.85)	.83 (.85)	.89 (.74)	.89 (.74)	.89 (.76)	.83 (.63)	-	-	.83 (.63)	-
7	.92 (.85)	.83 (.85)	.89 (.73)	.89 (.73)	.89 (.62)	.89 (.73)	.89 (.73)	.83 (.62)	.83 (.62)	-	-
8	.91 (.83)	1 (.73)	.83 (.73)	.83 (.73)	.89 (.73)	.83 (.63)	.83 (.63)	-	-	.83 (.63)	.83 (.63)
9	.91 (.82)	.83 (.72)	.83 (.72)	.89 (.62)	.89 (.62)	.89 (.72)	.83 (.62)	.83 (.62)	.83 (.62)	.83 (.62)	-
10	.90 (.80)	.83 (.62)	.83 (.62)	.83 (.62)	.89 (.62)	.83 (.62)	.83 (.62)	.83 (.62)	.83 (.62)	.83 (.62)	.83 (.62)

18.4 E^* - optimality $(p \geq 9)$

To establish the E^*-optimality of the design d^*, we shall first prove the following proposition.

Proposition 18.4.1 *If for any design we have three effects y_1, y_2, y_3 such that $y_i^t y_j = \pm 4$ for $i \neq j$, then for $p \geq 9$, the maximum variance for the estimate of an admissible effect is at least $\sigma^2/10$.*

PROOF: We note that the 16 effects form a group under multiplication. Write $G_1 = (1, y_1 y_2, y_1 y_3, y_2 y_3)$. Clearly, G_1 is a subgroup. Let G_2, G_3, G_4 denote the cosets of G_1. In the matrix $\boldsymbol{X}^t\boldsymbol{X}$, the diagonal sub-matrix corresponding to the elements of any G_i can be written as

$$
\boldsymbol{A} = \begin{pmatrix}
12 & 4 & 4 & \alpha \\
4 & 12 & \alpha & 4 \\
4 & \alpha & 12 & 4 \\
\alpha & 4 & 4 & 12
\end{pmatrix}
$$

where α is ± 4. Let p_i denote the number of admissible effects from G_i. Let $\boldsymbol{B}_i$ denote the matrix obtained from A by retaining only the rows and columns corresponding to the admissible effects from G_i. The diagonal elements of $\boldsymbol{B}_i^{-1} \cdot \sigma^2$ are lower bounds for the variances of the estimates of the corresponding effects. If any p_i is 3, the maximum diagonal element of $\boldsymbol{B}_i^{-1}$ is at least $1/10$ whereas if any p_i is 4, the maximum diagonal element of $\boldsymbol{B}_i^{-1}$ exceeds $1/10$. Since $p \geq 9$, at least one p_i must exceed 2. This completes the proof. ∎

It is easy to see that the results also hold if any $\boldsymbol{y}_i^t \boldsymbol{y}_j$ is ± 6 or ± 8. Thus, we can assume that $|\boldsymbol{y}_i^t \boldsymbol{y}_j| \leq 4$ and that the design does not have property Δ.

Henceforth, we shall assume that the designs under consideration do not have property Δ. To further restrict the class of competing designs we state the following proposition.

Proposition 18.4.2 *If a design has three admissible effects for which the corresponding sub-matrix of $\boldsymbol{X}^t\boldsymbol{X}$ has a row $(12, 4, 4)$, the maximum variance for the estimate of one of these effects exceeds $\sigma^2/10$.*

PROOF: Direct verification. ∎

We shall now consider only those designs which do not satisfy the conditions of Proposition 18.4.2.

To establish E^*-optimality of d^* for $p \geq 9$ we shall first consider competing designs where the effects are all even. Again, we shall assume that the competing designs do *not* have property Δ. It is easy to see that in this case we can find three mutually orthogonal effects $\boldsymbol{y}_1$, $\boldsymbol{y}_2$, $\boldsymbol{y}_3$ (not necessarily admissible). Write $G = (\mathbf{1}, \boldsymbol{y}_1\boldsymbol{y}_2, \boldsymbol{y}_1\boldsymbol{y}_3, \boldsymbol{y}_2\boldsymbol{y}_3)$ and form cosets w.r.t. G. We note that the elements of each coset are mutually orthogonal and that the product of any three elements of each coset is the remaining element of that coset. By Lemma 18.2.1, each coset must contain at least one element not orthogonal to 1. Denote these elements from the three cosets by $\boldsymbol{\xi}_1$, $\boldsymbol{\xi}_2$, $\boldsymbol{\xi}_3$. Now form sub-group $G^* = (\mathbf{1}, \boldsymbol{\xi}_1, \boldsymbol{\xi}_2, \boldsymbol{\xi}_1\boldsymbol{\xi}_2)$. Since the design does not have property Δ, 1 is orthogonal to $\boldsymbol{\xi}_1\boldsymbol{\xi}_2$. Each coset of G^* has the property that each element of the coset has one element orthogonal to it and two elements not orthogonal to it. If the number of admissible effects in any coset exceeds 2, by Proposition 18.4.2, the design cannot be E^*-optimal.

We now deal with the case where there are 8 odd effects and 8 even effects. We restrict attention to designs which do not have property Δ and do not satisfy conditions of Proposition 18.4.2.

As remarked in Section 18.2 we can form a subgroup $G_1 = (1, a, b, ab)$ containing mutually orthogonal even effects. It was also remarked that the coset G_2 containing the remaining effects has the property that each effect in $G_1(G_2)$ is connected with 1 or 3 effect in $G_2(G_1)$. If we wish to exclude designs which satisfy conditions of Proposition 18.4.2, we can have only one connection between an effect from G_1 to an effect in G_2. Thus, we can divide the eight even effects into four pairs such that the inner product of the effects in any pair is 4 whereas two effects not from the same pair are orthogonal. Similarly we can divide the eight odd effects into four pairs with a similar property. Inner product of any even effect with any odd effect is ± 2.

To establish E^*-optimality of d^* we successively eliminate classes of competing designs. Let x_1 and x_2 be two admissible connected effects of the same parity with inner product 4 and let x_3 and x_4 be two mutually orthogonal admissible effects of the opposite parity. Let $x_1^t x_3 = g$, $x_2^t x_3 = h$, $x_1^t x_4 = p$, $x_2^t x_4 = q$ where each of g, h, p, q is ± 2. Inverse of the covariance matrix for x_1, x_2 adjusted for x_3 and x_4 can be seen to have diagonal element greater than $\frac{1}{10}$ unless $gh + pq = 8$ i.e. unless $g = h$ and $p = q$. Thus, we can assume that if x_1 and x_2 have inner product 4 and if x_3 and x_4 are mutually orthogonal of the other parity $x_1^t x_3 = x_2^t x_3$ and $x_1^t x_4 = x_2^t x_4$.

Next, if we have four mutually orthogonal admissible effects x_3, x_4, x_5 and x_6, we must have $x_1^t x_i = x_2^t x_i$ for $i = 3, 4, 5$ or 6. In this case the inverse of the diagonal element of the covariance matrix adjusted for x_3, x_4, x_5 and x_6 has diagonal elements $\frac{1}{10}$ and hence such a design cannot be superior to d^*.

Next we consider the case where x_1 and x_2 have $x_1^t x_2 = 4$ and x_3, x_4, x_5, x_6, x_7 are admissible effects of the other parity. Each of the x_i's with $i \geq 3$ has another effect x_j with $j \geq 3$ such that $x_i^t x_j = 0$. Hence $x_1^t x_i = x_2^t x_i$, for $3 \leq i \leq 7$. Now, H, the part of K relating to $x_1, \cdots, x_7$ will have the following form

$$H = \begin{pmatrix} 12 & 4 & u \\ 4 & 12 & u \\ u^t & u^t & P \end{pmatrix}$$

where u is a 5×1 matrix with each element ± 2 whereas P is a 5×5 matrix corresponding to the effects x_3 to x_7. Since these are of the same parity, these contain one or two connected pairs. From this it can be deduced that the inner-product matrix of x_1, x_2 adjusted for x_3 to x_7 is $\begin{pmatrix} 12 - a & 4 - a \\ 4 - a & 12 - a \end{pmatrix}$ where $a \geq \frac{4}{3}$. From this it follows that the variance of the estimate of x_1 is at least $\sigma^2/10$.

We now consider partitioning all the admissible effects into r effects of one parity and $p - r$ effects of the other parity with r effects of one parity and $p - r$

effects of the other parity with $r \leq p - r$. From the above, we only need to consider the case where the effects in the first set are mutually orthogonal and $r \leq 3$. The inner product matrix for $x_1, \cdots, x_3$ adjusted for the others can be seen to have the inverse with a diagonal element not less than $1/10$ and this completes the proof of E^*-optimality of d^*.

18.5 D-optimality $(p \geq 9)$

We first show that a design with property Δ cannot have det $H > 2^{28} \cdot 10^{(p-8)}$ ($= \det H$ for d^*). Let d be a design with effects x_1, x_2, x_3 such that $x_1^t x_2 = x_1^t x_3 = 4$ and $x_2^t x_3 = \pm 4$. Now, $G = (1, x_1 x_2, x_1 x_3, x_2 x_3)$ is a sub-group. Form cosets $G_1 (= G)$, G_2, G_3 and G_4. The matrix of inner products of the elements of G_i is

$$L = \begin{pmatrix} 12 & 4 & 4 & \alpha \\ 4 & 12 & \alpha & 4 \\ 4 & \alpha & 12 & 4 \\ \alpha & 4 & 4 & 12 \end{pmatrix}$$

where $\alpha = \pm 4$. Let p_i denote the number of admissible effects from G_i and let H_i denote the $p_i \times p_i$ sub-matrix of L which corresponds to the admissible effects from G_i. It is clear that

$$\det H \leq \prod_{i=1}^{4} \det H_i \leq \prod_{i=1}^{4} \left(8^{p_i - 1} (8 + 4p_i) \right) = 8^{p-4} \prod_{i=1}^{4} (8 + 4p_i).$$

For a fixed value of $p = \sum p_i$, this is maximum when the p_i's are as nearly equal as possible. This happens when each p_i is 2 or 3. It is easy to verify that in that case $8^{p-4} \prod_{i=1}^{4} (8 + 4p_i)$ takes the value $2^{28} (10)^{p-8}$.

In view of the above, we shall restrict our attention to designs which do not have property Δ.

Again, we first consider the case where all the effects are even. Since the design does not have property Δ we can form cosets G_1, G_2, G_3, G_4 as indicated in Section 18.2. Let S_i denote the set of admissible effects in G_i and let p_i denote the number of effects in S_i. Since $\sum p_i \geq 9$ we must have at least one $p_i \geq 3$. Without loss of generality we can assume that $p_1 \geq 3$ and $p_1 + p_2 \geq 5$.

Suppose there does not exist any admissible effect which is not connected with any effect in S_1. We now write $X^t X$ as

$$\begin{bmatrix} 12 I_{p_1} & B \\ B^t & V \end{bmatrix}.$$

It is easy to see that there are at least $p_2 + 2$ pairs of connected effects where one effect is from S_1 and the other is from S_2. Similarly, there are $p_3(p_4)$ connected pairs with one from S_1 and the other from $S_3(S_4)$. Thus, $\boldsymbol{B}$ contains at least $p_2 + p_3 + p_4 + 2 = p - p_1 + 2$ entries which are ± 4. It is clear that

$$\det \boldsymbol{X}^t \boldsymbol{X} \le (12)^{p_1} \prod_{i=1}^{p-p_1} (12 - \tfrac{4}{3} s_i) \tag{18.3}$$

where s_i is the number of elements in the i-th column of $\boldsymbol{B}$ which are ± 4. We note that $\sum s_i \ge p - p_1 + 2$ and that the right side of (18.3) is maximum when $p - p_1 - 2$ of the s_i's are unity and 2 of the s_i's are 2. Substituting these we get

$$\begin{aligned}
\det \boldsymbol{X}^t \boldsymbol{X} &\le 12^{p_1} (\tfrac{32}{3})^{p-p_1-2} (\tfrac{28}{3})^2 \\
&= (16 \times 8)^{p_1} (\tfrac{32}{3})^{8-2p_1} (\tfrac{32}{3})^{p-10} (\tfrac{28}{3})^2 .
\end{aligned}$$

We note that $\det \boldsymbol{X}^t \boldsymbol{X}$ for d^* is $(16 \times 8)^4 (10)^{p-8}$ and that for $p = 9$, 10 or 11, $(\tfrac{32}{3})^{p-10} (\tfrac{28}{3})^2 \le 10^{p-8}$. Thus, d^* is not inferior if

$$(16 \times 8)^{p_1} (\tfrac{32}{3})^{8-2p_1} \le (16 \times 8)^4 .$$

This is easily seen to hold for $p_1 = 3$ or 4.

We now turn to the case where an effect from S_3 say is not connected with any effect from S_1. This implies $p_1 = 3$. Form set S_1^* by adding this effect to S_1 and partition $\boldsymbol{X}^t \boldsymbol{X}$ by partitioning the effects into S_1^* and its complement. This gives

$$\boldsymbol{X}^t \boldsymbol{X} = \begin{bmatrix} 12\boldsymbol{I}_4 & \boldsymbol{L} \\ \boldsymbol{L}^t & \boldsymbol{U} \end{bmatrix} .$$

Again,

$$\det \boldsymbol{X}^t \boldsymbol{X} \le (12)^4 \prod_{i=1}^{p-4} (12 - \tfrac{4}{3} s_i) \tag{18.4}$$

where s_i is the number of elements in the i-th column of $\boldsymbol{L}$ which are ± 4. If $\sum_i s_i = p - 4 + a$, the right side of (18.4) is maximum when $p - 4 - a$ of the s_i's are unity and the rest are 2. Thus we get

$$\det \boldsymbol{X}^t \boldsymbol{X} \le (12)^4 (\tfrac{32}{3})^{p-4-a} (\tfrac{28}{3})^a .$$

The right side of the above is not greater than $(16 \times 8)^4 (10)^{p-8}$ if

$$(\tfrac{32}{3})^{p-8-a} (\tfrac{28}{3})^a \le 10^{p-8} .$$

This is satisfied if $p - 8 \leq 2a$. For $p = 9$ or 10, we need $a = 1$ whereas for $p = 11$ we need $a = 2$.

When $p = 11$, we must have $p_1 = p_2 = 3$ and $p_3 + p_4 = 5$. We have at least six pairs of connected effects where one is from S_1^* and the other is from S_2. Similarly, we have at least $p_3 - 1$ pairs of connected effects with one from S_1^* and the other from S_3 which is not included in S_1^*. Also we have $p_4 - 1$ connected pairs with one from S_1^* and the other from S_4. Thus, $\boldsymbol{L}$ has at least 9 elements which are ± 4.

It is similarly seen that when $p = 9$ or 10, $\boldsymbol{L}$ has at least $p - 3$ elements which are ± 4.

Thus, no design with all even effects can be superior to d^*.

We now discuss the case where some effects are odd and the others are even. Let there be p_1 of one type and $p_2(\geq p_1)$ of the other type. Clearly $p_1 \leq 5$ and $p_2 \geq 5$. We divide p_1 effects into two sets say S and T such that (i) no effect in S is non-orthogonal to more than one effect in S whereas (ii) every effect in T is non-orthogonal to at least two effects in S. The existence of such a partition is easily checked. Denote the number of effects in $S(T)$ by $p_{11}(p_{10})$. We first consider the case where $p_{10} = 0$. We partition $\boldsymbol{X}^t\boldsymbol{X}$ corresponding to the p_{11} effects in S and p_2 effects of the other type. In view of the property (i) of S, the p_{11} effects in S can be partitioned into γ_1 sets of 2 effects each and s_1 sets of single effects such that

(a) two effects from the set are not orthogonal

(b) any two effects not from the same set are orthogonal.

Similarly, the p_2 effects of the other parity can be partitioned into γ_2 sets of 2 effects each and s_2 sets of single effects in such a way that the property (a) above holds. The elements of $\boldsymbol{X}^t\boldsymbol{X}$ corresponding to two effects not of the same parity are all ± 2. (Values ± 6 or ± 10 give value for $\boldsymbol{X}^t\boldsymbol{X}$ smaller than that for d^*). From this it can be seen that $\det \boldsymbol{X}^t\boldsymbol{X} \leq D(\gamma_1, s_1, \gamma_2, s_2)$ where

$$D(\gamma_1, s_1, \gamma_2, s_2) = (8 \cdot 16)^{\gamma_1} 12^{s_1} 16^{\gamma_2} (8 - \frac{\gamma_1}{2} - \frac{s_1}{3})^{\gamma_2} (12 - \frac{\gamma_1}{2} - \frac{s_1}{3})^{s_2}. \quad (18.5)$$

We note that $2\gamma_1 + s_1 = p_{11} < 5$ and $2\gamma_2 + s_2 \geq 5$. Further, $\gamma_1 + s_1 \leq 4$ and $\gamma_2 + s_2 \leq 4$. It can be verified that $D(\gamma_1 - 1, s_1 + 2, \gamma_2, s_2) \geq D(\gamma_1, s_1, \gamma_2, s_2)$ and $D(\gamma_1, s_1, \gamma_2 - 1, s_2 + 2) \geq D(\gamma_1, s_1, \gamma_2, s_2)$.

Since $p_{11} = 2\gamma_1 + s_1 = \gamma_1 + (\gamma_1 + s_1) \leq \gamma_1 + 4$ we have $\gamma_1 \geq p_{11} - 4$. Thus, the smallest value for γ_{11} is 0 if $p_{11} = 4$ and is $p_{11} - 4$ if $p_{11} > 4$. Similarly, the smallest value for γ_2 is $p_2 - 4$. We define

$$D_1(p_{11}, p_2) = D(\gamma_1^0, s_1^0, \gamma_2^0, s_2^0) \quad (18.6)$$

where γ_1^0 and γ_2^0 are the smallest possible values for γ_1 and γ_2. Further, $s_1^0 = p_{11} - 2\gamma_1^0$ and $s_2^0 = p_2 - 2\gamma_2^0$.

Using these values of γ_1^0, s_1^0, γ_2^0 and s_2^0, it can be verified that

$$D_1(p_{11} + 1, p_2) \geq 10 D_1(p_{11}, p_2).\tag{18.7}$$

Further, if $p_{11} \leq 4$

$$D_1(p_{11}, p_2 + 1) \geq 10 D_1(p_{11}, p_2).\tag{18.8}$$

We note that for design d^*, det $\boldsymbol{X}^t\boldsymbol{X} = 2^{28}(10)^{p-8}$. This equals $D_1(4, 7)$. From (18.7) and (18.8) it follows that when $p_{10} = 0$, designs with (p_{11}, p_2) such that $p_{11} \leq 4$ and $p_2 > 7$ cannot be superior to d^* (w.r.t. D-optimality). Also direct verification using (18.6) gives $D_1(5, 5) \leq 2^{28} \cdot 10^2$ and $D_1(5, 6) \leq 2^{28} \cdot 10^3$. Thus, we have established D-optimality of d^* when $p_{10} = 0$.

For designs with $p_{10} > 0$, we consider adjusting all the other effects for these p_{10} effects. Using property (ii) of the set T it can be shown that

$$\det(\boldsymbol{X}^t\boldsymbol{X}) \leq \left(\frac{28}{3}\right)^{p_{10}} D_1(p_{11}, p_2).$$

Since $p_{10} + p_{11} \leq 5$, when $p_{10} > 0$, $p_{11} \leq 4$ and hence by (18.7) and (18.8) we have $D_1(p_{11}, p_2) \leq D_1(4, 7) \cdot 10^{p_{11}+p_2-11}$. Thus, we have

$$\begin{aligned}
\det(\boldsymbol{X}^t\boldsymbol{X}) \;&\leq\; (28/3)^{p_{10}} \cdot D_1(4, 7) \cdot 10^{p_{11}+p_2-11} \\
&=\; (28/30)^{p_{10}}\, 2^{28} \cdot 10^3 10^{p-11} \\
&\leq\; 2^{28} \cdot 10^{p-8}.
\end{aligned}$$

This completes the proof of D-optimality of d^* when $p \geq 9$.

18.6 E^* and D-optimality for $p = 6, 7, 8$

Before we look at the competing designs we establish the following property of d^*.

Proposition 18.6.1 *With design d^*, if $p = 6, 7$, or 8, we can arrange the effects in four sets so that*

  (i) no set contains more than two effects and

  (ii) estimates of two effects not from the same set are uncorrelated.

PROOF: As noted in Section 18.3, when $p = 11$, the effects can be divided into four sets

$$(F_1, F_2 F_3, F_1 F_4)\,, (F_2, F_1 F_3, F_2 F_4)\,, (F_3, F_1 F_2, F_3 F_4)\,, (F_4, \mu)$$

with property (ii) stated above. When $p = 6$, i.e. when we wish to estimate only one interaction we have $(2,1,1,2)$ split. When $p = 7$ and when the two interactions to be estimated have a common factor, we get $(2,2,1,2)$ split. If however, these two do **not** have a common factor we get 3 admissible effects in the same set. In such a case, we define new factors G_1, G_2, G_3 and G_4 by $F_1 = G_1G_2G_4,\ F_2 = G_4,\ F_3 = G_3G_4,\ F_4 = G_2G_4$ or equivalently by

$$G_1 = F_1F_4,\quad G_2 = F_2F_4,\quad G_3 = F_2F_3,\quad G_4 = F_2.$$

In terms of the new factors we have the sets

$$(G_1G_2G_4, G_1, G_3),\ \ (G_4, G_2, G_1G_2G_3),\ \ (G_3G_4, G_2G_3, G_1G_2),\ \ (G_2G_4, \mu).$$

We can now estimate $\mu, G_1, G_2, G_3, G_4, G_1G_2, G_3G_4$. We note that no set contains more than two effects.

When $p = 8$, we need to estimate three two factor interactions. We have the following mutually exclusive cases for the three interactions to be estimated.
(a) $F_1F_2,\ F_1F_3,\ F_1F_4$; (b) $F_1F_2,\ F_1F_3,\ F_2F_3$; (c) $F_1F_1,\ F_3F_4,\ F_2F_4$.

For cases (a) and (b) we have $(2,2,2,2)$ split. For case (c) we use the factors G_1, G_1, G_3G_4 to get the same split.

We now proceed to establish the E^*-optimality of d^* when $p = 7$ or 8. If there is a connected pair of effects, each of these will be estimated with variance not less that $(3/32)\,\sigma^2$. It is easy to verify that when $p \geq 6$, if all the effects are even, there must be a connected pair of admissible effects. Thus, we need only discuss the case where there are both odd and even effects and where the effects of the same parity are orthogonal. If there are at least four effects of one parity, an effect x of the opposite parity will have covariance ± 2 with each of these and hence the estimate of x adjusted for these four effects will have variance at least equal to $(12 - 16/12)^{-1}$ i.e. $3/32$. Thus, d^* is E^*-optimal when $p = 7$ or 8.

When $p = 6$, the following design is superior to d^* w.r.t. E^*-optimality.

F_1	1	1	1	1	1	1	-1	-1	-1	-1	-1	-1
F_2	1	1	1	-1	-1	-1	1	1	1	-1	-1	-1
F_3	1	1	-1	-1	1	-1	-1	1	1	-1	1	1
F_4	1	-1	1	1	1	-1	1	-1	1	-1	-1	1

If the effects to be estimated are $1, F_1, F_2, F_3, F_4$ and F_1F_3, it is easy to verify that the variance of the estimate for each effect is $\frac{3264}{35840}$ which is less than $\frac{3}{32}$ and hence $p = 6, d^*$ is not E^*-optimal.

We now turn to D-optimality. In the first place we note that for d^*,

$$\det(X^tX) = 12^4(12 - 16/12)^{p-4} = 12^{8-p}(128)^{p-4}.$$

We have assumed that d^* is so arranged that Proposition 18.6.1 holds. ∎

We shall use the following result which is easy to verify.

Lemma 18.6.1 *Let $y, x_1, x_2, \cdots, x_k$ be admissible effects such that $y^t x_i = \pm 4$. Then*

$$\det(H^t H) \leq (128)(12/2)^{k-1}$$

where, $H = [y, x_1, \cdots, x_k]$.

We first discuss the case where all the effects are even. We establish the following result.

Theorem 18.6.1 *If there are four admissible effects x_1, x_2, x_3 and x_4 such that every admissible effect is connected to at least one x_i, $\det(X^t X) \leq 128^{8-p} \cdot 128^{p-4}$.*

PROOF: Form classes by connecting the remaining admissible effects to one of the x_i's. Let s denote the number of classes with exactly one element. For each class other than the ones with single elements the determinant of the information matrix can be seen not to exceed $(128)\left(\frac{21}{2}\right)^{h-2}$ where h is the number of effects in the class. From this it follows that

$$
\begin{aligned}
\det(X^t X) &\leq 12^s (128)^{4-s} \left(\frac{21}{2}\right)^{p-8+s} \\
&\leq 12^{8-p} (128)^{4-s} \left(12 \times \frac{21}{2}\right)^{p-8+s} \\
&\leq 12^{8-p} (128)^{p-4}.
\end{aligned}
$$

When all the effects are even we have to consider two possibilities. There may be five mutually orthogonal admissible effects. Call these y_1, y_2, y_3, y_4 and y_5. In that case, by Lemma 18.2.1, 1 is connected to the product of any four of these. From this it can be deduced that the product of any three of these is connected to each of the other two. Further, the product of all five is connected to each of these. Since these products taken three at a time and the product of all five constitute all the remaining eleven effects, it follows that each admissible effect is connected with at least two of the y_i's. Now choose $x_1 = y_1, x_2 = y_2, x_3 = y_3$ and $x_4 = y_4$. An application of Theorem 6.3 shows that $\det(X^t X) \leq 12^{8-p}(128)^{p-4}$. If there are at most four mutually orthogonal admissible effects, Theorem 18.6.1 applies directly. ∎

We now turn to the case where a competing design has both odd effects and even effects. There can be at most four mutually orthogonal admissible effects. If there are precisely four, $\det(X^t X) \leq 12^4 \left(12 - \frac{16}{12}\right)^{p-4}$. If there are only three mutually orthogonal admissible effects, the determinant is 12^3 times the determinant of the matrix reduced for these three effects. For an admissible effect of the same parity, the diagonal element of the reduced matrix is at most $32/3$. Let s denote the number of such effects. For an effect of different parity the diagonal element is 11. Let g denote the number of admissible effects of the other parity. Suppose this set contains t mutually orthogonal effects. We must have $t \leq 3$. If $g > t$, the reduced matrix must have at least one off-diagonal element in this $t \times (g - t)$ part of the reduced matrix whose absolute value is not less than 3. We now have

$$\det(X^t X) \leq 12^3 \left(\frac{32}{3}\right)^s 11^t \left(11 - \frac{9}{11}\right).$$

Since $t \leq 3$ and $11 > \frac{32}{3} > \left(11 - \frac{9}{11}\right)$ we have

$$
\begin{aligned}
\det(X^t X) &\leq 12^3 \cdot 11^3 \cdot (32/3)^{p-6} \\
&= 11^3 \cdot 12^{8-p} \cdot (12 \cdot 32/3)^{p-6} \cdot 12 \\
&\leq 12^{8-p} \cdot 128^{p-4}.
\end{aligned}
$$

If $g = t = 3, p = 6$, $\det(X^t X) \leq 12^3 \cdot 11^3 < 12^2 \cdot 128^2$. This completes the proof of D-optimality of d^* when $p = 6, 7$ or 8.

18.7 Summary and Concluding Remarks

We have presented a design for a 2^4 factorial in 12 runs and have shown that it is optimal. The optimality is established using the structure of competing designs. Some simple arguments from regression analysis are also used.

Acknowledgement. The work of the second author was partially supported by a research grant from the Natural Sciences and Engineering Research Council of Canada. Their support is gratefully acknowledged.

References

1. Box, G. E. P. and Bisgaard, S. (1993). George's Column. What can you find out from twelve experimental runs? *Quality Engineering*, **5**, 663–668.

2. Lin, D. K. J. and Draper, N. R. (1992). Projection properties of Plackett and Burman designs, *Technometrics*, **34**, 423–428.

3. Plackett, R. L. and Burman, J. P. (1946). The design of optimum multi-factorial experiments, *Biometrika*, **33**, 305–325.

4. Wang, J. C. and Wu, C. F. J. (1995). Hidden projection property of Plackett-Burman and related designs, *Statistica Sinica*, **5**, 235–250.

19

Robustness of D-optimal Experimental Designs for Mixture Studies

David W. Bacon and Rodney Lott

Queen's University, Kingston, Ontario, Canada

Abstract: The principal objective of mixture experiments is to assess the influences of the proportions and amounts of mixture ingredients, along with the effects of relevant processing variables, on performance characteristics of a mixture. Experimental designs, called "mixture designs", have been proposed to guide such investigations. This paper explores the effectiveness of some commonly used designs, including D-optimal designs, in providing predictions with relatively uniform variance from alternative model forms.

Keywords and phrases: D-optimal, experimental designs, mixture models

19.1 Introduction

A mixture can be defined as a product formed by combining ingredients. Common examples are alloys, plastics and food products. The focus of an experimental study of a mixture is often to determine the influences of the proportions and amounts of the mixture ingredients, and the effects of relevant processing variables, on the performance of the mixture.

Models for these relationships must satisfy two requirements:

1. the relative concentration of each of the q ingredients in the mixture must be nonnegative.
$$X_i \geq 0 \qquad i = 1, 2, \ldots, q \tag{19.1}$$

2. the relative concentrations of the ingredients in the mixture must sum to 1.
$$\sum_{i=1}^{q} X_i = 1 \tag{19.2}$$

19.2   Linear Empirical Model Forms

The most commonly used mixture models are the Scheffé polynomials [Scheffé (1958)]. A first degree Scheffé polynomial model can be written as

$$E(Y) = \sum_{1 \leq i \leq q} \beta_i X_i \qquad (19.3)$$

where $E(Y)$ denotes the expected value of the response variable of interest. A second degree Scheffé polynomial model can be written as

$$E(Y) = \sum_{1 \leq i \leq q} \beta_i X_i + \sum_{1 \leq i < j \leq q} \beta_{ij} X_i X_j \qquad (19.4)$$

Scheffé polynomial models of higher degrees can be developed in analogous fashion. A special case of the third degree Scheffé polynomial was suggested by Gorman and Hinman (1962), who called it the special cubic model:

$$E(Y) = \sum_{1 \leq i \leq q} \beta_i X_i + \sum_{1 \leq i < j \leq q} \beta_{ij} X_i X_j + \sum_{1 \leq i < j < k \leq q} \beta_{ijk} X_i X_j X_k \qquad (19.5)$$

When the value of the response variable is expected to change dramatically near a boundary of the constrained mixture region, Draper and St. John (1977) suggested to include inverse terms in the relative concentrations. Such models are referred to as extended Scheffé polynomials. A first degree model of this form is

$$E(Y) = \sum_{1 \leq i \leq q} \beta_i X_i + \sum_{1 \leq i \leq q} \beta_{-1} X_i^{-1} \qquad (19.6)$$

and a second degree model of this form is

$$E(Y) = \sum_{1 \leq i \leq q} \beta_i X_i + \sum_{1 \leq i < j \leq q} \beta_{ij} X_i X_j + \sum_{1 \leq i \leq q} \beta_{-1} X_i^{-1} \qquad (19.7)$$

Other forms of empirical linear models used to describe the performance of mixtures include ratio models [Cornell (1990)], log contrast models [Aitchison and Bacon-Shone (1984)], Cox models [Cox (1971)], and Becker models [Becker (1968)].

19.3   Experimental Designs for First Order Linear Mixture Models

Simplex designs (also called extreme vertices designs) are frequently used for experimental investigations in which a first degree Scheffé model is believed to provide an adequate representation of the relationship between a response variable of interest and the relative concentrations of the mixture ingredients. A simplex design for a mixture of three ingredients, each of whose relative concentrations is permitted to vary from 0 to 1, is as follows:

X_1	X_2	X_3
1	0	0
0	1	0
0	0	1

19.4 Experimental Designs for Second Order Linear Mixture Models

Two classes of mixture designs are widely used to produce information about second degree Scheffé polynomial models: simplex lattice designs and simplex-centroid designs. A simplex latti the following form:

X_1	X_2	X_3
1	0	0
0	1	0
0	0	1
0.5	0.5	0
0.5	0	0.5
0	0.5	0.5

A simplex-centroid design for a mixture of three ingredients has the following settings:

X_1	X_2	X_3
1	0	0
0	1	0
0	0	1
0.5	0.5	0
0.5	0	0.5
0	0.5	0.5
0.33	0.33	0.33
0.67	0.17	0.17
0.17	0.67	0.17
0.17	0.17	0.67

19.5   Experimental Design Criteria

As Box and Draper (1959, 1963) have pointed out, there can be many objectives in designing an experiment. The primary objective of interest in this study was to investigate the uniformity of across the feasible mixture space using designs that are D-optimal for alternative mixture models. D-optimality was chosen as the design criterion in this study because if its widespread use in existing commercial experimental design software. In addition, for linear models, a D-optimal design for a specified model form is identical to a G-optimal design for that model form [Kiefer and Wolfowitz (1960)].

19.6   Robustness of D-Optimal Mixture Designs

Robustness of an experimental design is its effectiveness for its intended use when one or more of the assumptions on which it is based is not true. One measure of the robustness of a design that is optimal for a specific model form is its performance when a different model form is specified. In most experimental investigations it is unusual for an appropriate form of empirical model for the process under study to be known in advance. A common practice is to test the fits of a number of alternative model forms after the data have been obtained. An experimental design that provides satisfactory information about a number of alternative model forms is therefore an attractive choice for such situations. The measure of design performance used in this study is the uniformity of the variance of the predicted response across a region of interest in the mixture space. The results of four cases are reported. Additional comparisons have been made by Lott (1995).

D-optimal designs for four cases were constructed for the following constrained mixture spaces:

for a mixture with two ingredients, $0.2 \leq X_i \leq 0.8,$ $i = 1,2$
for a mixture with three ingredients, $0.2 \leq X_i \leq 0.6,$ $i = 1,2,3$

Graphical displays of the contours of the normalized variance, $var(\hat{y})/\sigma^2$, of the predicted response have been constructed using the following transformed concentrations for the mixture ingredients:

$$X_i^* = \frac{X_i - L_i}{1 - \sum_{j=1}^{q} L_j} \qquad (19.8)$$

where L_i denotes the lower constraint on X_i. Consequently, the limits for each transformed concentration X_i^* are 0 and 1.

19.6.1 Case 1

The first case involves two mixture ingredients. The design under test is a three run D-optimal design for a second degree Scheffé polynomial model:

$$E(Y) = \beta_1 X_1 + \beta_2 X_2 + \beta_{12} X_1 X_2 \tag{19.9}$$

X_1	X_2
0.2	0.8
0.5	0.5
0.8	0.2

The model form used to test this design is a ratio model for two mixture ingredients:

$$E(Y) = \alpha_0 + \alpha_1 \frac{X_2}{X_1} \tag{19.10}$$

The normalized variance of the predicted response for this case is shown as the solid curve in Figure 19.1. For comparison, the dashed curve in Figure 19.1 shows the corresponding behaviour of the variance of the predicted response for model (19.10) from a three run design which is D-optimal for model (19.10):

X_1	X_2
0.2	0.8
0.8	0.2
0.8	0.2

As can be seen, the two variance patterns are very similar, indicating that little precision is lost by using the D-optimal design for a second degree Scheffé polynomial model in this case.

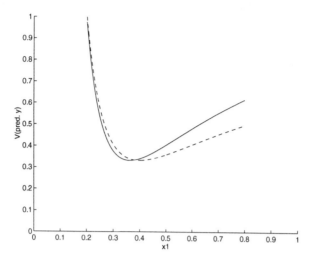

Figure 19.1: Normalized variance of the predicted response from a fitted ratio
model for two mixture ingredients:
———— from a three run D-optimal design for a second degree Scheffé polynomial
model
—— from a three run D-optimal design for a ratio model

19.6.2　Case 2

This case involves three mixture ingredients. The design is a six run D-optimal
design for a second degree Scheffé polynomial:

$$E(Y) = \beta_1 X_1 + \beta_2 X_2 + \beta_3 X_3 + \beta_{12} X_1 X_2 + \beta_{13} X_1 X_3 + \beta_{23} X_2 X_3 \quad (19.11)$$

X_1	X_2	X_3
0.2	0.6	0.2
0.6	0.2	0.2
0.2	0.2	0.6
0.4	0.2	0.4
0.2	0.4	0.4
0.4	0.4	0.2

The model form used to test this design is a ratio model for three mixture
ingredients:

$$E(Y) = \alpha_0 + \alpha_1 \frac{X_2}{X_1} + \alpha_2 \frac{X_2}{X_3} \quad (19.12)$$

A six run D-optimal design for model (19.12) is:

X_1	X_2	X_3
0.2	0.6	0.2
0.2	0.296	0.504
0.504	0.296	0.2
0.2	0.6	0.2
0.2	0.296	0.504
0.504	0.296	0.2

The solid curve in Figure 19.2 displays the normalized variance of the predicted response for this case, and the dashed curve the corresponding behaviour of the variance of the predicted response for model (19.12) from a six run design which is D-optimal for model (19.12). Once again there is only marginal loss from using the D-optimal design for a second degree Scheffé polynomial model in this case. In fact, the variance pattern from this design is more symmetric across the defined mixture region.

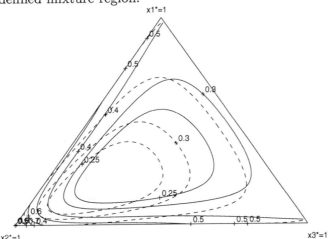

Figure 19.2: Normalized variance of the predicted response from a fitted ratio model for three mixture ingredients: ———— from a six run D-optimal design for a second degree Scheffé polynomial model
—— from a six run D-optimal design for a ratio model

19.6.3 Case 3

Case 3 also uses the ratio model for three mixture ingredients:

$$E(Y) = \alpha_0 + \alpha_1 \frac{X_2}{X_1} + \alpha_2 \frac{X_2}{X_3} \tag{19.13}$$

The design under test is a seven run D-optimal design for a special cubic model:

$$E(Y) = \beta_1 X_1 + \beta_2 X_2 + \beta_3 X_3 + \beta_{12} X_1 X_2 + \beta_{13} X_1 X_3 + \beta_{23} X_2 X_3 + \beta_{123} X_1 X_2 X_3 \tag{19.14}$$

X_1	X_2	X_3
0.2	0.6	0.2
0.6	0.2	0.2
0.2	0.2	0.6
0.4	0.2	0.4
0.2	0.4	0.4
0.4	0.4	0.2
0.3333	0.3333	0.3333

A seven run D-optimal design for model (19.13) is:

X_1	X_2	X_3
0.2	0.6	0.2
0.2	0.296	0.504
0.504	0.296	0.2
0.2	0.296	0.504
0.2	0.6	0.2
0.2	0.6	0.2
0.504	0.296	0.2

The solid and dashed curves in Figure 19.3 show the normalized variance contours of the predicted responses for model (19.13) from the seven run D-optimal design for a special cubic model and from the seven run D-optimal design for model (19.13), respectively.

As in Case 2, the symmetry of the variance pattern of the predicted response in this case is improved by using the seven run D-optimal design for a special cubic polynomial model rather the D-optimal design with the same number of runs for the model under consideration. The value of the normalized variance is understandably smaller near the locations of the four "additional" points in the D-optimal design for the ratio model.

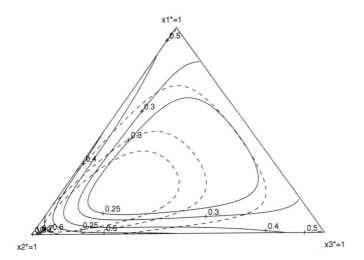

Figure 19.3: Normalized variance of the predicted response from a fitted ratio
model for three mixture ingredients:
    ———— from a seven run D-optimal design for a special cubic model
    — — from a seven run D-optimal design for a ratio model

19.6.4   Case 4

Case 4 is similar to Case 3, with three mixture ingredients and the seven run
D-optimal design shown in Case 3 for a special cubic model (19.14). This time,
however, the test model is an extended first degree Scheffé polynomial model:

$$E(Y) = \beta_1 X_1 + \beta_2 X_2 + \beta_3 X_3 + \beta_{-1}\frac{1}{X_1} + \beta_{-2}\frac{1}{X_2} + \beta_{-3}\frac{1}{X_3} \qquad (19.15)$$

A seven run D-optimal design for model (19.15) is:

X_1	X_2	X_3
0.2	0.6	0.2
0.6	0.2	0.2
0.2	0.2	0.6
0.4	0.2	0.4
0.2	0.4	0.4
0.4	0.4	0.2
0.2	0.6	0.2

The normalized variance contours of the predicted responses for model (19.15)
from the seven run D-optimal design for a special cubic model and from the
seven run D-optimal design for model (19.15) are shown as the solid and dashed
curves, respectively, in Figure 19.4. This time the patterns produced by the two
designs are very different from each other. Although symmetry exists across

the defined mixture region from both designs (aside from the aberration near X_2^* created by the one "additional" point in the seven run D-optimal design for model (19.15), the predicted variance from the seven run D-optimal design for the special cubic model is clearly more uniform.

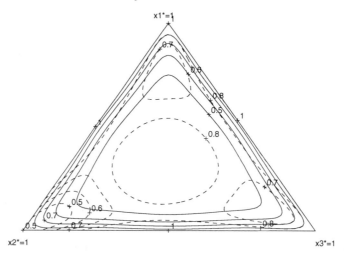

Figure 19.4: Normalized variance of the predicted response from a fitted extended first degree Scheffé polynomial model for three mixture ingredients:
———— from a seven run D-optimal design for a special cubic model
——— from a seven run D-optimal design for an extended first degree Sheffé polynomial model

19.7   Some General Observations

The precision of a predicted response from any mixture design is clearly non-uniform across the defined mixture region. Replicating the minimum run D-optimal design for a selected model improves the uniformity of the precision of the predicted response across the defined mixture region.

Simplex-lattice and simplex-centroid designs appear to be D-optimal or nearly D-optimal for many commonly used empirical mixture model forms. The D-optimal design for a second degree Scheffé polynomial model provides predicted response variance patterns for several commonly used empirical mix-ture model forms that are very close to those produced by D-optimal designs having equivalent numbers of runs for those models.

An efficient approach to model development is one in which experiments can be conducted sequentially, with the design of each successive experiment based upon all of the information obtained to that point. This approach is particularly effective for Sche ffé polynomial models, because the D-optimal design for a first degree Scheffé polynomial model is a subset of the D-optimal design for a second degree Scheffé polynomial model.

References

1. Aitchison, J. and Bacon-Shone, J. (1984). Log contrast models for experiments with mixtures, *Biometrika*, **71(2)**, 323–330.

2. Becker, N. G. (1968). Models for the response of a mixture, *Journal of the Royal Statistical Society, Series B*, **30(2)**, 349–358.

3. Box, G. E. P. and Draper, N. R. (1959). A basis for the selection of a response surface design, *Journal of the American Statistical Association*, **54**, 622–654.

4. Box, G. E. P. and Draper, N. R. (1963). The choice of a second order rotatable design, *Biometrika*, **50**, 335–352.

5. Cornell, J. A. (1990). *Experiments with Mixtures: Designs, Models, and the Analysis of Mixture Data*, Second Edition, John Wiley & Sons, Inc., New York.

6. Cox, D. R. (1971). A note on polynomial response functions for mixtures, *Biometrika*, **58**, 155–159.

7. Draper, N. R. and St. John, R. C. (1977). A mixtures model with inverse terms, *Technometrics* **19**, 37–46.

8. Gorman, J. W. and Hinman, J. E. (1962). Simplex lattice designs for multicomponent systems, *Technometrics* **4(4)**, 463–487.

9. Kiefer, J. and Wolfowitz, J. (1960). The equivalence of two extremum problems, *Canadian Journal of Mathematics*, **12**, 363–366.

10. Lott, R. (1995). Experimental designs for mixture studies, *M.Sc. Thesis*, Department of Chemical Engineering, Queen's University, Kingston, Ontario, Canada.

11. Scheffé, H. (1958). Experiments with mixtures, *Journal of the Royal Statistical Society, Series B*, **20 (2)**, 344–360.

20

A Bivariate Plot Useful in Selecting a Robust Design

Albert Prat and Pere Grima

Universitat Politècnica de Catalunya, Barcelona, Spain

Abstract: A sound engineering practice for improving quality and productivity is to design quality into products and processes. The ideas of G.Taguchi about parameter design were introduced in the US some ten years ago. Despite some strong controversy about some aspects of it, they play a vital role in the concept of robustness in the design of industrial products and processes. In this paper, we present a new methodology for designing products and processes that are robust to variations in environmental or internal variables. First, a tentative model for the response as a function of the design and noise factors is assumed. This model is then estimated using a single design matrix and the expected value and variance of the response are calculated over the space of the design factors. Finally, the best setting of the parameters values can be located in a newly developed bivariate plot where the distance to the target is plotted against the variance of the response.

Keywords and phrases: Design of experiments, distance-variance plot, quality improvement, robust product, variation reduction

20.1   Introduction

Faced with the old view of quality control by means of final inspection or the classical one of Statistical Process Control, we can designate as the modern method one that guarantees better economy and quality of a product. This would apply to the design of products such that their features are maintained at a desired level despite adverse factors in their production and utilization.

These products are said to be robust. In this paper, we present a new methodology for the selection of values for the parameters used in the design of robust products.

20.2   Background

Techniques for the robust design of products were first introduced in Japan by the engineer Genichi Taguchi (1984, 1986). The method bearing his name is well known and has been used widely. One of the steps involved in using the Taguchi method is the selection of design factors (parameters). This in essence consists of setting up a plan of experimentation where the value of some quality characteristic (response) is measured for a known combination of design factor values and also the so-called noise factors (that influence the response but cannot be easily controlled at a fixed value). From the results obtained during experimentation, one can determine the set of values for the design factors that elicit satisfactory values of the response, independent of the values taken by noise factors.

The method proposed by Taguchi for obtaining the optimal design factor value is through the use of product matrices in the design of the experimental plan and the utilization of signal-to-noise ratios in the analysis of the results [See Hunter (1985), Kacker (1985), Ross (1988), Taguchi and Phadke (1984)].

Despite being well-received, the contributions of Taguchi towards product design have some controversial aspects. Although in some cases the number of experiments that one plans to carry out is justifiably necessary for obtaining precise information, in most other cases, experimenting on each of the conditions of the product matrix would require an effort and dedication of resources that may be unnecessary and therefore could be reduced. [Shoemaker et al. (1991)]. Also the statistics and the associated procedures used are not very intuitive. It has been demonstrated [Box and Fung (1986)] that the statistical techniques used for analyzing the results obtained are not very adequate. For related discussion see also Kacker et al. (1991), Maghsoodloo (1990) and Tort-Martorell (1985).

20.3   Description of the Problem

We consider 3 types of variables that influence the quality characteristics of a product:

i) Those corresponding to parameters related to the design of the product, whose values are maintained constant at the desired levels. This type of variables is known as "constant design factors" or, simply, "design factors".

ii) Those outside the design of the product that show a certain variability around their average value. Typical examples of this type of variables are: the ambient temperature, the degree of humidity, the voltage of the electricity supply, etc. These variables are known as "external noise factors".

iii) The variables corresponding to parameters related to the design of the product (as in type i), whose nominal value can be selected by the designer. In practice, however, they show a certain variability around the value selected. An example of this type of variable is the value of the resistance that is placed in a circuit: the designer chooses its value, for example 10 Ω, but in practice it will not be exactly 10, but between 9.5 and 10.5 Ω. These variables are known as "design factors affected by internal noise" or simply "internal noise factors".

In the design of a product it will not always be necessary to consider the two types of noise factors. In some cases the external noise factors will suffice, and it is possible to consider all the design factors as constant. In others, it will be sufficient to consider solely the variability of some design factors.

The issue that we wish to addess is the determination of the nominal values of the design factors so that: 1) the response shows minimum variability around its average value, neutralizing as far as possible the variability transmitted by the noise factors and, 2) the average value of the response is as near as possible to its objective value. For solving this problem we outline below a 4-stage methodology in which it is assumed that the functional relationship linking response with the factors that influence it is not known, and it is therefore necessary to deduce an approximation experimentally. If this relationship is known, stages i) and ii) should be skipped, since they attempt to find an approximation to the unknown functional relationship.

20.4 Proposed Methodology

The proposed methodology is based on the concepts developed by G. Taguchi (1984, 1986) for the design of robust products, and on later analysis and proposals on this subject, especially the ones by Box and Jones (1990). It consists of the following steps:

i) Assume a tentative model for the response.

ii) Estimate the model parameters by running a factorial or fractional factorial design. If the model is quadratic, a central composite design is used.

iii) Analyze the estimated model, deducing the expressions for the variance and the expected value of the response.

iv) Calculate the expected value and the variance of the response for a large set of combinations of design factor values. Draw a bivariate plot of the optimal distance to the target versus the variance of the response and obtain the optimum combination of values for the design factors.

The use of a spreadsheet program (on a PC) facilitates the calculations and the plotting of the data. Short descriptions of the steps mentioned above are given in the next sections.

20.4.1   Hypothesis about the model for the response

The objective is to establish a hypothesis about the existing relation between the response and the factors (design factors, and noise factors) which must satisfy the following requirements: i) Explain satisfactorily the behaviour of the response, and ii) Facilitate easy estimation of the parameters and the subsequent analysis. Below we discuss 2 typical cases:

1. The noise factors are exclusively of the "external noise" type: in this case, the transmission of variability of the noise factors to the response can be neutralized (as far as possible) using the interaction of the noise factors with the design factors (Figure 20.1). Thus it is sufficient to use models of the type:

$$y = \beta_0 + \sum_{i=1}^{n} \beta_i x_i + \sum_{i=1}^{n-1} \sum_{k=i+1}^{n} \beta_{ik} x_i x_k \ + \ \sum_{j=1}^{m} \gamma_j z_j + \sum_{j=1}^{m-1} \sum_{l=j+1}^{m} \gamma_{jl} z_j z_l$$

$$+ \ \sum_{i=1}^{n} \sum_{j=1}^{m} \delta_{ij} x_i z_j + \varepsilon \qquad (20.1)$$

where y: response, x_i: design factors (constants), z_i: noise factors (random variables) and ε: not explained by the model.

2. The noise factors are exclusively of the internal type: In this case, the transmission of variability to the response can be reduced through the mutual interactions of the design factors, but also by using the possible non-linear relationship between the factors affected by variability and the response (Figure 20.2). In this case, it is advantageous to use models of the type:

$$y = \beta_0 + \sum_{i=1}^{k} \beta_i x_i + \sum_{i=1}^{k} \beta_{ii} x_i^2 + \sum_{i=1}^{k-1} \sum_{j=i+1}^{k} \beta_{ij} x_i x_j + \varepsilon \qquad (20.2)$$

where $x_1, \ldots, x_k$ represent design factors affected by internal noise. See Box and Jones (1990) and Jones (1990) for a detailed discussion on this area.

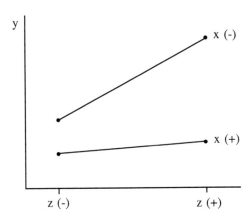

Figure 20.1: The interaction of x with z allows us to choose the values of x (in this case it will be the one coded with '+') which reduces the effect of the variability of z

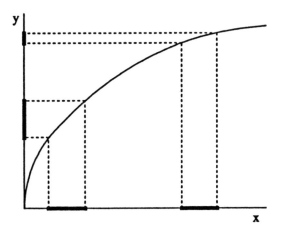

Figure 20.2: The quadratic relation between x and y allows us to reduce the transmission of variability to y by choosing the right values of x (in this case, the greatest possible value)

20.4.2   Estimation of the model parameters

The model parameters can be estimated by running a fractional design. If the model is of first order, a design with 2 levels will be sufficient, but if it is quadratic, it would be necessary to use a central composite design [Box and Draper (1987) and Myers (1976)] or a three level design. In either case, the experimental conditions are formulated throughout using only one design matrix, that we will call a combined matrix, which includes both the control factors and the noise factors. This set-up allows for greater flexibility over the Taguchi product matrix for the estimation of the important effects without confounding. The following example shows this advantage.

Example 1. Let us assume that we have 3 design factors A, B, C and 3 external noise factors O, P and Q. One possible combined matrix is the one that has as generators: $C = AB$ for the inner orthogonal array, and $Q = OP$ for the outer orthogonal array, giving the following defining relation:

$$I = ABC = OPQ = ABCOPQ$$

that corresponds to a fourth of a fraction of a complete factorial design of 2^6 and requires the same number of runs as in the product matrix of Taguchi. From the last defining relation we can see that the main effects cannot be estimated unless we assume that all two factor interactions as non significant. If this assumption cannot be made but we need to estimate the main effects, then the number of experiments required in a Taguchi product matrix is much larger.

However, a combined matrix with the following generators: $P = ABC$ and $Q = BCO$, leads to:

$$I = ABCO = BCOQ = APOQ$$

and consequently we can estimate all the main effects free of two factor interactions. This feature is especially useful when it is not required to estimate all the model parameters. For instance, if we are only interested in minimizing the response variance, all that is required is to estimate the effects of noise factors and the interactions with the control factors. This will be explained in more detail in the next section.

Shoemaker et al. (1989) and Box and Jones (1990) have discussed in detail the advantages of the combined matrix. They also proposed the possibility of using special designs like Addelman or Box-Behnken type designs.

20.4.3 Model analysis: Response variance and expected value

First order models, with only external noise factors

This is a model of the type shown in equation (20.1) and minimising the variability of the response is achieved by minimising the following expression:

$$V(y) = \sum_{j=1}^{m} \left(\gamma_j + \sum_{i=1}^{n} \delta_{ij} x_i \right)^2 V(z_j) + \sum_{j=1}^{m-1} \sum_{l=j+1}^{m} \gamma_{jl}^2 V(z_j z_l) + V(\varepsilon)$$

Considering that ε is independent of any factor, and that all factors included in the model have been previously coded to have values between -1 and $+1$, and making the hypothesis that the factors that can't be controlled (z) have a uniform distribution in that interval, we get the following expression:

$$V(y) = \frac{1}{3} \sum_{j=1}^{m} \left(\gamma_j + \sum_{i=1}^{n} \delta_{ij} x_i \right)^2 + \frac{1}{9} \sum_{j=1}^{m-1} \sum_{l=j+1}^{m} \gamma_{jl}^2 + V(\varepsilon) \qquad (20.3)$$

Example 2. Let us assume that we want to minimise the variability of a quality characteristic (response) that can be expressed by the following model:

$$y = 11 + 2x - 1.5z + 3xz$$

In this case, we have to minimise the following expression:

$$V(y) = (-1.5 + 3x)^2 V(z)$$

The value of x that minimises $V(y)$ is $x = 0.5$.

Example 3. Consider the following model:

$$y = 15 - 5x_1 + 3x_2 - z_1 + z_2 + 2x_1 x_2 - 2z_1 z_2 + 2.5x_1 z_1 + 2.5x_1 z_2 + 2x_2 z_1 - x_2 z_2$$

To minimise the variability of y, we need to minimise the following expression:

$$
\begin{aligned}
V(y) &= (-1 + 2.5x_1 + 2x_2)^2 V(z_1) + (1 + 2.5x_1 - x_2)^2 V(z_2) + 4V(z_1 z_2) \\
&= (1/3)[(-1 + 2.5x_1 + 2x_2)^2 + (1 + 2.5x_1 - x_2)^2 + 4/9
\end{aligned}
$$

Setting the derivatives of $V(y)$ with respect to x_1 and x_2 to zero we obtain: $x_1 = -0.13$ and $x_2 = 0.67$. These will be the optimal values of x_1 and x_2 because any other pair of values would produce a greater variability in the response y.

In the previous examples, the values of the controllable factors which minimise the variability of the response were found. But this was done without taking into account the mean value of the response. However, the response will have an optimal value (that we shall call τ). We will also be interested in minimising the distance between the mean value of the response and its optimal value. Therefore, we have to minimise the following expression:

$$\tau - E(y) = \tau - \beta_0 + \sum_{i=1}^{n} \beta_i x_i + \sum_{i=1}^{n-1} \sum_{k=i+1}^{n} \beta_{ik} x_i x_k$$

where: $E(\varepsilon) = 0$ and $E(z_i) = 0$.

Example 4. Given the simple model in Example 2:

$$y = 11 + 2x - 1.5z + 3xz$$

we want to minimise the distance between the mean value of the response and its optimal value (assuming a value of $\tau = 10$). For this we should find the value of x that minimises (for absolute values) the following expression:

$$D(\tau) = 10 - 11 - 2x$$

It can be easily seen that the optimal value of x is -0.5 because for this value we obtain $D(\tau, \underline{x}) = 0$.

In general, the distance to the optimum will be:

$$D(\tau, \underline{x}) = \tau - E(y) \tag{20.4}$$

Second order models, with only internal noise factors

In this case the model corresponds to the type shown in equation (20.2) and hence we have:

$$V(y) = V\left(\beta_0 + \sum_{i=1}^{k} \beta_i x_i + \sum_{i=1}^{k} \beta_{ii} x_i^2 + \sum_{i=1}^{k-1} \sum_{j=i+1}^{k} \beta_{ij} x_i x_j\right) + V(\varepsilon)$$

When there is only one design factor, a linear approximation of the variation transmitted to the response takes the form:

$$V(y) = V(x)(dy/dx)^2 + V(\varepsilon)$$

In general, with k design factors we obtain:

$$V(y) = \sum_{i=1}^{k} V(x_i)(\partial f/\partial x_i)^2 + V(\varepsilon) \tag{20.5}$$

where the value of $V(\varepsilon)$ may be ignored as it only adds a constant to $V(y)$, and does not affect the conclusions obtained.

Since $E(\varepsilon) = 0$ and the x_i (design factors) can be considered random variables independent of ε, with a distribution: $x_i \sim N(\mu_i, \sigma_i)$, the mathematical expectation for y at any defined point is:

$$E(y) = \beta_0 + \sum_{i=1}^{k} \beta_i E(x_i) + \sum_{i=1}^{k} \beta_{ii} E(x_i^2) + \sum_{i=1}^{k-1} \sum_{j=i+1}^{k} \beta_{ij} E(x_i) E(x_j) \, ,$$

and:

$$E(x^2) = V(x) + E^2(x) \, .$$

Then:

$$E(y) = \beta_0 + \sum_{i=1}^{k} \beta_i E(x_i) + \sum_{i=1}^{k} \beta_{ii} Var(x_i) + \sum_{i=1}^{k} \beta_{ii} E(x_i)^2 + \sum_{i=1}^{k-1} \sum_{j=i+1}^{k} \beta_{ij} E(x_i) E(x_j)$$

20.4.4 Choosing the optimum values for design factors: the distance-variance plot

As shown previously, the value that minimises the response variability is different from the one that minimises its distance to the optimum.

There is no general rule that can facilitate the selection of an optimal value for x, because in some cases we would like to minimise the response variability even if there is a large distance to the target and in other cases do the reverse.

Looking at the model used in examples 2 and 4, a way of selecting the value of x consists of first calculating the values of $E(y)$ and $V(y)$ for every possible value of the control factor. Then, on examination of the calculated values, it is possible to select the value of x that best fits the requirements of the designer.

A simple method that can be generalised for more complicated models for the selection of the optimal value of x, is given below. It is possible to calculate and display graphically all the relevant data by using any spreadsheet programme running on a PC. The steps involved are:

1. In column A of the spreadsheet, enter the values of x between the intervals -1 and $+1$ with increments of 0.1 (this increment could be reduced).

2. In column B, calculate the variance of the response for each value of x, using formula (20.3).

3. In column C, calculate the distance to the target using formula (20.5).

An illustration of the method is shown in Table 20.1. The model parameters and the optimal value for the response (τ) are entered in column G, so that if any of the data is changed, $V(y)$ and the distance to the target are automatically re-calculated.

From this spreadsheet tabulation, a bivariate plot of the distance to the target and the variance can be generated as shown in Figure 20.3. Note that every data point plotted, corresponds to a value of x. From Figure 20.3, one can observe that there is a value of x for which the variance equals zero and the distance to the target is equal to one. There is also an x for which the distance equals zero and the variance equals 0.75. The plot in Figure 20.3 has all the information required in order to select the data point that corresponds to the values of the design factors that best fit the objectives of the designer.

The identification of which value of x corresponds to each one of the plotted data points is simple. In this case, it can be done by counting in increments of 0.1 from the data points that correspond to $x = -1$ and $x = +1$. In other cases, they can be estimated either from their coordinates or by using software packages such as STATGRAPHICS that allow for the identification of any characteristic of a plotted point by just placing the cursor on it.

Additionally, as shown above, this kind of plot can be used for any number of design factors and noise factors. Taking the model used in Example 3, there are two design factors that are both between the interval -1, $+1$ with an increment of 0.1. It is possible to have $21 \times 21 = 441$ combinations of these for which we can calculate the response variance and the distance to the optimum using the spreadsheet programme (see Table 20.2).

The distance-variance plot obtained from this computation is shown in Figure 20.4. From the figure we get $\tau = 24$. The plotted data point that corresponds to a minimum variance has been highlighted with a box around it and the values of x for this are $x_1 = -0.1$ and $x_2 = 0.6$.

20.5   Conclusions

The most important aspects of this method may be summarized in five points:

i. When the functional relationship between the response and the factors is not known it can be found through the experimentation using a single design matrix that includes both, noise and design factors. This provides sensible estimates of the effects and a reduction in the number of experiments [In agreement with the proposals of Shoemaker et al. (1991)].

ii. The proposed method does not include abstract concepts, and is easily understood, used and interpreted even by people with little knowledge of mathematics or statistics.

iii. The calculations are simple and a spreadsheet program for personal computers (e.g. LOTUS) is the only tool used in the application of this method.

iv. The information is presented as a graph (it is always a bivariate diagram) that clearly summarizes the information available in order to make the most suitable decision in every different case.

v. The method is valid for the design of robust products subject to either internal or external noise.

Table 20.1: Spreadsheet calculation to obtain the values for variance and the distance to the target for examples 2 and 4

	A	B	C	D	E	F	G	H
1	X	Var(Y)	Dist.					
2	−1	6.75	2			cte	11	
3	−0.9	5.88	1.8			x	2	
4	−0.8	5.07	1.6			z	−1.5	
5	−0.7	4.32	1.4			xz	3	
6	−0.6	3.63	1.2					
7	−0.5	3	1			tau	10	
8	−0.4	2.43	0.8					
9	−0.3	1.92	0.6					
10	−0.3	1.47	0.4					
11	−0.1	1.08	0.2					
12	0	0.75	0					
13	0.1	0.48	−0.2					
14	0.2	0.27	−0.4					
15	0.3	0.12	−0.6					
16	0.4	0.03	−0.8					
17	0.5	0	−1					
18	0.6	0.03	−1.2					
19	0.7	0.12	−1.4					
20	0.8	0.27	−1.6					
21	0.9	0.48	−1.8					
22	1	0.75	−2					

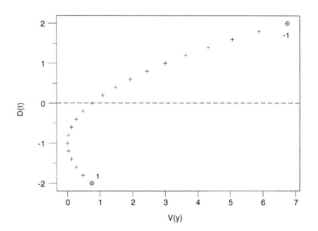

Figure 20.3: Distance-Variance plot for examples 2 and 4

Table 20.2: The first row of the table of values for the model in example 3

	A	B	C	D	E	F	G
1	$x1$	$x2$	$V(y)$	Dist.			
2	-1	-1	10.6	5		cte	15
3	-1	-0.9	9.9	4.9		$x1$	-5
4	-1	-0.8	9.3	4.8		$x2$	3
5	-1	-0.7	8.7	4.7		$z1$	-1
6	-1	-0.6	8.1	4.6		$z2$	1
7	-1	-0.5	7.5	4.5		$x1x2$	2
8	-1	-0.4	7.0	4.4		$z1z2$	-2
9	-1	-0.3	6.5	4.3		$x1z1$	2.5
10	-1	-0.2	6.1	4.2		$x1z2$	2.5
11	-1	-0.1	5.7	4.1		$x2z1$	2
12	-1	0	5.3	4		$x2z2$	-1
13	-1	0.1	4.9	3.9			
14	-1	0.3	4.6	3.8		tau	24
15	-1	0.4	4.3	3.7			

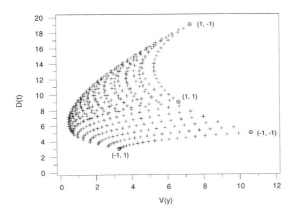

Figure 20.4: Distance-variance plot for example 3

References

1. Box, G. E. P., Hunter W. G. and Hunter J. S. (1978). *Statistics for experimenters*, New York: John Wiley.

2. Box, G. E. P. and Fung, C. A. (1986). Studies in quality improvement: minimizing transmitted variation by parametrer design, *Center for Quality and Productivity Improvement*, Report **8**, University of Wisconsin-Madison.

3. Box, G. E. P. and Draper N. R. (1987). *Empirical Model Building and Response Surfaces*, New York: John Wiley.

4. Box G. E. P. and Jones S. (1990). Designing products that are robust to the environment, *Center for Quality and Productivity Improvement*, Report **28**, University of Wisconsin-Madison.

5. Hunter, J. S. (1985). Statistical design applied to product design, *Journal of Quality Technology*, **17**, 210–221.

6. Jones, S. P. (1990). Designs for minimizing the effect of envirinmental variables, *Unpublished Ph.D. dissertation*, University of Wisconsin, Department of Statistics.

7. Kackar, R. N. (1985). Off-Line quality control, parametrer design, and the Taguchi method, *Journal of Quality Technology*, **17**, 176–188.

8. Kackar, R. N., Lagergren, E. S. and Filliben. J. J. (1991). Taguchi's fixed-element arrays are fractional factorials, *Journal of Quality Technology*, **23**, 107–115.

9. Maghsoodloo, S. (1990). The exact relation of Taguchi's signal-to-noise ratio to his quality loss function, *Journal of Quality Technology*, **22**, 57–67.

10. Myers, R. H. (1976). *Response surface methodology*, Boston: Allyn and Bacon, Inc.

11. Ross, P. J. (1988). *Taguchi techniques for quality engineering*, New York: McGraw-Hill.

12. Shoemaker, A. C., Tsui K. L. and Wu, C. F. J. (1991). Economical experimentation methods for robust design, *Technometrics*, **33**, 415–427.

13. Taguchi G. and Phadke M. S. (1984). Quality engineering through design optimization, *Presented at IEEE GLOBECOM, Atlanta*.

14. Taguchi, G. (1986). *Introduction to quality engineering: designing quality into products and processes*, Tokyo: Asian Productivity Organization.

15. Taguchi, G., Elsayed, E. A. and Hsiang, T. (1989). *Quality engineering in production systems*, New York: McGraw-Hill.

16. Tort-Martorell, J. (1985). Diseños factoriales fraccionales, aplicación al control de calidad mediante el diseño de productos y procesos, *Unpublished Ph.D. dissertation*, Universitat Politècnica de Catalunya, Spain.

21

Process Optimization Through Designed Experiments: Two Case Studies

A. R. Chowdhury, J. Rajesh and G. K. Prasad

Indian Statistical Institute, Bangalore, India

Abstract: This paper contains two case studies in process optimization using designed experiments. The first case study shows an application of Design of Experiments to reduce the defect level in a Frame Soldering Process and the second case study describes the reduction in plating thickness variation in an Electro-Plating Process.

Keywords and phrases: Analysis of variance, electroplating, factorial experiment, frame soldering

21.1 Introduction

Design of Experiments was introduced in the 1920's by Sir Ronald A. Fisher (1947) as an efficient procedure for planning data collection to study variation from different sources. Subsequently it has been successfully used to optimize agricultural yields, textile production, etc. [see Box et al. (1978), Cochran and Cox (1957), Montgomery (1984)]. Taguchi (1987) popularized its applications in engineering areas and in particular in the automobile industry [see also Phadke (1989)]. In this paper we present two case studies on process optimization using designed experiments. In Section 21.2 we consider the reduction of defect level in a frame soldering process. Section 21.3 considers variation reduction in an electroplating process. Some concluding remarks are given in Section 21.4.

21.2 Case Study 1 - Frame Soldering Process Improvement

A company is engaged in manufacturing radio frequency component, namely tuner. The tuners have to undergo the entire assembly process, which involves frame soldering process, where various feed through capacitors are soldered in the frame. During assembly the capacitors are loosely placed in their respective positions on the frame. After application of flux, the frame is allowed to pass through the oven by a conveyor belt. The oven has three heaters. The temperature in each of them is maintained within 215-235°C. This method of frame soldering is preferred over hand soldering because it offers high rate of production with uniform quality of solder. There are two types of defects observed in frame soldering, viz. Burning and Dry Solder. For the past two years, the company was facing in-house quality problems regarding defects in frame soldering. The non-conformance percentage for various models was found to be varying from 0.03% to 0.5%. Apart from this, there were customer complaints to the level of 0.45%. Hence, it was decided to investigate the reasons for high level of non conformance.

The objective of this study was to arrive at suitable levels of the process parameters involved in the frame soldering process so as to minimize the incidence of defects.

The existing practice involves lot of trials by changing the process parameters which requires a long time to arrive at a reasonably satisfactory level of performance. The main drawback of this procedure is the lack of insight into the possible interaction of different parameters and therefore varying only one parameter at a time while trying to keep others at a constant level could lead to the incidence of defects.

Thus, a systematic and scientific experimental approach was necessary and the same was adopted to get a comprehensive picture for the factors/interactions affecting the quality of frame soldering and thereby arriving at a suitable combination of the process parameters.

21.2.1 Experimentation

After discussion with the concerned technical personnel of the frame soldering process, following factors and levels were chosen.

Controllable Factors:

	Level 1	Level 2	Level 3
Temperature of Heater 1 (A)	220°C	217°C	-
Temperature of Heater 2 (B)	220°C	217°C	-
Temperature of Heater 3 (C)	220°C	217°C	-
Conveyor Speed (E)	6 Volts	7 Volts	8 Volts

Uncontrollable Factor:

Ambient Temperature (D)

This is actually room temperature. This is difficult to control. Hence, it was considered as a noise factor and two levels are taken into account. One is low (morning), which is room temperature at 9 am and the other is high (afternoon) which is the room temperature at 2 pm.

21.2.2 Design and analysis

Since many interactions were suspected for a frame soldering experiment, a full factorial design was used involving 48 (2 x 2 x 2 x 3 x 2) runs. Total number of defects (Burning and Dry Solder) in 60 frames per run were counted and recorded as response.

To facilitate analysis of data on the number of defects, a transformation $\sqrt{x} + 1$, was considered, where x denotes the number of defects in 60 frames under each experimental condition. Since ambient temperature was a noise factor and it was necessary to reduce the defect rate, signal to noise (S/N ratio) for the "Smaller the Better" type, was calculated for the transformed data under morning and afternoon conditions for each of the 24 combinations of the controllable factors. The experimental data thus transformed were analyzed using the analysis of variance (ANOVA) technique and is shown in Table 21.1. The average response plots are shown in Figure 21.1.

From the ANOVA table and the average response plots, the following important conclusions were drawn.

- The significant factors were B, E and the interaction BC.

- Therefore the best combination appeared to be **B2 C1 E3**, which implies

  Temperature of Heater 2: 217°C, Temperature of Heater 3: 217°C, Conveyor Speed: 8 Volts.

Although A was not significant, it was decided to use the second level of A (A2) due to technical considerations; hence **A2 = 217°C** was chosen.

Source	Degrees of Freedom	Sum of Squares	Mean Squares	F-ratio	Contribution Ratio (p%)
A	1	6.671	6.671	2.72	1.8
B	1	31.263	31.263	12.76 **	12.4
C	1	4.196	4.196	1.71	0.7
E	2	98.812	49.406	20.16**	41.4
AB	1	5.131	5.131	2.09	1.1
AC	1	9.365	9.365	3.82	3.0
AE	2	3.613	1.806	-	-
BC	1	40.825	40.825	16.67 **	16.5
BE	2	3.617	1.809	-	-
CE	2	3.793	1.897	-	-
Pooled Residual	(15)	(36.744)	(2.45)		23.1
Total	23	233.007			100.0

Table 21.1: Analysis Variance (Based on S/N Ratio)

**Significant at 1% level of significance

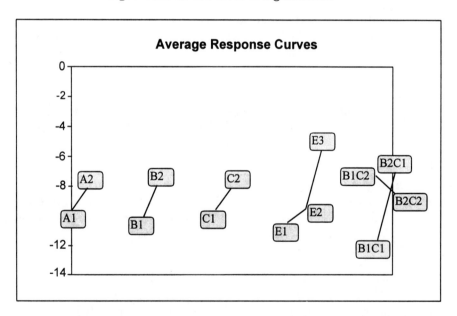

Figure 21.1: Average response plots

Confirmation trials

Using the best levels that had been determined, confirmatory trial runs were carried out with the same type of frames. One hundred frames were soldered under the best combination of the experiment and all them were found to be defect free. The confirmatory trial has yielded satisfactory result with a defect rate of zero and the production department was advised to operate the machine at the suggested levels of process parameters. The company has incorporated the best combination of process parameters of frame soldering process in their work instruction and implemented the same. After one month of implementation the defect rate dropped from (9-12) to (0-2)%.

21.3   Case Study 2 - Electro-plating Process

A Printed Circuit Board (PCB) is the base electronic component with electrical interconnections on which several components are mounted in a compact manner to give the desired electrical output. In the electroplating (pattern plating) process, the circuitry was plated with copper. The copper deposition rates were observed to be varying a great extent. The adverse effects of improper plating were poor solderability, low ductility and low mechanical strength. They were very much essential for the repeated component replacements and long life of PCBs. Therefore, it was important to maintain variation due to plating thickness as low as possible.

The specification width of thickness inside the hole was 25 to 35 microns. Though PCBs were not rejected at pattern plating stage, there were many solderability problems reported from the customers end. The existing electroplating (pattern plating) process performance was studied through nested ANOVA. The process performance was not satisfactory with high levels of mean (32 microns) and standard deviation (8 microns) of plating thickness.

To identify and examine the nature and sources of overall plating thickness variation, the data were collected through a hierarchical design as shown in Figure 21.2. The pattern plating operation could be carried out in any of the ten tanks numbered 16 to 25, which were classified into four stages as shown in Figure 21.2. Pre-plating operations were carried out in the tanks numbered 1 to 15. In Figure 21.2 Stage I and Stage II had three tanks and Stage III and Stage IV had two tanks each. From each tank two panels were considered and the thickness was measured at three positions viz., top, center and bottom as shown in Figure 21.2. A panel consisted of one or more circuits, which would be routed to the required shape during routing operation. In each position two observations were recorded. After collecting the data in the above format, analysis was carried out through nested ANOVA. The details are given in Table 21.2.

From the analysis, it was concluded that the variation due to stages and the variation due to tanks were not significant, but the variation from panel to panel and the variation between positions within the panels were highly significant.

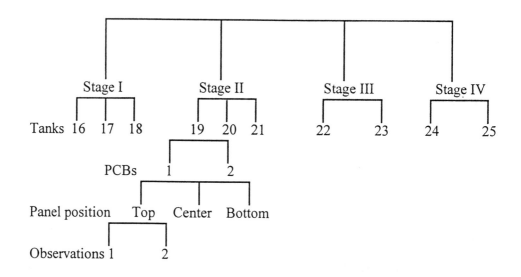

Figure 21.2: Hierarchical design

Table 21.2: Nested ANOVA				
Source	Degrees of Freedom	Sum of Squares	Mean Squares	F-Ratio
Stage	3	351.48	117.16	1.62
Tank	6	246.27	41.04	1.89
Panel to Panel	10	911.33	91.13	20.66*
Within Panel	40	176.56	4.41	1.93 *
Error	60	137.34	2.29	
Total	119	1822.98		

*Significant at 5% level of significance

The results of Table 21.2 were reviewed with the R&D manager and the concerned personnel. The primary objective was to minimize the variation from panel to panel and the variation within the panels. During a brain storming session the various factors influencing the plating thickness were listed. These were:

1. Plating time	2. Current density
3. Bath temperature	4. Bath concentration
5. Air and mechanical agitation	6. Anode to cathode
7. Throwing power (surface to hole plating, 1:1)	surface ration (2:1)
8. Anode area and position	9. Claming position

Out of these factors, the last two were considered to be critical since the remaining factors were fixed according to the standard given by the manufacturer of the proprietary chemicals used, and the cycle time of the total process. It was decided to carry out an experiment with these two factors to find out the combination that minimizes plating thickness variation with a mean closer to the target (30 microns).

21.3.1   Experimentation

Anode area and position

The anode to cathode ratio was 2:1, that is, for two anode bars one cathode flight bar is designed. As the current was passed to the anode, the copper pellets in the bags would get dissolved in the electrolytic copper sulphate solution and would be deposited over the cathode (the panels). To produce uniform copper plating on the panels the distance between anode bags to panels was a concerning factor. Higher the distance, less was the copper deposition and lower the distance, more was the deposition. Hence to produce uniform plating thickness an approximate uniform distance must be maintained between the panels and the anode bags which was not possible in the existing arrangement with four anode bags. Therefore, it was necessary to increase the number of anode bags to have uniform plating thickness over the panels. As the flight bar carrying the panels was divided into three segments arrangement with five bags might not be helpful in reducing the panel to panel variation. So either six anode bags with two bags in each segment or seven anode bags with three at the center segment with two bags on both sides could be helpful. Therefore, the levels for this factor were four bags, six bags and seven bags. This factor was considered because we wanted to reduce panel to panel variation.

Clamping position

Since our objective was also to reduce within panel variation, clamping position was considered. The clamping position in the panel would form a low current density area and the areas away from the clamps would form a high current density portion. At high current density, the plating would be more and it would be less at the low current density portion. This variation in the plating could be reduced by properly positioning the clamps. The possible clamping positions were single clamping, double clamping and side clamping.

The factors and levels are shown in Table 21.3.

With two factors each at three levels the number of tests required for a full factorial experiment was $3^2 = 9$. The physical layout of the experiment is shown in Table 21.4.

Table 21.3: Factors and levels			
Factor	Level-1	Level-2	Level-3
Anode bags (A)	4 bags	6 bags	7 bags
Clamping position (B)	Side	Single	Double

Table 21.4: Physical layout of the experiment									
Test	1	2	3	4	5	6	7	8	9
A (# of bags)	4	4	4	6	6	6	7	7	7
B	Side	S	D	Side	S	D	Side	S	D

S: Single, D: Double

A full factorial experiment was conducted as per the plan. The actual number of trials was only three as all the levels of clamping position could be covered in the same flight bar in each of the trials. The trials were conducted by changing the number of anode bags and using different clamping positions. Because of the configuration of the clamping position, the loading of the panels was done as follows:

Side Clamping - six panels (six replicates);
Single Clamping - six panels (six replicates);
Double Clamping - three panels (three replicates)

Thickness was measured at ten positions in each of the panels.

21.3.2 Analysis

The mean thickness (X) and the standard deviation ($s.d$) for each replication was calculated. The details are given in Table 21.5. For each test the following responses were considered and they were analyzed through ANOVA.

(i) $\log_{10}(s.d)_{ij}^2$ (ii) $20\log_{10}[Xij/(s.d_{ij})]$
(iii) $-10\log_{10}(1/10)\sum|X_{ijk} - 30|$

where, $X_{ijk} = k^{th}$ observation of plating thickness in j^{th} replication of the ith trial. $(s.d)_{ij}$ = standard deviation of the ith trail and j^{th} replication and X_{ij} = mean thickness of ith trial and j^{th} replication.

In response (iii) 30 indicates the target value (the mean of the specifications). The ANOVA was conducted with the above three responses and details are given in the Tables 21.6–21.8 and in Figures 21.3–21.5.

Table 21.5: Mean and s.d (in microns)					
	Side clamping		Single clamping		Double clamping
	Mean	s.d	Mean	s.d	Mean	s.d
4 Anode bags	32.67	4.82	28.19	3.14	9.73	3.88
	38.30	7.80	21.44	1.57	31.67	2.95
	31.70	5.72	26.75	1.60	26.50	4.63
	31.41	8.63	21.26	1.18		
	26.61	7.58	18.24	2.92		
	33.51	7.91	26.16	2.80		
6 Anode bags	28.20	3.74	25.95	2.36	29.29	2.92
	26.64	6.10	26.46	2.98	25.71	2.27
	32.90	4.09	29.65	2.68	24.20	2.68
	36.72	5.63	29.82	2.92		
	28.52	5.30	26.48	2.73		
	33.93	7.62	25.53	2.41		
7 Anode bags	25.47	5.16	18.86	2.38	25.92	2.09
	26.54	4.17	24.02	1.62	30.93	2.08
	24.42	1.76	27.58	1.81	31.38	2.78
	30.21	3.94	29.77	2.33		
	28.76	7.03	30.86	3.48		
	30.89	7.74	25.20	2.32		

Table 21.6: ANOVA using $\log_{10}(s.d)_{ij}^2$ as response				
Source	Degrees of Freedom	Sum of Squares	Mean Squares	F-Ratio
A	2	0.48	0.24	3.34*
B	2	5.32	2.66	36.94*
AB	4	0.57	0.142	1.98
Error	36	2.59	0.072	
Total	44	8.82		

*significant at 5% level of significance.

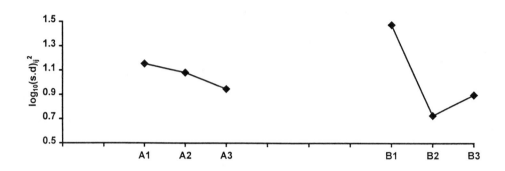

Figure 21.3: Average response plots with $\log_{10}(s.d)^2_{ij}$ as response

Table 21.7: ANOVA using $20\log_{10}[Xij/(s.d_{ij})]$ as response				
Source	Degrees of Freedom	Sum of Squares	Mean Squares	F-Ratio
A	2	25.65	12.82	1.86
B	2	355.56	177.78	25.80*
AB	4	26.58	6.64	0.964
Error	36	248.05	6.89	
Total	44	655.86		

*significant at 5% level of significance.

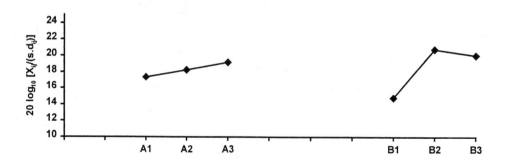

Figure 21.4: Average response plots with $20\log_{10}[Xij/(s.d_{ij})]$ as response

Source	Degrees of Freedom	Sum of Squares	Mean Squares	F-Ratio
A	2	60.27	30.13	2.61**
B	2	130.91	65.45	5.67*
AB	4	29.44	7.36	0.64
Error	36	415.88	6.89	
Total	44	655.86		

Table 21.8: ANOVA using $-10\log_{10}(1/10)\sum|X_{ijk}-30|$ as response

*significant at 5% level of significance,
**significant at 10% level of significance.

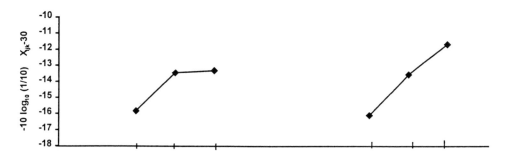

Figure 21.5: Average response curves with $-10\log_{10}(1/10)\sum|X_{ijk}-30|$ as response

From ANOVA tables and average response plots, it was concluded that the variation due to clamping position was highly significant. The effect of anode bags was significant in two cases (in one case it was significant at only 10%). The interaction effect was not significant. From Table 21.6, it was concluded that seven anode bags and double clamping as the best combination since it gave a mean of 29 microns and *s.d* of 2.3 microns. Hence the optimal combination was *seven anode bags - double clamping*.

Confirmation trials were performed to check the effectiveness of the optimal combination in the actual production. The results were highly encouraging. The optimal combination gave a *s.d* of 2.5 to 3.0 microns and an average of 29 to 30 microns. The optimal combination has been implemented in the actual production. The side clamps were unavoidable because they gave support to the sides of the panels at both ends of a flight bar. Therefore, the flight bar design, was changed in such a way that dummy panels were used at both ends of the flight bar.

21.4 Summary and Conclusions

In this paper, we discussed two case studies in design of experiments. The first one was on a frame soldering process and the second one on an electroplating process. Both resulted in substantial process improvement, confirming the power of experimental design for process optimization.

Acknowledgments. The authors are grateful to Mr. R. C. Sarangi, Director, INDAL and Prof. A. K. Choudhury of Indian Statistical Institute for their encouragement, support and guidance.

References

1. Box, G. E. P., Hunter, W. G. and Hunter J. S. (1978). *Statistics for Experimenters - An Introduction to Design, Data Analysis and Model Building*, John Wiley and Sons, Inc. New York.

2. Cochran, W. G. and Cox, G. M. (1957). *Experimental Designs*, John Wiley and Sons Inc., New York.

3. Fisher, R. A. (1947). *The Design of Experiments*, Oliver and Boyd, 5th Ed., Edinburgh.

4. Montgomery, D. C. (1984). *Design and Analysis of Experiments*, John Wiley and Sons Inc., New York.

5. Phadke, M. (1989). *Quality Engineering Using Robust Design*, Prentice Hall, Englewood Cliffs.

6. Taguchi, G. (1987). *Systems of Experimental Design*, Volumes 1 and 2, UNIPUB/Kraus International Publications, New York.

22

Technological Aspects of TQM

C. Hirotsu

University of Tokyo, Tokyo, Japan

Abstract: A Beyond Analysis of Variance technique is introduced as an off-line technology for Total Quality Management. It takes into account the characteristics of factors involved in an experiment for a more satisfactory analysis of data. It also efficiently utilizes the intrinsic natural ordering appearing frequently in relation to time, temperature, dose and so on.

Keywords and phrases: Controllable factor, indicative factor, interaction, isotonic inference, noise factor

22.1   Introduction

Recently in the Total Quality Management (TQM) activities in Japan a philosophical aspect is likely to be more emphasized rather than the technological aspect which used to be the basis of the Statistical Quality Control (SQC). It is partly because the manufacturing processes in the factories are now mostly stable reducing the percent defectives of products to satisfactory level. It should, however, be noted that the off-line techniques, especially for early stages of developing new products, are even more necessary than it used to be and that the classical Analysis of Variance (ANOVA) techniques cannot fully afford it. We therefore need to develop new methodologies responding to the requirements from the recent TQM activities. In this paper we address the following topics,

(i) Taking nonnumerical prior information into the analysis,

(ii) Modelling other than the normal means,

(iii) Testing and modelling the generalized interactions.

22.2 Use of Nonnumerical Information

22.2.1 Natural orderings in parameters

The gum hardness data in Table 22.1 come from a one-way layout experiment with five levels of added amount of sulfur. There is an obvious upward trend in the responses in accordance with the increased amount of sulfur. It is, however, often too restrictive to assume a linear regression model and more reasonably only a monotone relationship

$$\mu_1 \leq \cdots \leq \mu_a \tag{22.1}$$

is assumed in the usual ANOVA model

$$y_{ij} = \mu_i + \varepsilon_{ij}, \ \ i = 1, \cdots, a; \ \ j = 1, \cdots, n_i.$$

Then the usual ANOVA techniques are not efficient enough and a specific method of analyzing the ordered parameters is required.

Table 22.1: Gum hardness data

Added amount	Data	Total
5	55 57	112
15	61 63	124
25	65 60	125
35	65 62	127
45	64 66	130

The idea is naturally extended to the two-way problems, where essentially two cases arise. In one case there are natural orderings in both of rows and columns and the two-way ordered alternative

$$\mu_{i+1j+1} - \mu_{i+1j} - \mu_{ij+1} + \mu_{ij} \geq 0, \ \ i = 1, \cdots, a-1; \ \ j = 1, \cdots, b-1 \tag{22.2}$$

is of interest. This alternative implies that the differences $\mu_{ij} - \mu_{i'j}$ are trending upwards as the level j of the column becomes larger for any $1 \leq i' < i \leq a$. In another case there is a natural ordering only in columns, say, and the hypothesis

$$\mu_{ij+1} - \mu_{i'j+1} \geq \mu_{ij} - \mu_{i'j}, \ \ i, i' = 1, \cdots, a(i \neq i'); \ \ j = 1, \cdots, b-1 \tag{22.3}$$

is of interest. This is the two-sided ordered alternative since rows are permutable and free from ordering.

As a nonnormal case we give an example of comparing multinomial distributions.

Table 22.2: Taste testing data

Foods	1(Terrible)	2	3	4	5(Excellent)
1	9	5	9	13	4
2	7	3	10	20	4
3	14	13	6	7	0
4	11	15	3	5	8
5	0	2	10	30	2

The data of Table 22.2 are the result of the taste testing of five foods evaluated in five ordered categories [Bradley, Katti and Coons (1962)]. The purpose of the experiment is to identify the most preferable food among the five. In comparing rows we consider the ordered alternative

$$p_{ij+1}/p_{i'j+1} \geq p_{ij}/p_{i'j}, \quad i, i' = 1, \cdots, a(i \neq i'); \quad j = 1, \cdots, b-1, \qquad (22.4)$$

which implies that the food i is more preferable than i' or vice versa. The hypothesis (22.4) is exactly the hypothesis (22.3) in $\log p_{ij}$ instead of μ_{ij}. The analysis of sensory data like this will become more important in evaluating customer satisfaction which is the key issue of the TQM.

These natural orderings frequently occur also in relation to passage of time, including the monotonic change of the percent defectives, increasing hazard rates and so on. A nonparametric analysis of a dose response relationship is another important example. For taking these nonnumerical information into the analysis a general theory is given in Section 22.4.

22.3 Classification of Factors

In data analysis the character of a factor plays an important role. We define here five factors and explain how they are incorporated in analyzing data.

(i) Controllable factor : Reproducible and can be specified by the experimenter. The purpose of the experiment will mostly be to determine the optimal level of this factor.

(ii) Indicative factor: Reproducible but uncontrollable by the experimenter. A typical example is the region in the adaptability experiment of the rice varieties.

(iii) Concomitant factor: Reproducible but observable only after the experiment. Measurable environmental effects such as temperatures are typical examples.

(iv) Variation factor: Reproducible in the laboratory but not in the actual field of application. Individuals in a clinical trial is a typical example.

(v) Block factor: Nonreproducible factor introduced to reduce systematic background variation. Blocks in a randomized block design are typical examples.

The controllable, indicative and concomitant factors are considered to be of fixed effects. The variation factor may be treated as a fixed indicative factor within an experiment but works as if it were a random noise on extending the result to the actual field. The block factor can be treated either as fixed or random and may not either way cause serious effects excepting the recovery of interblock information since no interaction is assumed in relation with other factors in the experiment. These considerations of factors play an important role especially in analyzing interactions.

In case of the controllable × controllable interactions, the purpose of the experiment will be to determine the optimal combination of the levels of the two factors and some simple interaction models such as Mandel(1969) and Johnson and Graybill(1972) should be useful to point it out. The usual ANOVA techniques in the texts seem to assume only this situation. If one factor is indicative, the purpose will be to select an optimal level of the controllable factor for each level of the indicative factor. Then the procedure of grouping the homogeneous levels of the indicative factor is desirable so that a single level of the controllable factor can commonly be applicable to as many levels of the indicative factor as possible. If a variation factor is involved, the purpose will be to find out an optimal level of the controllable factor which makes the objective characteristics robust in the possible range of the variation factor. The last example of Section 22.2 gives yet another case of interaction analysis. Originally this is a problem of one-way layout comparing treatments based on the ordered categorical responses and mathematically, the analysis of the generalized interactions as shown by Eq. (22.4) is required. In all these situations the row- and /or column-wise multiple comparisons procedure should be useful for modelling and analyzing interaction instead of a global F test or the multiple comparison of 1 df interaction contrasts such as $\mu_{ij} - \mu_{i'j} - \mu_{ij'} + \mu_{i'j'}$ or $\mu_{ij} - \bar{\mu}_{i.} - \bar{\mu}_{.j} + \bar{\mu}_{..}$. In the last example of Section 22.2 it is obvious that the row-wise comparisons are essential for defining a most preferable food.

22.4   A General Theory for Testing Ordered Alternatives

Suppose $\mathbf{y}$ is distributed as a-variate normal with mean $\boldsymbol{\mu}$ and known covariance matrix Ω. Noting that order restrictions like (22.1) and (22.2) can be expressed in the matrix notation as $A'\boldsymbol{\mu} \geq \mathbf{0}$, we consider generally testing the null hypothesis $H_0 : A'\boldsymbol{\mu} = \mathbf{0}$ against the restricted alternative $H_1 : A'\boldsymbol{\mu} \geq \mathbf{0}$, where the inequality is element-wise with at least one inequality strong. Then the essentially complete class of the tests is formed from all those convex acceptance regions that are increasing in every element of

$$(AA')^{-1}A'\Omega^{-1}\{\mathbf{y} - E(\mathbf{y} \mid H_0)\}, \tag{22.5}$$

where $E(\mathbf{y} \mid H_0)$ is the conditional expectation of $\mathbf{y}$ given the sufficient statistics under the null hypothesis H_0 [Hirotsu (1982)].

In a more general situation where $\mathbf{y}$ is distributed according to some distribution with the unknown parameter vector $\boldsymbol{\mu}$ and possibly some nuisance parameters , the asymptotic equivalent of Eq. (22.5) is

$$(A'A)^{-1}A'\boldsymbol{\nu}(\hat{\boldsymbol{\mu}}_0) \tag{22.6}$$

where $\boldsymbol{\nu}(\hat{\boldsymbol{\mu}}_0)$ is the efficient score vector evaluated at the maximum likelihood estimator $\hat{\boldsymbol{\mu}}_0$ under the null hypothesis H_0; see Hirotsu (1982) and Cohen & Sacrowitz (1991) for details. The equations (22.5) and (22.6) can sometimes be very simple and give reasonable test statistics.

As an important special case we consider the simple ordered alternative (22.1) in the one-way layout. Then we can apply (22.5) with $A' = D'_a$ an $a - 1$ by a differential matrix with the ith row $(0 \cdots 0 -1\ 1\ 0 \cdots 0)$, $\mathbf{y} = (\bar{y}_{1.}, \cdots, \bar{y}_{a.})'$ and $\Omega = (\sigma^2/n)I$, where n is the repetition number, $\sigma^2 = \mathrm{var}(y_{ij})$, I an identity matrix and we follow the usual dot bar notation so that $\bar{y}_{i.} = (y_{i1} + \cdots + y_{in})/n$. The ith standardized element of $(D'_a D_a)^{-1} D'_a (\sigma^2/n)^{-1}\mathbf{y}$ is

$$y_i^* = \sigma^{-1}\{ni(a-i)/a\}^{\frac{1}{2}}(\bar{Y}_i^* - \bar{Y}_i), \quad i = 1, \cdots, a-1,$$

with

$$\bar{Y}_i = \frac{\bar{y}_{1.} + \cdots + \bar{y}_{i.}}{i}, \quad \bar{Y}_i^* = \frac{\bar{y}_{i+1.} + \cdots + \bar{y}_{a.}}{a - i},$$

which we denote by matrix notation as

$$\mathbf{y}^* = (y_1^*, \cdots, y_{a-1}^*)' = \sigma^{-1}\sqrt{n}P_a^{*'}\mathbf{y}.$$

According to the complete class lemma the two proposed statistics are:

$$\begin{aligned}
\text{the cumulative chi-squared} \quad &: \quad \chi^{*2} = \Sigma y_i^{*2}, \text{ and}\\
\text{the maximal component of } \chi^{*2} \quad &: \quad \max t = \max y_i^*.
\end{aligned}$$

The χ^{*2} is for the two-sided problem whereas the max t is for the original one-sided problem. Of course max t is easily extended to the two-sided problems based on the maximum of y_i^{*2}.

There is a very good chi-squared approximation for the null and alternative distributions of the χ^{*2} [Hirotsu (1979a)]. Also available is a very elegant and efficient algorithm for obtaining the p value of the max t method based on the

Markov property of the subsequent components $y_i^*, i = 1, \cdots, a - 1$ [Hawkins (1977)].

The essential characteristics of the χ^{*2} and max t are the directional properties deduced from the strong positive correlations among $y_i^{*'}s$. In particular the χ^{*2} is expanded in a series of independent chi-squared variables,

$$\chi^{*2} = \frac{a}{1 \cdot 2}\chi_{(1)}^2 + \frac{a}{2 \cdot 3}\chi_{(2)}^2 + \cdots + \frac{a}{(a-1) \cdot a}\chi_{(a-1)}^2,$$

where $\chi_{(l)}^2$ is the 1df chi-squared detecting the departure from the null model in the direction of Chebyshev's lth order orthogonal polynomial [Hirotsu (1986), Nair (1986)]. The statistic max y_i^{*2} is characterized as the likelihood ratio test statistic for the change of means in an independent sequence of normal variables.

The power of the χ^{*2} and max t has been discussed on several occasions, see Hirotsu (1978, 1979b, 1992), and Hirotsu et al. (1992), for example.

22.5 Testing Ordered Alternatives in Interaction Effects

The hypothesis (22.4) can be expressed as $(D_a' \otimes D_b')\boldsymbol{\mu} \geq \mathbf{0}$, with $\boldsymbol{\mu}$ the vector of μ_{ij} arranged in dictionary order and $\otimes$ a Kronecker product. Then ideas in Section 22.4 lead to the statistics,

the cumulative chi-squared type : $\chi^{**2} = n \parallel (P_a^{*'} \otimes P_b^{*'})\mathbf{y} \parallel^2$,

the maximal contrast type : maximal component of $\sqrt{n}(P_a^{*'} \otimes P_b^{*'})\mathbf{y}$,

where $\mathbf{y}$ is the vector of $\bar{y}_{ij.}$ arranged in dictionary order, n the repetition number in each cell and the variance σ^2 is assumed to be unity for brevity. The maximal contrast type statistic may be called the max max t and will be useful for the row- and/or the column-wise multiple comparisons discussed in Section 22.2.

A very good chi-squared approximation is available for the distribution of χ^{**2} and also available is a very efficient exact algorithm for evaluating the p value for max max t.

When there is a natural ordering only in the columns the quadratic type statistic is naturally extended to defining

$$\chi^{*2} = \parallel (P_a' \otimes P_b^{*'})\mathbf{y} \parallel^2,$$

where P_a' is an $a - 1$ by a orthogonal matrix satisfying $P_a'P_a = I_{a-1}, P_aP_a' = I_a - \boldsymbol{j}_a\boldsymbol{j}_a'$ with $\boldsymbol{j}_a = (1 \ 1 \ \cdots \ 1)'$.

The maximal contrast type statistics are defined in two ways,

$$T_1 = \max_{\|\boldsymbol{a}\|=1, \boldsymbol{a}'\boldsymbol{j}=0} \parallel (\boldsymbol{a}' \otimes P_b^{*'})\mathbf{y} \parallel^2 \tag{22.7}$$

and

$$T_2 = \max_j \| (P'_a \otimes \boldsymbol{p}^{*'}_{j(b)}) \mathbf{y} \|^2 . \tag{22.8}$$

Under the null hypothesis of additivity, T_1 is distributed as the largest root of the Wishart matrix $W(P^{*'}_b P^*_b, a - 1)$ if $a \geq b$ and T_2 is evaluated by the multivariate chi-squared distribution with correlations $\boldsymbol{p}^{*'}_{j(b)} \boldsymbol{p}^*_{j'(b)}$ and possessing the Markov property among the subsequent components. T_1 and T_2 are respectively useful for row- and column-wise multiple comparisons. The statistic T_1 is used in Hirotsu (1991) for comparing treatments based on repeated measures and T_2 is used in Hirotsu (1992) for defining a block interaction effects model for a two-way table.

22.6 Analysis of Generalized Interactions

For the two-way contingency table with ordered categories such as Table 22.2 the asymptotic equivalent of those statistics as developed in Section 22.5 can be derived according to the general complete class lemma. The result is stated in the following.

Let z_{ij} be $\sqrt{y_{..}} y_{ij} / \sqrt{y_{i.} y_{.j}}$ and define R' and C' to stand for P'_a and P'_b which satisfy $R'R = I_{a-1}, RR' = I_a - \boldsymbol{r}\boldsymbol{r}'$ and $C'C = I_{b-1}, CC' = I_b - \boldsymbol{c}\boldsymbol{c}'$, respectively, where $\boldsymbol{r} = ((y_1./y_{..})^{1/2}, \cdots, (y_a./y_{..})^{1/2})'$ and $\boldsymbol{c} = ((y_{.1}/y_{..})^{1/2}, \cdots, (y_{.b}/y_{..})^{1/2})'$. $R^{*'}$ and $C^{*'}$ are similarly defined to $P^{*'}_a$ and $P^{*'}_b$ so as to be orthogonal to $\boldsymbol{r}$ and $\boldsymbol{c}$ instead of $\boldsymbol{j}_a$ and $\boldsymbol{j}_b$, respectively. Then all the statistics in Section 22.5 can be defined through $R', R^{*'}, C', C^{*'}$ instead of $P'_a, P^{*'}_a, P'_b, P^{*'}_b$ and have asymptotically equivalent optimal properties. Those statistics are very useful for industrial, agricultural, and clinical trial data and some applications are given in Hirotsu (1978, 1982, 1983a,b, 1992 and 1993).

For higher-way layouts and contingency tables we can proceed similarly following the general complete class lemma. An example of a three-way contingency table will be given in Section 22.7. For a highly fractional factorial experiment with ordered categorical responses one should refer to Hamada and Wu (1990) and the discussions on their article.

22.7 Applications

22.7.1 Gum hardness data

For the data of Table 22.1 the p value by the cumulative chi-squared method is 0.02 suggesting a monotone tendency whereas the usual F test gives the p

value of 0.051 and fails to detect the effects of the added amount of sulfur at significance level 0.05. It should be noted that in this application the variance σ^2 has been replaced by its unbiased estimate, the sample variance, and a necessary modification of the distribution theory has been made.

22.7.2 Taste testing data

For the data of Table 22.2 Snell (1964) fitted a logistic distribution with different location but common scale parameters and obtained a goodness of fit chi-squared 49.1 with 12 df. McCullagh (1980) improves the chi-squared value into 21.3 with 8 df by allowing the scale parameters different and showed that Food 5 has the largest location and the smallest scale parameter values and thus most preferable.

Applying the row-wise multiple comparison procedure based on the asymptotic equivalent of (22.7) the classification of rows into three subgroups $I_1 = (1,2), I_2 = (3,4)$ and $I_3 = (5)$ has been obtained in Hirotsu (1990). For the column-wise comparisons the asymptotic equivalent of (22.8) has been applied to obtain the homogeneous subgroups $J_1 = (1,2)$ and $J_3 = (3,4,5)$. These results suggest a block interaction model

$$p_{ij} = p_{i.}p_{.j}\lambda_{u\vartheta} \, , \; i \in I_u, \; u = 1,2,3 \; ; \; j \in J_\vartheta, \; \vartheta = 1,2 \, . \tag{22.9}$$

The standardized block interaction effects are given in Table 22.3, from which we conclude that Food 5 is rated high, 3 and 4 are low, and 1 and 2 are intermediate.

Table 22.3: Estimating block interaction pattern

	Subgroups of rating	
Subgroups of foods	(1,2)	(3,4,5)
(1,2)	-2.21	2.21
(3,4)	6.47	-6.47
(5)	-5.09	5.09

The goodness of fit chi-squared for the model (22.9) is obtained as 24.9 with 14 df and seems to improve the fittings by Snell (1964) and MuCullagh (1980). It is, however, still large for its df, suggesting the need to analyze the residuals. Hirotsu (1990) strongly recommended the two-step procedure first analyzing the systematic effects by a systematic statistic such as the cumulative chi-squared statistic and then analyze the residuals to detect short-term deviations. For the details of the residual analysis one should refer to Hirotsu (1990).

22.7.3 Three-way ordinal categorical data

The data of Table 22.4 are the number of cancer patients classified by their age, the presence of metastasis into lymph node and the soaking grade. We are interested in analyzing the effects of age on the metastasis and soaking. Since there is a natural ordering in each of three factors the linear score test and the cumulative chi-squared test are applied based on the loglinear model for the parameters in a multinomial distribution.

Table 22.4: Number of cancer patients

	Age i							
	1. ~ 39		2. $40 \sim 49$		3. $50 \sim 59$		4. $60 \sim$	
	Metastasis j							
Soaking grade k	1. $-$	2. $+$	1. $-$	2. $+$	1. $-$	2. $+$	1. $-$	2. $+$
1. slight	9	4	10	5	6	2	8	4
2. medium	5	3	9	9	13	9	5	3
3. severe	12	5	15	16	11	18	9	9

(i) Testing the three-way interaction

We first test the effect of age on the association between the metastasis and the soaking. This is done by testing the null hypothesis $H_{\alpha\beta\gamma}$ for three factor interaction in the loglinear model for the expectation parameter,

$$\log m_{ijk} = \mu + \alpha_i + \beta_j + \gamma_k + (\alpha\beta)_{ij} + (\alpha\gamma)_{ik} + (\beta\gamma)_{jk} + (\alpha\beta\gamma)_{ijk}.$$

The maximum likelihood estimator (mle) $\hat{m}_{ijk}$ under $H_{\alpha\beta\gamma}$ are calculated by the usual iterative scaling procedure, from which we obtain the efficient score vector $\boldsymbol{\vartheta}_{\alpha\beta\gamma}$ evaluated at $H_{\alpha\beta\gamma}$,

$$\boldsymbol{\vartheta}_{\alpha\beta\gamma} = (y_{111} - \hat{m}_{111},\ y_{112} - \hat{m}_{112},\ \cdots,\ y_{423} - \hat{m}_{423})'.$$

Then the covariance matrix of $\boldsymbol{\vartheta}_{\alpha\beta\gamma}$ is obtained as

$$V(\boldsymbol{\vartheta}_{\alpha\beta\gamma}) = \mathrm{diag}(\hat{\mathbf{m}}) - \mathrm{diag}(\hat{\mathbf{m}})J\{J'\mathrm{diag}(\hat{\mathbf{m}})J\}^{-1}J'\mathrm{diag}(\hat{\mathbf{m}}),$$

where J is a $24 \times (24 - 6)$ matrix such that $J'\mathbf{y}$ is a set of linearly independent sufficient statistics under $H_{\alpha\beta\gamma}$.

The linear score statistic and its variance are defined by

$$
\begin{aligned}
S_w &= \{\mathbf{w}_1' \otimes (1\ 0) \otimes \mathbf{w}_3'\}(\mathbf{y} - \hat{\mathbf{m}}), \\
v_{(S_w)} &= \{\mathbf{w}_1' \otimes (1\ 0) \otimes \mathbf{w}_3'\}V(\boldsymbol{\vartheta}_{\alpha\beta\gamma})\{\mathbf{w}_1 \otimes (1\ 0)' \otimes \mathbf{w}_3\}.
\end{aligned}
$$

Then the standardized score statistic is obtained as 0.94 for the score vectors $\mathbf{w}_1' = (1, 2, 3, 4), \mathbf{w}_3' = (1, 2, 3)$, for example. This is not significant at level 0.05 as evaluated by the chi-squared with 1 df.

The cumulative chi-squared statistic $\chi^{*2} = 4.20$ is obtained as the standardized sum of squares of the elements of

$$\{T_4' \otimes (1\ 0) \otimes T_3'\}(\mathbf{y} - \hat{\mathbf{m}})$$

where

$$T_h' = \begin{bmatrix} 1 & 0 & 0 & \cdots & 0 \\ 1 & 1 & 0 & \cdots & 0 \\ \multicolumn{5}{c}{\dotfill} \\ 1 & 1 & \cdots & 1 & 0 \end{bmatrix}_{h-1 \times h}$$

is the matrix for accumulation. The constants for the chi-squared approximation $d\chi_f^2$ for χ^{*2} are $d = 1.90, f = 3.15$ and $\chi^{*2} = 4.20$ is not significant as evaluated by $1.90\chi_{3.15}^2$.

(ii) Testing $H_{\alpha\beta} : (\alpha\beta)_{ij} \equiv 0$ assuming $H_{\alpha\beta\gamma}$

Next we test the effect of the age on metastasis assuming no three factor interaction. The mle under $H_{\alpha\beta\gamma} \cap H_{\alpha\beta}$ is $\hat{m}_{ijk} = y_{i.k}\, y_{.jk}/y_{..k}$ and the efficient score vector evaluated under the null model is obtained as

$$\boldsymbol{\vartheta}_{\alpha\beta} = (y_{11.} - \hat{m}_{11.}, \cdots, y_{42.} - \hat{m}_{42.})'.$$

The $(2(i-1)+j,\ 2(i'-1)+j')$th element of the covariance matrix $V(\boldsymbol{\vartheta}_{\alpha\beta})$ of $\boldsymbol{\vartheta}_{\alpha\beta}$ is

$$\sum_{k=1}^{3} \frac{y_{i.k}(\delta_{ii'}y_{..k} - y_{i'.k})y_{.jk}(\delta_{jj'}y_{..k} - y_{.j'k})}{y_{..k}^2(y_{..k} - 1)},$$

from which we obtain the variance

$$\{\mathbf{w}_1' \otimes (1\ 0)\}V(\boldsymbol{\vartheta}_{\alpha\beta})\{\mathbf{w}_1 \otimes (1\ 0)'\}$$

for the linear score statistic $\{\mathbf{w}_1' \otimes (1\ 0)\}\boldsymbol{\vartheta}_{\alpha\beta}$. The standardized statistic is 0.98 and it is not significant at level 0.05.

The cumulative chi-squared statistic $\chi^{*2} = 2.68$ is obtained as the standardized sum of squares of the elements of $\{T_4' \otimes (1\ 0)\}\boldsymbol{\vartheta}_{\alpha\beta}$. The constants for the chi-squared approximation are determined as $d = 1.356$ and $f = 2.21$ and χ^{*2} is not significant as evaluated by $1.356\chi_{2.21}^2$.

(iii) Testing $H_{\alpha\gamma} : (\alpha\gamma)_{ik} \equiv 0$ assuming $H_{\alpha\beta\gamma}$

This situation is similarly dealt with as the previous situation (ii) and we obtain the linear score statistic 0.56 and the cumulative chi-squared $\chi^{*2} = 2.04$ with constants $d = 1.74$ and $f = 3.44$ for the chi-squared approximation. They are not significant.

Throughout (i) $\sim$ (iii) we conclude that there is no significant effect of age on the metastasis and soaking. Finally we test the association between the metastasis and the soaking throughout the age.

(iv) Testing $H_{\beta\gamma} : (\beta\gamma)_{jk} \equiv 0$ in the pooled table throughout age

By pooling the data throughout age we obtain Table 22.5. This is a 3 by 2 table with ordered categories and any of the cumulative chi-squared and the max χ^2 can be applied. The cumulative chisquared is obtained as $\chi^{*2} = 7.01$ and the constants for chi-squared approximation are $d = 1.290, f = 1.55$. The χ^{*2} is significant at level 0.05 since $1.290\chi^2_{1.55}(0.05) = 6.48$. We therefore conclude that there is a significant association between metastasis and soaking throughout the age. There is an apparent tendency in Table 22.5 that the soaking grade is relatively more severe in the presence of metastasis.

Table 22.5: Pooled data throughout age

	metastasis	
Soaking grade	$-$	$+$
slight	33	15
medium	32	24
severe	47	48

22.8   Concluding Remarks

The statistical procedures introduced here extend the usual ANOVA techniques in the following sense

  (i) Order restrictions are effectively adopted in the analysis,

 (ii) The characteristics of factors are effectively involved in the analysis,

(iii) Various techniques for the row- and/or the column-wise multiple comparisons for interaction effects are developed,

 (iv) The analysis of generalized interactions are developed,

  (v) The analysis of residuals following the analysis by a systematic statistic is introduced.

   Especially some of the isotonic inferences on interaction effects are newly introduced. The procedure may therefore be called the Beyond Analysis of Variance (BANOVA) Techniques.

References

1. Bradley, R. A., Katti, S. K., and Coons, I. J. (1962). Optimal scaling for ordered categories, *Psychometrika*, **27**, 355–374.

2. Cohen, A. and Sackrowitz, H. B. (1991). Tests for independence in contingency tables with ordered categories, *Journal of Multivariate Analysis*, **36**, 56–67.

3. Hamada, M. and Wu, C. F. J. (1990). A critical look at accumulation analysis and related methods (with discussion), *Technometrics*, **32**, 119–130.

4. Hawkins, D. M. (1977). Testing a sequence of observations for a shift in location, *Journal of the American Statistical Association*, **72**, 180–186.

5. Hirotsu, C. (1978). Ordered alternatives for interaction effects, *Biometrika*, **65**, 561–570.

6. Hirotsu, C. (1979a). An F-approximation and its application, *Biometrika*, **66**, 577–584.

7. Hirotsu, C. (1979b). The cumulative chi-squares method and Studentized maximal contrast method for testing an ordered alternative in a one-way analysis of variance model, *Reports of Statistical Application Research, JUSE*, **26**, 12–21.

8. Hirotsu, C. (1982). Use of cumulative efficient scores for testing ordered alternatives in discrete models, *Biometrika*, **69**, 567–577.

9. Hirotsu, C. (1983a). An approach to defining the pattern of interaction effects in a two-way layout, *Annals of the Institute of Statistical Mathematics*, **A35**, 77–90.

10. Hirotsu, C. (1983b). Defining the pattern of association in two-way contingency tables, *Biometrika*, **70**, 579–589.

11. Hirotsu, C. (1986). Cumulative chi-squared statistic as a tool for testing goodness of fit, *Biometrika*, **73**, 165–173.

12. Hirotsu, C. (1990). Discussion on Hamada & Wu's paper, *Technometrics*, **32**, 133–136.

13. Hirotsu, C. (1991). An approach to comparing treatments based on repeated measures, *Biometrika*, **78**, 583–594.

14. Hirotsu, C. (1992). *Analysis of Experimental Data, Beyond Analysis of Variance* (in Japanese), Tokyo: Kyoritsu-shuppan.

15. Hirotsu, C. (1993). Beyond analysis of variance techniques : some applications in clinical trials, *International Statistical Review*, **61**, 183–201.

16. Hirotsu, C., Kuriki, S. and Hayter, A. J. (1992). Multiple comparison procedure based on the maximal component of the cumulative chi-squared statistic, *Biometrika*, **79**, 381–392.

17. Johnson, D. E. and Graybill, F. A. (1972). An analysis of a two-way model with interaction and no replication, *Journal of the American Statistical Association*, **67**, 862–868.

18. Mandel, J. (1969). The partitioning of interactions in analysis of variance, *Journal of Research of the National Bureau of Standards*, **B73**, 309–328.

19. McCullagh, P. (1980). Regression models for ordinal data, *Journal of the Royal Statistical Society*, **B42**, 109–142.

20. Nair, V. N. (1986). On testing against ordered alternatives in analysis of variance model, *Biometrika*, **73**, 493–499.

21. Snell, E. J. (1964). A scaling procedure for ordered categorical data, *Biometrics*, **20**, 592–607.

23

On Robust Design for Multiple Quality Characteristics

Jai Hyun Byun and Kwang-Jae Kim

Gyeongsang National University, Chinju, Korea
Pohang Institute of Technology, Pohang, Korea

Abstract: In product and process design we have multiple quality characteristics very frequently. It is not easy to find optimal design parameter setting when there are multiple quality characteristics, since there will be conflict among the selected levels of the design parameters for each individual quality characteristic. In this paper we propose a modified desirability function approach and devise a scheme which gives a systematic way of solving mutiple quality characteristic problem.

Keywords and phrases: Decision making, desirability function, injection molding, multiple quality characteristics, robust design

23.1 Introduction

Parameter design, also called robust design, is one of the three design phases wherein the best nominal values of the product or process parameters are determined [Taguchi (1978)]. The basic steps of parameter design for identifying optimal settings of design parameters are as follows: (1) identify controllable factors and noise factors, (2) construct the design and noise matrices, (3) conduct the parameter design experiment and evaluate the signal-to-noise (S/N) ratio for each test run of the design matrix, (4) determine new settings of the design parameters using the S/N ratio values, and (5) conduct confirmation experiments [Kackar(1985)].

An assumption implicitly employed in the standard procedure described above is that there exists only one performance or quality characteristic whose mean and variance are simultaneously considered by the S/N ratio in steps (3) and (4). This assumption is not always justified. A common problem in product or process design is the selection of optimal factor levels which essentially involves simultaneous consideration of multiple conflicting quality characteristics. In reality, the customer's perception of the quality of a product is determined by multiple performance characteristics [Hauser and Clausing (1988)].

The purpose of this paper is to extend Taguchi's parameter design into the multiple quality characteristic case using the desirability function approach. The approach proposed in this paper can easily be understood and applied by practitioners.

23.2   Desirability Function Approach

Suppose there are m quality characteristics $\mathbf{y} = (y_1, y_2, \ldots, y_m)$ which are determined by a set of process parameters $\mathbf{x} = (x_1, x_2, \ldots, x_p)$. A general multiple response problem can then be defined as

$$y_i = f_i(x_1, x_2, \ldots, x_p), \quad j = 1, 2, \ldots, m,$$

where $\mathbf{f} = (f_1, f_2, \ldots, f_m)$ represents the functional relationship between the process parameters and responses. The exact form of $\mathbf{f}$ is usually unknown, but can be estimated using model building techniques such as regression or response surface methodology.

One popular approach to dealing with multiple characteristics has been the use of a dimensionality reduction strategy, which converts a multiple characteristic problem into one with a single aggregate characteristic and solves it using conventional techniques for a single characteristic problem.

The early work by Harrington (1965) has been extended and applied by a number of authors in conjunction with classical response surface methodology. The desirability function approach first transforms estimated response on each quality characteristic ($\hat{y}_j$) to a scale-free value d_j, called desirability. d_j is a value between 0 and 1 and increases as the desirability of the corresponding response increases. The individual desirabilities (d_j's) are combined into an overall desirability value D. Then the objective is to find the optimal solution $\mathbf{x}^*$ which maximizes the value of D, i.e., the parameter setting which achieves the optimal compromise among multiple quality characteristics. Harrington suggests various types of transformations from $\hat{y}_j$ to d_j, including a uniform transformation (i.e., d_j is set at 1 within the specification limits, and 0 otherwise) and a nonlinear transformation emphasizing midspecification quality. As a transformation from individual d_j's to the overall desirability D, Harrington suggests the use of the geometric mean of d_j's, i.e.,

$$D = (d_1 d_2 \ldots d_m)^{1/m}, \tag{23.1}$$

and treats it as though it were a measure of a single quality characteristic.

Derringer and Such (1980) modify Harrington's approach by employing a different transformation scheme from $\hat{y}_j$ to d_j. For a larger-the-better (LTB) type quality characteristic, they consider the transformation given by

$$d_j = \begin{cases} 0, & \hat{y}_j \leq y_j^{\min}, \\ \left(\dfrac{\hat{y}_j - y_j^{\min}}{y_j^{\max} - y_j^{\min}}\right)^r, & y_j^{\min} < \hat{y}_j < y_j^{\max}, \\ 1, & \hat{y}_j \geq y_j^{\max}, \end{cases} \tag{23.2}$$

where $y_i^{\min}$ and $y_j^{\max}$ denote the minimum acceptable value and the highest value of $\hat{y}_j$, respectively. The value of r, specified by the user, would indicate the degree of stringency of the characteristic. A larger value of r is used to make the desirability curve steeper if it is very desirable for $\hat{y}_j$ to be close to $y_j^{\max}$. Equation (23.2) can be easily modified for the smaller-the-better (STB) or nominal-the-best (NTB) (two-sided) type quality characteristics. Once D is constructed using individual d_j's as in (23.1), they employ an existing univariate search technique to maximize D over the process parameter ($\mathbf{x}$) domain. Later, Derringer (1994) modifies the overall desirability function D. The new form of D is still based on the concept of a (weighted) geometric mean;

$$D = (d_1^{w_1} d_2^{w_2} \ldots d_m^{w_m})^{1/\sum w_j}, \tag{23.3}$$

where w_j's are relative weights among m quality characteristics, $j = 1, 2, \ldots, m$. If all w_j's are set to 1, (23.3) is reduced to (23.1).

Harrington (1965) argues that the individual d_j can and should be self-weighting, which is manifested by the larger (or smaller) value of r for more (or less) important quality characteristic in (23.2). On the other hand, Derringer (1994) uses a different approach by incorporating the weights directly in the functional form of D, as given in (23.3). This would reduce the cognitive burden required on users by separating the weights from the shape of desirability functions. However, the value of D given in (23.3) (or D^* at $\mathbf{x}^*$) still does not allow a clear interpretation, except the principle that a higher value of D is preferred. As an example, if the overall desirability D at $\mathbf{x}_1$ ($D(\mathbf{x}_1)$) is higher than D at $\mathbf{x}_2$ ($D(\mathbf{x}_2)$), then $\mathbf{x}_1$ is considered a better design point than $\mathbf{x}_2$, but it is generally impossible to assign a physical meaning to the desirability values $D(\mathbf{x}_1)$ and $D(\mathbf{x}_2)$ as well as to the difference $D(\mathbf{x}_1) - D(\mathbf{x}_2)$.

23.3 Robust Design with Multiple Characteristics

A robust design procedure for multiple characteristics is proposed in this section. Suppose there are m quality characteristics, n control factor level combinations, and r noise factor level combinations for each of the n control factor level combinations. $y_{ijk}(i = 1, 2, \ldots, n; \quad j = 1, 2, \ldots, m; \quad k = 1, 2, \ldots, r)$ denotes the response value of the j-th quality characteristic obtained at the i-th factor level combination and the k-th noise level combination.

In this paper, the classical S/N ratio is used as the performance measure. However, any other kind of performance measure can be employed as appropriate without additional effort or complication. [For a discussion on the proper use of Taguchi's S/N ratio and other performance statistics, see Nair (1992)].

As can be seen in (23.2), it is necessary to specify the minimum acceptable value and the highest value for the performance measure of each characteristic. Since S/N ratio serves as the performance measure in this discussion, hereafter the notation y_j in (23.2) will be replaced by SN_j in order to avoid confusion. For the j-th quality characteristic, the minimum acceptable value $SN_j^{\max}$ is specified in such a way that any value of SN_j lower than $SN_j^{\max}$ would result in an unacceptable product. If the desirability increases as SN_j increases, there is no theoretical upper limit on SN_j. For this case, $SN_j^{\max}$ is set at the value of SN_j such that any higher value than this has little practical advantage.

It is usually difficult to find the minimum acceptable value and the highest value for the S/N ratio a priori because there does not exist an absolute range of the S/N ratio values in general. Therefore we suggest that $SN_j^{\min}$ and $SN_j^{\max}$ be specified based on the given experimental data in the following

$$
\begin{aligned}
SN_j^{\min} &= \min_{1 \leq i \leq n}\{SN_{ij}\}, \\
SN_j^{\max} &= \max_{1 \leq i \leq n}\{SN_{ij}\},
\end{aligned}
$$

where SN_{ij} is the S/N ratio of the j-th quality characteristic at the i-th factor level combination. Then, the desirability of the j-th quality characteristic at the i-th factor level combination, d_{ij}, is expressed as

$$
d_{ij} = \begin{cases}
0, & SN_{ij} \leq SN_j^{\min} \\
\left(\dfrac{SN_{ij}-SN_j^{\min}}{SM_j^{\max}-SN_j^{\min}}\right)^r, & SN_j^{\min} < SN_{ij} < SN_j^{\max} \\
1, & SN_j^{\max} \leq SN_{ij}
\end{cases} \tag{23.4}
$$

S/N ratio for the multiple quality characteristic case are typically computed as

$$
SN_{ij} = \begin{cases}
-10 \log\left[\frac{1}{r}\sum_{k=1}^{r} \frac{1}{y_{ijk}^2}\right], & \text{for an LTB type characteristic,} \\
-10 \log\left[\frac{1}{r}\sum_{k=1}^{r} y_{ijk}^2\right], & \text{for an STB type characteristic,} \\
10 \log\left[\frac{\bar{y}_{ij}^2}{s_{ij}^2}\right], & \text{for an NTB type characteristic,}
\end{cases} \tag{23.5}
$$

where, $\bar{y}_{ij} = \dfrac{1}{r}\sum_{k=1}^{r} y_{ijk}$ and $s_{ij}^2 = \dfrac{1}{r-1}\sum_{k=1}^{r}(y_{ijk} - \bar{y}_{ij})^2$. Then the procedure of analyzing multiple characteristics data is as follows.

1. Calculate SN_{ij}, $i = 1, \ldots, n$, $j = 1, \ldots, m$.

2. Calculate individual desirability d_{ij} using (1.4), $i = 1, \ldots, n$, $j = 1, \ldots, m$.

3. For the i-th factor level combination, obtain the overall desirability D_i as follows:

$$D_i = \min_{1 \leq j \leq m} \{d_{oj}^{w_i}\}, \qquad i = 1, \ldots, n, \qquad (23.6)$$

   where w_j is the relative importance of the j-th characteristic, and w_j's are scaled so that $\sum_{j=1}^{m} w_j = m$.

4. Estimate the effect of each control factor on the overall desirability. Identify important control factors on the overall desirability, and determine the optimal levels of the factors.

5. For an NTB type characteristic, select a factor as an adjustment factor. An adjustment factor should have a small effect on the overall desirability, but an important effect on the mean response of the characteristic.

6. Conduct confirmation experiments and plan future actions.

   In Step 3, values for the relative importance of multiple characteristics can be assessed by simply assigning weights directly or by using an eigenvector method which is a very widely used weight elicitation method among practitioners [Saaty (1977) and Yager (1997)]. If one cannot find an adjustment factor by analyzing data in Step 5, it should be identified using engineers' knowledge on the nature of the particular problem [Phadke (1989)]. In many situations an adjustment factor can be found prior to the design optimization [Phadke and Dehnad (1988)].

23.4   An Illustrative Example

A robust design optimization method is applied to an injection molded housing for a new copier by Greenall (1989). Four response variables are considered in order to optimize the process capability : overall housing length (y_1), overall housing width (y_2), flatness (y_3), and surface roughness (y_4). After a brainstorming session 7 control factors are considered likely to have an effect : injection pressure (A), injection speed (B), mold temperature (C), melt temperature (D), holding pressure (E), cool time (F), and hold time (G). An L_{18} orthogonal array is used as the design matrix, and at each of the 18 conditions the process is run until it is stabilized and then 10 parts are molded.

For length, width, and flatness standard measurement techniques are used. Surface quality is measured as an ordered categorical data : bad, acceptable, and good. Subjective scores are given to the categories 1, 2, and 3 to "bad", "acceptable", and "good" surface, respectively. As advocated by Hamada and Wu (1990) and Nair (1990), this scoring scheme performs reasonably well in analyzing the ordered categorical data. Since y_1, y_2, and y_3 are NTB type characteristics, S/N ratio are computed using (23.5). For y_4, mean response is analyzed, since dispersion tests are rarely applicable in the multifactor setting [Hamada and Wu (1990)]. In this example, r is set at 1 and the relative weights of the four characteristics are assumed to be the same. S/N ratios, individual desirabilities, and overall desirability are given in Table 23.1.

Factor effects on the overall desirability are given in Table 23.2. Factors B, C and F seem to be important. The optimal parameter setting would be $B_3C_3F_3$ or $B_3C_3F_2$.

Factor effects on the mean responses of the NTB type characteristics (y_1, y_2, and y_3) are summarized in Table 23.3. Table 23.4 indicates which factors are important for the mean responses of y_1, y_2, and y_3, and overall desirability. Using this information, one can identify factors that can serve as the adjustment factors. First, one needs to select factors which are unimportant to the overall desirability. They are factors A, D, E, and G in this case. Secondly, identify factors which are important to the mean response of only one quality characteristic : A for y_1, and G for y_3. Since A and G are influential on two different characteristics, A is chosen as an adjustment factor for y_1 and G for y_3. Finally, factor E is chosen for y_2, since E is the only remaining factor which has a major effect on y_2. It is crucial to conduct confirmation experiments to verify that the suggested optimum conditions indeed result in the projected overall desirability.

23.5   Summary and Concluding Remarks

A common problem in product or process design is to determine the optimal parameter setting when there exist multiple quality characteristics, which may be conflicting with each other. Most of the work on parameter design so far, however, has been concerned with the single quality characteristic case. This paper extends Taguchi's parameter design into the multiple quality characteristic case using the desirability function approach.

The experimental and computational procedure of our approach is virtually unaffected by the number of quality characteristics considered. Moreover, our approach can easily accommodate any combination of larger-the-better, smaller-the-better, or nominal-the-best type characteristics because it employs the signal-to-noise ratio, not the actual observations, in the desirability function development.

Table 23.1: Injection molded housing experiment : S/N ratios, individual desirabilities, and overall desirability

Experiment Number	SN_{i1}	d_{i1}	SN_{i2}	d_{i2}	SN_{i3}	d_{i3}	$SN_{i4}^\dagger$	d_{i4}	D_i
1	27.05	0.535	22.95	0.640	3.39	2.272	1.00	0.000	0.000
2	25.66	0.454	27.76	0.915	3.22	0.265	1.00	0.000	0.000
3	22.92	0.293	23.26	0.658	15.20	0.759	1.33	0.165	0.165
4	34.97	1.000	29.26	1.000	21.04	1.000	1.00	0.000	0.000
5	23.76	0.342	24.37	0.721	13.40	0.685	1.00	0.000	0.000
6	26.75	0.518	11.71	0.000	15.96	0.791	1.33	0.165	0.000
7	22.73	0.282	18.75	0.401	13.98	0.709	1.50	0.250	0.250
8	21.33	0.200	20.69	0.512	8.58	0.486	1.00	0.000	0.000
9	21.08	0.185	18.70	0.398	5.67	0.366	2.33	0.665	0.185
10	25.27	0.431	27.16	0.880	3.52	0.278	1.00	0.000	0.000
11	17.98	0.003	18.02	0.360	1.38	0.190	1.00	0.000	0.000
12	18.40	0.028	25.55	0.789	15.10	0.755	1.67	0.335	0.028
13	17.93	0.000	23.74	0.685	12.47	0.647	3.00	1.000	0.000
14	29.62	0.686	22.50	0.615	12.47	0.647	1.33	0.165	0.165
15	21.48	0.208	22.78	0.631	11.13	0.592	3.00	1.000	0.208
16	26.39	0.496	20.88	0.523	-3.22	0.000	1.83	0.415	0.000
17	22.60	0.274	19.65	0.452	-2.74	0.020	1.33	0.165	0.020
18	30.52	0.739	14.70	0.170	8.76	0.494	2.00	0.500	0.170

† SN_{i4} is the mean response of the surface quality (y_4) which is measured as an ordered categorical data, whereas SN_{i1}, SN_{i2} and SN_{i3} are the usual S/N ratios.

Table 23.2: Factor effects on overall desirability

1	0.0667	0.0322	0.0417	0.0688	0.0975	0.0080	0.0630
2	0.0657	0.0622	0.0308	0.0747	0.0393	0.0867	0.0797
3	—	0.1042	0.1260	0.0550	0.0617	0.1038	0.0558

† Factor A has two levels. All other factors have three levels.

Table 23.3: Mean responses

Characteristic Level	y_2 1	y_2 2	3	1	y_2 2	3	2	y_3 2	3
Factor A	69.4	64.4	–	63.6	70.0	–	9.0	7.0	–
B	65.5	68.0	67.2	67.1	67.4	65.8	7.3	8.7	8.0
C	68.4	58.8	73.5	69.6	70.7	60.1	7.4	-1.4	18.0
D	65.2	64.6	71.0	66.5	65.3	68.6	7.6	7.7	8.8
E	61.7	68.2	70.8	57.2	68.7	74.5	8.9	9.8	5.3
F	61.6	68.9	70.2	57.3	71.8	71.3	13.9	3.8	6.3
G	67.2	66.6	66.9	68.0	64.1	68.3	2.8	10.3	11.0

Table 23.4: Important factors for mean responses and overall desirability

Response	Mean y_1	Mean y_2	Mean y_3	Overall Desirability
Factor A	O[†]			
B				O
C	O		O	O
D	O			
E	O	O		
F	O	O		O
G			O	

† O means that the factor is important to the corresponding mean response or overall desirability.

Acknowledgment. This paper is supported by *Non-Directed Research Fund,* Korea Research Foundation, Korea, 1997.

References

1. Derringer, G. (1994). A balancing act: optimizing a product's properties, *Quality Progress*, 51–58.

2. Derringer, G. and Suich, R. (1980). Simultaneous optimization of several response variables, *Journal of Quality Technology*, **12**, 214–219.

3. Greenall, R. (1989). A Taguchi optimization of the manufacturing process for an injection molded housing, *In Taguchi Methods : Applications in the World Industry*, A. Bendell, J. Disney, and W. A. Pridmore (eds.), IFS Publications/Springer-Verlag, 295–313.

4. Hamada, M. and Wu, C. F. J. (1990). A critical look at accumulation analysis and related topics, *Technometrics*, **32**, 119–130.

5. Harrington, E. (1965). The desirability function, *Industrial Quality Control*, **21**, 494–498.

6. Hauser, J. and Clausing, D. (1988). The house of quality, *Harvard Business Review*, **66**, 63–73.

7. Kackar, R. (1985). Off-Line quality control, parameter design, and Taguchi method, *Journal of Quality Technology*, **17**, 165–188.

8. Nair, V. N. (1990). Discussion of a critical look at accumulation analysis and related topics, *Technometrics*, **32**, 151–152.

9. Nair, V. N., Ed. (1992). Taguchi's parameter design: A Panel Discussion, *Technometrics*, **34**, 127–161.

10. Phadke, M. S. (1989). *Quality Engineering Using Robust Design*, Prentice Hall, Englewood Cliffs.

11. Phadke, M. S. and Dehnad, K. (1988). Two step optimization for robust product and process design, *Quality and Reliability Engineering International*, **4**, 105–112.

12. Saaty, T. (1977). A scaling method for priorities in hierarchical structures, *Journal of Mathematical Psychology*, **15**, 234–281.

13. Taguchi, G. (1978). Off-Line and on-line quality control systems, *Proceedings of International Conference on Quality Control*, Tokyo, Japan.

14. Yager, R. (1977). Multiple objective decision-making using fuzzy sets, *International Journal of Man-Machine Studies*, **9**, 375–382.

24

Simultaneous Optimization of Multiple Responses Using a Weighted Desirability Function

Sung H. Park and Jun O. Park

Seoul National University, Seoul, Korea

Abstract: The object of multiresponse optimization is to determine conditions on the independent variables that lead to optimal or nearly optimal values of the response variables. Derringer and Suich (1980) extended Harrington's (1965) procedure by introducing more general transformations of the response into desirability functions. The core of the desirability approach condenses a multivariate optimization into a univariate one. But because of the subjective nature of this approach, inexperience on the part of the user in assessing a product's desirability value may lead to inaccurate results. To compensate for this defect, a weighted desirability function is introduced which takes into consideration the variances of the responses.

Keywords and phrases: Coefficient of variation, desirability function, simultaneous optimization

24.1 Introduction

In many experimental situations, multivariate optimization is necessary but complex compared with univariate optimization. Interrelationship that may exist among the responses can render univariate investigation meaningless. For example, if we desire to optimize several response functions simultaneously, it would be futile to obtain separate individual optimum. The main difficulty stems from the fact that two or more response variables are under investigation simultaneously, and the meaning of optimum becomes unclear since there is no unique way to order multivariate values of a multiresponse function. Furthermore, the optimal condition for one response may be far from optimal or even lead to physically impractical conditions for the remaining responses.

The object of the multiresponse optimization is to determine conditions on the input variables that lead to optimal or nearly optimal values of the response variables. In an effort to find optimal conditions on several responses, the desirability function approach will be introduced in the next section. The other methods used in the literature are briefly outlined as follows.

(1) Graphical superimposition method

In case there are only two or three input variables, this method is not only easy to understand and use, but also simple and straightforward. First of all, for each response, a response contour is obtained by fitting, in general, the second-order response model. Then by superimposition of response contours, we arrive at optimal conditions. Even though this procedure is practically very useful, it is difficult to identify one set of conditions or one point in the experimental region as being optimal, when the number of input variables exceeds three.

(2) Primary and secondary function

Myers and Carter (1973) introduced an algorithm for determining conditions on the input variables that maximize or minimize a primary response function subject to having an equality constant on a secondary response function. In other words, the secondary response function imposes certain constraints on the optimization of the primary response function. Biles (1975) extended this approach to include several secondary response functions within specified ranges. Biles's procedure employs a modification of the method of steepest ascent described by Box and Wilson (1951). In many cases, it is necessary to optimize the responses simultaneously rather than to optimize one response with the other constraints. Therefore, this method may be adapted to the restricted cases. Recently Vining and Myers (1990) proposed a dual response approach which combines Taguchi and response surface methods. Lin and Tu (1995) developed Vining and Myers's approach using the MSE(Mean Squared Error) criterion.

(3) Distance function approach

Khuri and Conlon (1981) presented several distance functions that measure the overall closeness of the response functions to achieve their respective optimal values at the same set of operating conditions. Multiresponse optimization is thus reduced to minimizing an appropriate distance function with respect to the input variables. This approach permits the user to account for the variances and covariances of the estimated responses and for the random error variation associated with the estimated ideal optimum.

(4) P_M and P_V measures

Park, Kwon, and Kim (1995) studied simultaneous optimization of multiple responses for robust design. They suggested two measures referred to as P_M and P_V. The P_M measure can be used without prior knowledge about the estimated mean responses. The P_V measure is reasonable when we have prior knowledge about the mean responses. P_V is simple and easy to compute and allows the user to make a decision on the range of the estimated mean responses. However, because of inappropriate decision on the range of mean responses, it may lead to inaccurate results.

24.2 Desirability Function Approach

24.2.1 Desirability function

Suppose each of the k response variables is related to the p independent variables by

$$y_i = f_i(x_1, x_2, \cdots, x_p) + \epsilon_i, \qquad i = 1, 2, \cdots, k$$

where f_i denotes the functional relationship between y_i and $x_1, x_2, \cdots, x_p$. If we make the usual assumption that $E(\epsilon_i) = 0$ and $Var(\epsilon_i) = \sigma_i^2$ for each i, then $E(y_i) = \eta_i = f_i(x_1, x_2, \cdots, x_p)$, $i = 1, 2, \cdots, k$, where η_i is represented by second order models within a certain region of interest in general.

The desirability function involves transformation of each estimated response variable $\hat{y}_i(= \hat{\eta}_i)$ to a desirability value d_i, where $0 \le d_i \le 1$. The value of d_i increases as the desirability of the corresponding response increases. The individual desirabilities are then combined using the geometric mean G,

$$G = (d_1 \times d_2 \times \cdots \times d_k)^{\frac{1}{k}}.$$

When k is large, the variation of d_i has much influence on G. So Park (1981) suggested the harmonic mean H,

$$H = \frac{k}{\frac{1}{d_1} + \frac{1}{d_2} + \cdots + \frac{1}{d_k}} = \frac{k \prod_{j=1}^{k} d_j}{\sum_{i=1}^{k} \prod_{j \ne i}^{k} d_j}.$$

This single value of G or H gives the overall assessment of the desirability of the combined response levels. If any $d_i = 0$ (that is, if one of the response variables is unacceptable), then G or H is 0 (that is, the overall product is unacceptable).

24.2.2 One-sided transformation (maximization of $\hat{y}_i$)

Let d_i be the ith individual desirability function, which is usually defined by

$$d_i = \begin{cases} 0 & \text{if} \quad \hat{y}_i \leq y_{i*}, \\ \left[\frac{\hat{y}_i - y_{i*}}{y_i^* - y_{i*}}\right]^r & \text{if} \quad y_{i*} \leq \hat{y}_i \leq y_i^*, \\ 1 & \text{if} \quad \hat{y}_i \geq y_i^*, \end{cases} \qquad (24.1)$$

where y_{i*} is the minimum acceptable value of $\hat{y}_i$ and y_i^* is the satisfactory value of $\hat{y}_i$ for $i = 1, 2, \cdots, k$ and r is an arbitrary positive constant. A large value of r would be specified if it were desirable for the value of $\hat{y}_i$ to increase rapidly above y_{i*}. On the other hand, a small value of r would be specified if having values of $\hat{y}_i$ considerably above y_{i*} were not of critical importance. For this reason the desirability function approach permits the user to make subjective judgements on the importance of each response. This is attractive to an experienced user. However, because of the subjective nature of the desirability approach, inexperience on the part of the user in assessing a product's desirability value may result in improper results. That is, the choice of r value contains user's subjective judgements and an inappropriate r value may result in an improper optimum condition.

Minimization of $\hat{y}_i$ is equivalent to maximization of $-\hat{y}_i$. Therefore, minimization of $\hat{y}_i$ is not elaborated here.

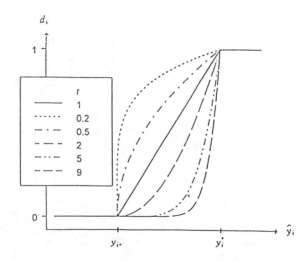

Figure 24.1: Transformation (24.1) for various values of r

24.2.3 Two-sided transformation

When the response variable y_i has both a minimum acceptable value and a maximum acceptable value, the individual desirability function is defined by

$$d_i = \begin{cases} \left[\dfrac{\hat{y}_i - y_{i*}}{c_i - y_{i*}}\right]^s & \text{if} \quad y_{i*} \le \hat{y}_i \le c_i, \\[2ex] \left[\dfrac{\hat{y}_i - y_i^*}{c_i - y_i^*}\right]^t & \text{if} \quad c_i \le \hat{y}_i \le y_i^*, \\[2ex] 0 & \text{if} \quad \hat{y}_i \le y_{i*} \quad \text{or} \quad \hat{y}_i \ge y_i^*, \end{cases} \qquad (24.2)$$

where c_i is the target value for the ith response, and s and t are arbitrary positive constants. In this situation, y_{i*} is the minimum acceptable value of $\hat{y}_i$ and y_i^* is the maximum acceptable value. The values of s and t in the two-sided transformation play the same role as that of r does in the one-sided transformation.

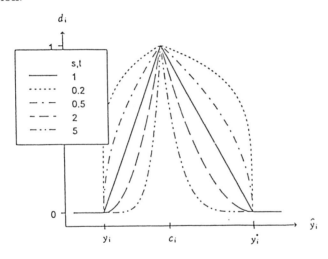

Figure 24.2: Transformation (24.2) for various values of s and t

Since $\hat{y}_i$ is a continuous function of x_i, $i = 1, 2, \cdots, p$, both G and H are continuous functions of d_i's, respectively. Therefore, it follows that both G and H are continuous in the $x_i, i = 1, 2, \cdots, p$. As a result, existing univariate search techniques can be used to maximize G or H over the independent variables domain.

24.3   Weighted Desirability Function Approach

The desirability functions (24.1) and (24.2) assume that, when y_{i*} and y_i^* are determined by the user, each y_i has the same degree of importance whether y_i's have different degrees of importance or different variances. Such assumption is not practical in real situations. Therefore, we propose here the weighted desirability function

$$\text{WG} = (d_1^{w_1} \times d_2^{w_2} \times \cdots \times d_k^{w_k})^{\frac{1}{k}}, \quad \sum_{i=1}^{k} w_i = k, \quad w_i > 0, \quad i = 1, 2, \cdots, k$$

where w_i is the weight for y_i, and the average of w_i's is 1. If y_i's are all equally important or they have equal variances, we may use $w_i = 1$ for each y_i. If we have no prior information for the importance of each y_i, a good choice of w_i is to make w_i proportional to the coefficient of variation of y_i,

$$\text{CV}_i = \frac{\sqrt{\text{MSE}_i}}{\bar{y}_i}$$

where $\bar{y}_i$ is the sample mean of the ith response, and $\sqrt{\text{MSE}_i}$ is the estimate of σ_i obtained from fitting a regression model. We may say that WG redistributes each weight in G according to each CV_i. Note that WG decreases as its weight w_i increases since d_i varies in $[0, 1]$. If any $d_i = 0$, WG is 0 like G. That is, if one of the response is unacceptable, then the overall product is unacceptable.

Each weighted desirability function is,

$$d_i^{w_i} = \begin{cases} 0 & \text{if} \quad \hat{y}_i \leq y_{i*}, \\ \left[\frac{\hat{y}_i - y_{i*}}{y_i^* - y_{i*}}\right]^{rw_i} & \text{if} \quad y_{i*} \leq \hat{y}_i \leq y_i^*, \\ 1 & \text{if} \quad \hat{y}_i \geq y_i^*, \end{cases} \tag{24.3}$$

in one-sided transformation and

$$d_i^{w_i} = \begin{cases} \left[\frac{\hat{y}_i - y_{i*}}{c_i - y_{i*}}\right]^{sw_i} & \text{if} \quad y_{i*} \leq \hat{y}_i \leq c_i, \\ \left[\frac{\hat{y}_i - y_i^*}{c_i - y_i^*}\right]^{tw_i} & \text{if} \quad c_i \leq \hat{y}_i \leq y_i^*, \\ 0 & \text{if} \quad \hat{y}_i \leq y_{i*} \quad \text{or} \quad \hat{y}_i \geq y_i^*, \end{cases} \tag{24.4}$$

in two-sided transformation. The constants r, s and t play the same role as those of G. The performance of this weighted desirability function is illustrated in the following example.

24.4 Example

In the development of a tire tread compound, the optimal combination of three ingredient(independent) variables - hydrated silica level x_1, silane coupling agent level x_2, and sulfur level x_3 - was sought. The properties to be optimized and the constraint levels to be achieved were as follows:

PICO abrasion Index, y_1	$120 < y_1$	$y_{1*} = 120$	
200% Modulus, y_2	$1000 < y_2$	$y_{2*} = 1000$	
Elongation at Break, y_3	$400 < y_3 < 600$	$y_{3*} = 400$	$y_3^* = 600$
Hardness, y_4	$60 < y_4 < 75$	$y_{4*} = 60$	$y_4^* = 75$

For y_1 and y_2, the one-sided transformation given by (24.3) was used and for y_3 and y_4, the two-sided transformation given by (24.4) was used. In the example given in Derringer and Suich (1980), they employed the rotatable central composite design with six center points in three variables. Table 24.1 shows the data which were fitted to the second degree polynomial models,

$$\hat{y}_i = b_0^i + \sum_{s=1}^{3} b_s^i x_s + \sum_{s=1}^{3} \sum_{t=s}^{3} b_{st}^i x_s x_t, \qquad i = 1, 2, 3, 4$$

where b_0^i, b_s^i's and b_{st}^i's are regression coefficient estimates for the ith response variable. The resultant fitted equations are

$$
\begin{aligned}
\hat{y}_1 =\ & 139.12 + 16.49x_1 + 17.88x_2 + 10.91x_3 - 4.01x_1^2 - 3.45x_2^2 - 1.57x_3^2 \\
& + 5.13x_1x_2 + 7.13x_1x_3 + 7.88x_2x_3 \\
\hat{y}_2 =\ & 1261.11 + 268.15x_1 + 246.50x_2 + 139.48x_3 - 83.55x_1^2 - 124.79x_2^2 \\
& + 199.17x_3^2 + 69.38x_1x_2 + 94.13x_1x_3 + 104.38x_2x_3 \\
\hat{y}_3 =\ & 400.38 - 99.67x_1 - 31.40x_2 - 73.92x_3 + 7.93x_1^2 + 17.31x_2^2 + 0.43x_3^2 \\
& + 8.75x_1x_2 + 6.25x_1x_3 + 1.25x_2x_3 \\
\hat{y}_4 =\ & 68.91 - 1.41x_1 + 4.32x_2 + 1.63x_3 + 1.56x_1^2 + 0.06x_2^2 - 0.32x_3^2 \\
& - 1.63x_1x_2 + 0.13x_1x_3 - 0.25x_2x_3.
\end{aligned}
$$

The $\bar{y}_i$, $\sqrt{\text{MSE}_i}$, $\text{CV}_i \times 100$ and w_i for each y_i are given in Table 24.2.

We use the grid-search method to find an optimum formulation. Table 24.3 shows the difference in the optimum condition and the estimated value of each response between G and WG.

Figures 24.3–24.6 show the performance of the $d_i^{w_i}$, $i = 1, 2, 3, 4$. We assume that $r = 1$ and $s = t = 1$ in (24.3) and (24.4).

Table 24.1: Experimental design

x_1	x_2	x_3	y_1	y_2	y_3	y_4
-1	-1	1	102	900	470	67.5
1	-1	-1	120	860	410	65
-1	1	-1	117	800	570	77.5
1	1	1	198	2294	240	74.5
-1	-1	-1	103	490	640	62.5
1	-1	1	132	1289	270	67
-1	1	1	132	1270	410	78
1	1	-1	139	1090	380	70
-1.633	0	0	102	770	590	76
1.633	0	0	154	1690	260	70
0	-1.633	0	96	700	520	63
0	1.633	0	163	1540	380	75
0	0	-1.633	116	2184	520	65
0	0	1.633	153	1784	290	71
0	0	0	133	1300	380	70
0	0	0	133	1300	380	68.5
0	0	0	140	1145	430	68
0	0	0	142	1090	430	68
0	0	0	145	1260	390	69
0	0	0	142	1344	390	70

Table 24.2: Each weight proportional to CV

i	$\overline{y}_i$	$\sqrt{MSE_i}$	$CV_i \times 100$	w_i
1	133.1	5.61	4.22	0.45
2	1255.0	328.69	26.19	2.82
3	417.5	20.55	4.92	0.53
4	69.8	1.27	1.82	0.20

Table 24.3: Different optimal conditions under G and WG

	G	WG
	$x_1=-0.050$ $x_2=0.145$ $x_3=-0.868$	$x_1=-0.158$ $x_2=0.437$ $x_3=-0.879$
$120 < y_1$	129.5	130.38
$1000 < y_2$	1300.0	1300.02
$400 < y_3 < 600$	465.7	471.00
$60 < y_4 < 75$	68.0	69.62

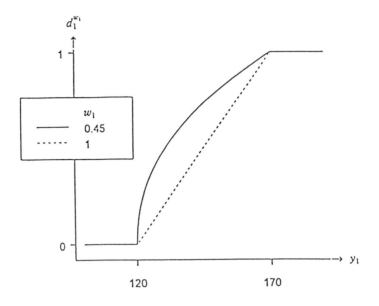

Figure 24.3: The performance of d_1

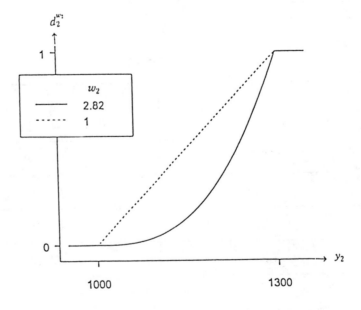

Figure 24.4: The performance of d_2

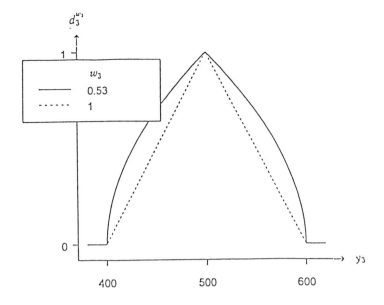

Figure 24.5: The performance of d_3

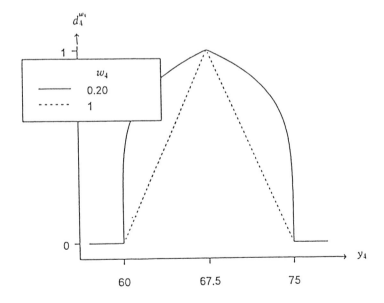

Figure 24.6: The performance of d_4

24.5 Conclusion

Since the desirability function condenses a multivariate optimization problem into a univariate one and WG is continuous in x_i, $i = 1, 2, \cdots, p$ as G is, we have to consider only the univariate techniques to find the maximum of the weighted desirability function. In this paper, the grid-search method was used to find the maximum value.

One can see the different optimum conditions resulting from WG and G. The estimated value of the response for which the CV_i is large is moved to the desirable point(maximum value in one-sided transformation or target value in two-sided transformation) in WG. On the other hand, the estimated value of the response whose CV_i is small is moved in the opposite direction. Therefore, if the response whose CV_i is relatively large is more important than others, it is useful to employ WG instead of G.

References

1. Biles, W. E. (1975). A response surface method for experimental optimization of multiresponse process, *Industrial and Engineering Chemistry Process Design and Development*, **14**, 152–158.

2. Box, G. E. P. and Wilson, K. B. (1951). On the experimental attainment of optimum conditions, *Journal of Royal Statistical Society.* **13**, 1–45.

3. Derringer, G. and Suich, R. (1980). Simultaneous optimization of several response variables, *Journal of Quality Technology*, **12**, 214–219.

4. Harrington, E. C., Jr. (1965). The Desirability Function, *Industrial Quality Control*, **21**, (10), 494–498.

5. Khuri, A. I. and Conlon, M. (1981). Simultaneous optimization of multiple response represented by polynomial regression function, *Technometrics*, **23**, 363–375.

6. Lin, D. J. and Tu, W. (1995). Dual Response Surface Optimization, *Journal of Quality Technology*, **27**, 34–39.

7. Myers, R. H. and Carter, W. H., Jr. (1973). Response surface techniques for dual response systems, *Technometrics*, **15**, 301–317.

8. Park, S. H. (1981). Simultaneous Optimization Techniques for Multipurpose Response Functions, *Journal of Military Operations Research Society of Korea*, **7**, 118–138.

9. Park, S. H., Kwon. Y. M. and Kim. J. J. (1995). *Simultaneous optimization of multiple responses for robust design, Total Quality Management,* Chapman & Hall, London, 381–390.

10. Vining, G. G. and Myers, R. H. (1990). Combining Taguchi and Response Surface Philosophies : A Dual Response Approach , *Journal of Quality Technology,* **22**, 38–45.

25

Evaluating Statistical Methods Practiced in Two Important Areas of Quality Improvement

Subir Ghosh and Luis A. Lopez

University of California, Riverside, CA
National University of Colombia, Colombia, SA

Abstract: This paper presents evaluations and comparisons of statistical methods practiced in factor screening as well as in analysis with missing data. In quality performance evaluation studies using factorial experiments, the method of a normal probability plot is routinely used for screening out factors with insignificant effect on the quality characteristic. We demonstrate that the method of a normal probability plot for factor screening can be misleading. A screened out factor may turn out to be significant in the presence of certain interactions. Three methods of analysis in the presence of missing data for explanatory or response variables are considered under the multiple linear regression model by ignoring the missing data rows, substituting the corresponding column means from the data rows with no missing data, and substituting the Yates/EM algorithm estimates for the missing data. Five illustrative examples are given. In all examples, the third method has turned out to be the winner.

Keywords and phrases: Coefficient of determination, estimation, factorial experiments, interactions, mean square error, missing data, normal probability plot, regression model

25.1 Introduction

Statistical methods such as design of experiments and multiple regression play important roles in industrial Quality Improvement activities. In this paper we discuss

(i) factor screening using normal probability plots in the context of designed experiments

(ii) analysis of multiple regression models in the presence of missing data.

The first issue is considered in Section 25.2 and the latter is in Section 25.3. Some concluding remarks are given in Section 25.4.

25.2 Misleading Probability Plots in Factor Screening

For a chemical product produced in a pressure vessel, the goal is to maximize the filtration rate (response variable). The following four factors are believed to influence the filtration rate.

A: Temperature, B: Pressure, C: Concentration of formal dehyde,
D: Stirring rate.

Two levels of each factor, low (−) and high (+), are chosen to perform a single replicate complete factorial experiment with sixteen runs [for the data see Montgomery (1991), page 291]. Numerical values of the estimated main effects and interactions are given in Table 25.1.

Table 25.1: Estimated factorial effects

Factorial Effects	Estimate
A	21.625
B	3.125
C	9.875
D	14.625
AB	0.125
AC	-18.125
AD	16.625
BC	2.375
BD	-1.125
CD	-0.375
ABC	1.875
ABD	4.125
ACD	-1.625
BCD	-2.625
ABCD	1.375

Fifteen estimated factorial effects given in Table 25.1 are first ordered and then used in drawing the normal probability plot of ordered effects given in Figure 9-8 on page 293 of Montgomery (1991). Effects that are on or near the line are claimed to be negligible and effects that are far from the line are declared to be important. From the plot, the main effects of A, C, and D and the AC and AD interactions are found to be important. Factor B is then screened out for subsequent analysis. Montogomery (1991) displayed the main effect plots for A, C, and D, and the interaction plots for AC and AD to support the conclusions from the normal probability plot. Moreover, the significance testing using the full model with three factors A, C, D concludes that the main effects A, C, and D, the interactions AC and AD are significant at 1% level and the interactions CD and ACD are not significant.

In our analysis it is demonstrated that factor B may not be as important as factors A, C, and D, but it should not be ignored particularly because of its main effect and interactions with factors A, C, and D. Practical considerations like cost, time, convenience, etc., may dominate over statistical benefits in decision making to ignore the main effect of B and the interaction effects BC, ABC, BCD, ACD and ABC. Even then strong recommendations should be made in keeping these effects in the model.

Following the recommendation from the normal probability plot, the factor B is first completely screened out. Then the model (Model I) with the general mean (μ), the main effects A, C, and D, and the interactions AC and AD is fitted to the data and the coefficient of determination (R^2) has turned out to be 96.6%. The complete model (Model II) with all interactions present is now fitted to observe the change in the value of R^2. The R^2 for Model II is 96.9% which is not much different from 96.6% of Model I. But again the maximum value of R^2 with three factors A, C, and D is 96.9% and no further improvement is possible. This is not a satisfactory situation.

The following three models (Models III, IV and V) are now considered without screening out factor B and in the presence of interactions. All three models have the general mean (μ), the main effects A, B, C, and D and the interactions AC and AD. For Model IV, there are interactions ABD, BCD, BC, ABC, and ACD and for Model V, there is an additional interaction ABCD. In Table 25.2 Models III, IV, and V are compared with respect to the values of the coefficient of cetermination (R^2), the adjusted coefficient of determination (R^2_{adj}), sum of squares due to error (SSE), mean square error (MSE), and degrees of freedom (DF) due to error. The main effect of B is not significant in Model III but significant at 5% level in both Models IV and V. In Model IV, factor B becomes significant when the interactions ABD, BCD, BC, ABC, and ACD are included in the model.

25.3   Analysis With Missing Data

In the study of the dependence of the response variable y on k explanatory variables $x_1, \ldots, x_k$, the observations on y and $x_1, \ldots, x_k$ are denoted by $(y_i, x_{i1}, \ldots, x_{ik})$, $i = 1, \ldots, n$. The observations $y_1, \ldots, y_n$ are assumed to be uncorrelated with mean

$$E(y_i) = \beta_0 + \beta_1 x_{i1} + \ldots + \beta_k x_{ik}, i = 1, \ldots, n, \qquad (25.1)$$

and variance σ^2. The parameters $\beta_0, \beta_1, \ldots, \beta_k$ and σ^2 are unknown constants. In matrix notation, (1) is expressed as

$$\mathbf{E(y)} = \mathbf{X}\boldsymbol{\beta}, \ \mathbf{V(y)} = \sigma^2 \mathbf{I} \qquad (25.2)$$

Table 25.2: Comparison of models III, IV, and V

	Model V			Model IV		Model III	
	Parameter	Estimate	p-value	Parameter	p-value	Parameter	p-value
	μ	70.1	0.000	μ	0.000	μ	0.000
	A	10.8	0.000	A	0.000	A	0.000
	B	1.56	0.020	B	0.026	B	0.168
	C	4.94	0.001	C	0.000	C	0.000
	D	7.31	0.000	D	0.000	D	0.000
	AC	-9.06	0.000	AC	0.000	AC	0.000
	AD	8.31	0.000	AD	0.000	AD	0.000
	ABD	2.06	0.009	ABD	0.011		
	BCD	-1.31	0.032	BCD	0.045		
	BC	1.19	0.041	BC	0.059		
	ABC	0.937	0.072	ABC	0.108		
	ACD	-0.812	0.099	ACD	0.149		
	ABCD	0.688	0.140				
R^2	99.9%			99.8%		97.3%	
R^2_{adj}	99.5%			99.1%		95.5%	
SSE	5.69			13.25		156.06	
DF	3			4		9	
MSE	1.90			3.31		17.34	

where the ith row of $\mathbf{X}$ is $(1, x_{i1}, \ldots, x_{ik})$ and $\boldsymbol{\beta} = (\beta_0, \beta_1, \ldots, \beta_k)'$. The matrix $\mathbf{X}$ is of dimension $(n \times p)$ and $\boldsymbol{\beta}$ is $(p \times 1)$ where $p = (1 + k)$.

Missing data are very common in practice and can in fact happen in different ways with y and $x_1, \ldots, x_k$. We compare the methods that are available for the missing data analysis. In particular, three methods are considered for such comparisons. The first method ignores the missing data rows completely, the second method substitutes the means of the complete data columns for the missing data, and the third method substitutes the Yates/EM algorithm estimates of the missing data. Some theoretical comparisons are made for the first two methods. Empirical comparisons are made for all three methods in Examples 1—5. Examples 1–5 represent the possible ways of missing data with y and $x_1, \ldots, x_k$.

Table 25.3: Comparison of models I and II

	Model II			Model I	
	Parameter	Estimate	p-value	Parameter	p-value
	μ	70.1	0.000	μ	0.000
	A	10.8	0.000	A	0.000
	C	4.94	0.003	C	0.001
	D	7.31	0.000	D	0.000
	AC	-9.06	0.000	AC	0.000
	AD	8.31	0.000	AD	0.000
	ACD	-0.81	0.521		
	CD	-0.56	0.642		
R^2		96.9%			96.6%
R^2_{adj}		94.1%			94.9%
SSE		179.50			195.10
DF		8			10
MSE		22.40			19.51

Suppose that there are $s_1 (\geq 0)$ missing data on y and $s_2 (\geq 0)$ missing data on $x_1, \ldots, x_k$. Assume that in the first q_1 elements of y and q_1 rows of X, all observations are available. In the remaining $(n - q_1)$ positions in y and the corresponding $(n - q_1)$ rows of X,

i. There are missing data in all of the first $q_2 (\geq 0)$ positions in y but no missing data in the corresponding q_2 rows of X,

ii. There are missing data in all of the next $q_3 (\geq 0)$ positions in y and also in some positions in each of the corresponding q_3 rows of X,

iii. all observations are available in the remaining $q_4 = (n - q_1 - q_2 - q_3)$ positions of y but there are missing data in some positions in each of the corresponding q_4 rows of X.

Note that $s_1 = q_2 + q_3$ and $s_2 < (q_3 + q_4)k$.

Illustrative special cases

In a study of the vending machine service routes in the distribution system of a soft drink bottler [see Montgomery and Peck (1992, p. 125)], the following response and explanatory variables are considered.

$y =$ The amount of time required by the route driver to
         service the vending machines in an outlet,

$x_1 =$ The number of cases of product stocked,

$x_2 =$ The distance walked by the route driver.

There are 25 observations on each of y, x_1 and x_2. Suppose now that there are two missing data, i.e., $s_1 + s_2 = 2$. The following situations are now considered.

Case 1.

Suppose that $s_1 = 2, s_2 = 0$, $q_1 = 23$, $q_2 = 2$ and $q_3 = q_4 = 0$. In other words, the last two observations on y, namely 19.83 and 10.75, are assumed to be missing.

Case 2.

Supppose that $s_1 = 0$, $s_2 = 2$, $q_1 = 23$, $q_2 = q_3 = 0$, and $q_4 = 2$. Assume that the last two observations on x_2, namely 635 and 150, are missing.

Case 3.

Suppose that $s_1 = 0, s_2 = 2$, $q_1 = 23, q_2 = q_3 = 0$, and $q_4 = 2$. Assume that in the last two observations on x_1 and x_2, 8 for x_1 and 150 for x_2 are missing.

Case 4.

Suppose that $s_1 = 1, s_2 = 1, q_1 = 23, q_2 = 1, q_3 = 0$, and $q_4 = 1$. Assume that in the last two observations on y and x_2, 19.83 for y and 150 for x_2 are missing.

Case 5.

Suppose that $s_1 = 1, s_2 = 1, q_1 = 23, q_2 = q_4 = 0$, and $q_3 = 1$. Assume that the last observation on both y and x_2, 10.75 for y and 150 for x_2 are missing.

Estimation

We denote the first q_1 elements in $\mathbf{y}$ by $\mathbf{y}_1$ and the first q_1 rows in $\mathbf{X}$ by X_1. Let $\bar{y}_c = (\mathbf{j}'_{q_1}\mathbf{y}_1/q_1)$ and $(1, \bar{x}_{1c}, \ldots, \bar{x}_{kc}) = (\mathbf{j}'_{q_1}X_1/q_1)$ where $\mathbf{j}'_{q_1}$ is vector with all elements unity. The next $(q_2 + q_3)$ positions in $\mathbf{y}$ have missing data and they are denoted by $\mathbf{y}_2(\mathbf{w}^{(2)}) = \bar{y}_c\mathbf{j}_{q_2} + \mathbf{w}^{(2)}$ and $\mathbf{y}_3(\mathbf{w}^{(3)}) = \bar{y}_c\mathbf{j}_{q_3} + \mathbf{w}^{(3)}$ where $\mathbf{w}^{(2)}$ and $\mathbf{w}^{(3)}$ are vectors of unknown constants. Note that $\mathbf{y}_2(\mathbf{0}) = \bar{y}_c\mathbf{j}_{q_2}$ and $\mathbf{y}_3(\mathbf{0}) = \bar{y}_c\mathbf{j}_{q_3}$ where the vector $\mathbf{0}$ has all elements zero. The elements in the last q_4 positions in $\mathbf{y}$ are denoted by $\mathbf{y}_4$.

Let

$$\mathbf{y}^{(1)} = \begin{bmatrix} \mathbf{y}_1 \\ \mathbf{y}_2(\mathbf{0}) \\ \mathbf{y}_3(\mathbf{0}) \\ \mathbf{y}_4 \end{bmatrix} \text{ and } \mathbf{y}^{(2)} = \begin{bmatrix} \mathbf{y}_1 \\ \mathbf{y}_2(\omega^{(2)}) \\ \mathbf{y}_3(\omega^{(3)}) \\ \mathbf{y}_4 \end{bmatrix}. \tag{25.3}$$

The q_2 rows of $\mathbf{X}$ corresponding to $\mathbf{y}_2(\boldsymbol{\omega}^{(2)})$ are denoted by $\mathbf{X}_2$. In the last $(q_3 + q_4)$ rows and the jth column of $\mathbf{X}$, represent each missing data by the mean of the jth column of $\mathbf{X}_1$ plus an unknown constant. Thus the last (q_3+q_4) rows of $\mathbf{X}$ are denoted by $\mathbf{X}_3(\mathbf{W}^{(3)})$ and $\mathbf{X}_4(\mathbf{W}^{(4)})$ where $\mathbf{W}^{(3)}$ and $\mathbf{W}^{(4)}$ are matrices with unknown constants in the missing data positions and zero elsewhere. Again, $\mathbf{X}_3(0)$ and $\mathbf{X}_4(0)$ represent the cases $\mathbf{W}^{(3)} = 0$ and $\mathbf{W}^{(4)} = 0$. Let

$$\mathbf{X}^{(1)} = \begin{bmatrix} \mathbf{X}_1 \\ \mathbf{X}_2 \\ \mathbf{X}_3(0) \\ \mathbf{X}_4(0) \end{bmatrix} \text{ and } \mathbf{X}^{(2)} = \begin{bmatrix} \mathbf{X}_1 \\ \mathbf{X}_2 \\ \mathbf{X}_3(\mathbf{W}^{(3)}) \\ \mathbf{X}_4(\mathbf{W}^{(4)}) \end{bmatrix}. \qquad (25.4)$$

Denote

$$\mathbf{b} = (\mathbf{X}_1'\mathbf{X}_1)^{-1}\mathbf{X}_1'\mathbf{y}_1 \text{ and } \mathbf{b}^{(1)} = (\mathbf{X}^{(1)'}\mathbf{X}^{(1)})^{-1}\mathbf{X}^{(1)'}\mathbf{y}^{(1)}. \qquad (25.5)$$

Notice that $\mathbf{b}$ and $\mathbf{b}^{(1)}$ are estimators of β in methods 1 and 2, respectively. Clearly $\mathbf{b}^{(1)}$ is a competitor of $\mathbf{b}$. In Section 25.2, it is shown that the estimator $\mathbf{b}$ is in fact an unbiased estimator of β and the estimator $\mathbf{b}^{(1)}$ is a biased estimator of β. However, the performance of $\mathbf{b}^{(1)}$ is better than $\mathbf{b}$ in terms of variance.

It can be checked that the quantity $(\mathbf{y}^{(2)} - \mathbf{X}^{(2)}\mathbf{b})'(\mathbf{y}^{(2)} - \mathbf{X}^{(2)}\mathbf{b})$ when minimized with respect to $\mathbf{W}^{(3)}$, $\mathbf{W}^{(4)}$, $\boldsymbol{\omega}^{(2)}$ and $\boldsymbol{\omega}^{(3)}$, the minimum value is attained at $\hat{\mathbf{W}}^{(3)}$, $\hat{\mathbf{W}}^{(4)}$, $\hat{\boldsymbol{\omega}}^{(2)}$ and $\hat{\boldsymbol{\omega}}^{(3)}$ satisfying $\mathbf{y}_2(\hat{\boldsymbol{\omega}}^{(2)}) = \mathbf{X}_2\mathbf{b}$, $\mathbf{y}_3(\hat{\boldsymbol{\omega}}^{(3)}) = \mathbf{X}_3(\hat{\mathbf{W}}^{(3)})\mathbf{b}$ and $\mathbf{y}_4 = \mathbf{X}_4(\hat{\mathbf{W}}^{(4)})\mathbf{b}$. This is known as the Yates method of estimation of the missing data [see page 25 in Little and Rubin (1987)]. There is of course the problem of identifiability of parameters $\mathbf{W}^{(3)}$, $\mathbf{W}^{(4)}$, $\boldsymbol{\omega}^{(2)}$ and $\boldsymbol{\omega}^{(3)}$. To illustrate this, it is to be noted that the elements of $\mathbf{w}^{(3)}$ and $\mathbf{W}^{(3)}$ may not be identifiable when $q_3 > 1$.

We now consider the missing data cases 1-5, estimate the missing values and present the results as Examples 1–5 respectively. In Examples 1–4, the missing data are estimated by the Yates method and the EM algorithm [see Little and Rubin (1987)]. In Example 5, the missing data are estimated by the EM algorithm. Three methods of dealing with the missing data are considered by ignoring the missing data rows, subsituting the corresponding column means of $(\mathbf{y}_1, \mathbf{X}_1)$ for the missing data, and subsituting the Yates / EM algorithm estimates of the missing data.

Example 1

The $\mathbf{b} = (2.50, 1.55, 0.016)'$ is based on 23 observations on y, x_1 and x_2. The Yates estimates of the missing data are $25.06(= 2.50+1.55 \times 8+0.016 \times 635)$ and $11.10(= 2.50+1.55 \times 4+0.016 \times 150)$. The $\bar{y}_c = 23.001$ and therefore the estimate of $\boldsymbol{\omega}^{(2)}$ is $(2.059, -11.901)$. The EM algorithm starts with the initial values (step

1) of the missing data as 23.001 and then finds the least squares estimates of β as $\mathbf{b}^{(1)} = (3.43, 1.57, 0.0143)'$ based on 25 observations on y, x_1 and x_2. The step 2 values of the missing data are then $25.0705 (= 3.43 + 1.57 \times 8 + 0.0143 \times 635)$ and $11.855 (= 3.43 + 1.57 \times 4 + 0.0143 \times 150)$. The step 3 values of the missing data are again obtained. The process converges at the step 5 with the least squares estimates of β to $\mathbf{b} = (2.50, 1.55, 0.016)'$. Therefore, the estimates of the missing data by the EM algorithm are exactly identical to the Yates estimates.

Example 2

The $\mathbf{b}$ remains the same as in Example 1. The Yates estimates of the missing data are $308.125 (= (19.83 - 2.50 - 1.55 \times 8)/0.016)$ and $128.125 (= (10.75 - 2.50 - 1.55 \times 4)/0.016)$. The EM algorithm starts with the initial values (step 1) of the missing data as $\bar{x}_{2c} (= 410.74)$ and then find the least squares estimates of β as $\mathbf{b}_1 = (2.16, 1.61, 0.0149)'$ based on 25 observations on y, x_1 and x_2. The step 2 values of the missing data are then 321.477 and 144.295. The process again converges at the step 6 with the least squares estimates of β to $\mathbf{b}$ and therefore the estimates of the missing data by the EM algorithm are exactly identical to the Yates estimates.

Example 3

The $\mathbf{b}$ remains the same as in Example 1. The Yates estimates of the missing data are $4.62581 (= (19.83 - 2.50 - 0.016 \times 635)/1.55)$ for x_1 and $128.125 (= (10.75 - 2.50 - 1.55 \times 4)/0.016)$ for x_2. The EM algorithm starts with the initial values (step 1) of the missing data as $\bar{x}_{1c} (= 9)$ and $\bar{x}_{2c} (= 410.74)$. The process again converges at the step 6 with the least squares estimates of β to $\mathbf{b}$ and therefore the estimates of the missing data by the EM algorithm are exactly identical to the Yates estimates.

Example 4

The $\underline{b}$ remains the same as in Example 1. The Yates estimates of the missing data are $25.06 (= 2.50 + 1.55 \times 8 + 0.016 \times 635)$ for y and $128.125 (= (10.75 - 2.50 - 1.55 \times 4)/0.016)$ for x_2. The EM algorithm starts with the initial values (step 1) of the missing data as $\bar{y}_c (= 23.001)$ and $\bar{x}_{2c} (= 410.74)$. The process again converges at the step 4 wth the least squares estimate of β to $\mathbf{b}$ and therefore the estimates of the missing data by the EM algorithm are exactly identical to the Yates estimators.

Example 5

The $\mathbf{b} = (2.36, 1.62, 0.0144)'$ based on 24 observations on y, x_1 and x_2. The Yates estimates of the missing data are not possible to obtain because there is a single equation in two unknowns. The EM algorithm starts with the initial values (step 1) of the missing data as $\bar{x}_{2c}(= 420.08)$ and $\bar{y}_c(= 22.869)$. It then finds the least squares estimates of β as $\mathbf{b} = (2.82, 1.52, 0.0161)'$ based on 25 observations on y, x_1 and x_2.

The step 2 values of the missing data are then $15.6633(= 2.82 + 1.52 \times 4 + 0.0161 \times 420.08)$ for y and $420.081(= (15.6633 - 2.82 - 1.52 \times 4)/0.0161)$ for x_2. The least squares estimate of β is $(2.41, 1.61, 0.0145)'$. The step 3 values of the missing data are then 14.9412 for y and 420.083 for x_2. The process converges at step 5 with the least squares estimate of β to $\mathbf{b} = (2.36, 1.62, 0.0144)'$. The EM algorithm estimates of the missing data are 420.083 for x_2 and 14.8892 for y.

It is observed in Examples 1–4 that the Yates estimates of the missing data are exactly identical to the EM algorithm estimates. In Example 5, the missing data on y and x_2 are in fact for the 25th observation. The Yates method is unable to estimate the missing data but the EM algorithm provides the estimates for it. The estimates $\mathbf{b}$ of β in Examples 1–5 are same for the Yates method and EM algorithm.

The following three data sets obtained by the three methods of dealing with the missing data are now considered for analysis.

(i) The available complete data in $(\mathbf{y}_1, \mathbf{X}_1)$. There are 23 observations in Examples 1–4 and 24 observations in Example 5.

(ii) The 25 observations on y, x_1, and x_2 obtained by substituting $\bar{y}_c, \bar{x}_{1c}$ and $\bar{x}_{2c}$ for the missing data on y, x_1 and x_2, respectively.

(iii) The 25 observations on y, x_1 and x_2 obtained by substituting the Yates / EM algorithm estimates of the missing data.

For the above three data sets, the comparisons are made of the numerical values of the mean square error (MSE) and the coefficients of determination with and without adjustment (R^2_{adj} and R^2). It is found that the analysis with the data set (iii) gives the smallest value of MSE and the largest values of R^2 and R^2_{adj}. Table 25.4 presents comparison of MSE's, R^2's and R^2_{adj}'s for the three data sets in Examples 1–5.

25.4 Conclusions and Remarks

The normal probability plot, of ordered effects requires the least squares estimation of all parameters in the model. In the example of Section 25.2, a single replicate of a complete factorial experiment is performed so that the estimation

Table 25.4: Comparison of MSE's, R^2's and R^2_{adj}'s for the three methods in Examples 1–5

Example	Data set	Number of Observations	Degrees of freedom	MSE	$R^2(\%)$	$R^2_{adj}(\%)$
1	i	23	20	10.5	96.3	95.9
	ii	25	22	15.8	93.8	93.3
	iii	25	22	9.5	96.4	96.0
2	i	23	20	10.5	96.3	95.9
	ii	25	22	10.4	96.0	95.7
	iii	25	22	9.5	96.4	96.0
3	i	23	20	10.5	96.3	95.9
	ii	25	22	12.0	95.4	95.0
	iii	25	22	9.5	96.4	96.0
4	i	23	20	10.5	96.3	95.9
	ii	25	22	10.5	96.0	95.7
	iii	25	22	9.5	96.4	96.0
5	i	24	21	11.1	95.9	95.5
	ii	25	22	13.3	94.8	94.4
	iii	25	22	10.6	95.9	95.5

of the general mean, all main effects and all interactions are possible. However, with the single replication, the significance testing of factorial effects is not possible. If more replications are available, then the significance testing is a helpful tool in determining the significance of factorial effects. In factor screening from a single replicate experiment, the method of probability plots originally suggested by Daniel (1959, 1976), is a useful tool. But the technique should be used with caution. As it is demonstrated in the example of Section 25.1, a screened out factor by the probability plot method may turn out to be significant in presence of certain interactions.

Search designs, probing designs and other methods discussed in Srivastava (1990), are powerful alternatives to the normal probability plot. Sequential assembly of fractions is a very useful technique of augmenting two or more fractions together in resolving ambiguities [Box, Hunter, and Hunter (1978, p. 396)] or in getting designs with higher revealing power [Srivastava (1975, 1984)]. Many researchers have contributed to this area. The methods presented in Ghosh (1987) are recommended in addition to the normal probability plot.

Three methods of dealing with the missing data are compared in this paper. In the first method, the estimator **b** of β is unbiased. In the second method, the estimator $\mathbf{b}^{(1)}$ of β is biased. The $\mathbf{b}^{(1)}$ has lower component-wise variance than **b**. However the first method performs better over the second method as demonstrated in Table 4 with lower values for MSE and higher values for $R^2(\%)$

and $R^2_{adj}(\%)$ in Examples 1–5. The third method has a clear edge over the second method and is slightly better than the first method. The EM algorithm always provides estimates of the missing data and such estimates are observed to be identical to the Yates estimates in Examples 1–4. In Example 5, the Yates estimates of the missing data do not exist but the EM algorithm does provide estimates of the missing data. The third method with the EM algorithm is recommended for the analysis of the missing data.

Acknowledgements. This work is sponsored by the Air Force Office of Scientific Research under grant F49620-95-1-0094.

References

1. Box, G. E. P., Hunter, W. G. and Hunter, J.S. (1978). *Statistics for Experimenters: An Introduction to Design, Data Analysis, and Model Building*, John Wiley, New York.

2. Daniel, C. (1959). Use of half-normal plots in interpreting factorial two level experiments, *Technometrics*, **1**, 311–342.

3. Daniel, C. (1976). *Applications of statistics to industrial experimentation*, John Wiley & Sons, New York.

4. Ghosh, S. (1987). Influential nonnegligible parameters under the search linear model, *Communications in Statistics, Theory and Methods*, **16**, 1013–1025.

5. Little, R. J. A. and Rubin, D. B. (1987). *Statistical Analysis With Missing Data*, John Wiley & Sons, New York.

6. Montgomery, D. C. (1991). *Design and Analysis of Experiments*, Third edition, John Wiley & Sons, New York.

7. Montgomery, D. C. and Peck, E. A. (1992). *Introduction to Linear Regression Analysis*, Second edition, John Wiley & Sons, New York.

8. Srivastava, J. N. (1975). Designs for searching nonnegligible effects. *A Survey of Statistical Designs and Linear Models* (J.N. Srivastava, ed.), North-Holland, Amsterdam, 507–519.

9. Srivastava, J. N. (1984). Sensitivity and revealing power: two fundamental statistical criteria other than optimality arising in discrete experimentation, *Experimental Designs, Statistical Models and Genetic Statistics Models, and Genetic Statistics* (K. Hinkelmann, ed.), Marcel Dekker, New York, 95–117.

10. Srivastava, J. N. (1990). Modern factorial design theory for experimenters and statisticians, *Statistical Design and Analysis of Industrial Experiments* (S. Ghosh, ed.), Marcel Dekker, New York, 311–406.

PART IV
STATISTICAL METHODS FOR RELIABILITY

26

Applications of Generalized Linear Models in Reliability Studies for Composite Materials

Makarand V. Ratnaparkhi and Won J. Park

Wright State University, Dayton, OH

Abstract: In this paper, the applications of generalized linear models for studying the stochastic behavior of the fatigue life of a composite material are discussed. The composite materials are used in many engineering structures such as aircraft. The physical parameters of different composite materials are estimated using experimental data. The generalized linear models, in general, provide a wide class of probility models that are useful for modeling the collected data. In particular, the concept of link function, is useful for modeling the physical parameters of composite materials. To illustrate the methodology developed in this paper we have included the analysis of fatigue data reported in Kim and Park (1980).

Keywords and phrases: Fatigue life, generalized linear models, normal, lognormal and gamma distributions, graphite/epoxy laminates, residual strength degradation

26.1 Introduction

Composite materials are routinely used in many engineering structures such as aircraft. The fatigue behavior of these materials is of interest in many studies. In particular, the estimates of the physical parameters of a composite material, based on the experimental data, are useful in the study of fatigue behavior. These estimates are obtained using a model that represents the data. A number of mathematical models, for example, differential equations for the residual strength degradation are formulated for estimating the parameters of different composite materials. For example, Yang (1977, 1978) has introduced models for residual strength degradation in graphite/epoxy specimens. These models seem to be of interest even today [Wu et al. (1996)].

The mathematical models, as mentioned above, are routinely used for estimation purposes. However, for predicting the fatigue life of a composite material we need stochastic models. The statistical distributions derived from the above mentioned differential equations are often useful for the prediction of the fatigue life. Ratnaparkhi and Park (1986) considered Yang's (1978) differential equation for obtaining the lognormal distribution as a model for the fatigue life of graphite/epoxy laminates. They also provided the ad-hoc method for estimating the parameters of this material.

The objective of this paper is to demonstrate the use of generalized linear models (GLIM) in the analysis of fatigue data. In particular, in Section 26.3, we have considered the lognormal distribution of Ratnaparkhi and Park (1986) as an error distribution within the framework of GLIM for analyzing the graphite/epoxy data of Kim and Park (1980). As an alternative to the lognormal distribution we have considered the gamma distribution with two different link functions for modeling the same data. The methodology developed in this paper can be also used for other composite materials.

The main results of this paper are presented in Section 26.3. For the derivation of these results we need the methods developed by Yang (1978) and Ratnaparkhi and Park (1986) for the analysis of fatigue data for graphite/epoxy laminates. Their results are presented in Section 26.2 for ready reference. A summary and some concluding remarks are included in Section 26.4.

26.2   Models for Fatigue Life of a Composite Material

The fatigue life of a composite material is a complex phenomenon. In our study we have considered only those factors that are reported in Yang (1978). First, we describe the experimental aspects of the data collection that are relevant for the statistical development of the model. The details of such experiments and related engineering aspects are discussed in a paper by Yang (1977) and its references.

26.2.1   Fatigue failure experiments

The stress levels of interest are decided for a composite material that is of interest. A specimen of a composite material is subjected to fatigue (stress) loading for a number of cycles. The residual strength of a material is recorded at the desired level of stress and cyclic loadings. The experiment is continued until the specimen breaks under the fatigue loadings. The fatigue life, the number of loading cycles at which the fatigue occurs, is recorded. The experiment is repeated for number of specimens and for different stress levels.

26.2.2   Notation and terminology

p – number of stress levels.
m_i – number of specimens tested at the i-th stress level.
n – number of fatigue loadings or cycles
N – number of cycles for fatigue failure. The random variable (r.v.) N denotes the fatigue life of a composite material.
S – applied stress
$g(S)$ – a non-negative function of S.
$Y(n)$ – the residual strength at cycle n.

26.2.3   Deterministic and probability models for the fatigue data

The residual strength $Y(n)$ is used as a measure of fatigue. The residual strength decreases with the increased number of fatigue loadings. The rate at which the residual strength decreases is given by the differential equation

$$dY(n)/dn = -g(S)/[c(Y(n))^{(c-1)}], c \neq 0 \qquad (26.1)$$

where c, a physical parameter of a composite material, is called the shape parameter of a composite material. For the engineering aspects of this equation refer to Yang (1977, 1978).

For studying the fatigue behavior of a material the classical $S - N$ curve is used routinely. The $S - N$ curve is a basic method of representing the fatigue data. It is a plot of predetermined values of stress S against the number of cycles (N) until failure for each stress value. A log scale is almost always used for N. For predicting the fatigue failure the variable N is considered as a random variable depnding on S. In many studies the distribution of the r.v. N is assumed to be lognormal [Dowling (1993, p. 343)]. In many applications, the characteristic life (N_0) of a material is expressed as

$$kS^b N_0 = 1 \qquad (26.2)$$

where k and b are the physical parameters of the composite material.

Now we consider the probabilistic model corresponding to the equation (26.1). First we note that N represents the fatigue life of a composite material. To derive the distribution of N we need the solution of the equation (26.1) which also provides a relationship between $Y(n)$ and N.

The solution of (26.1) is

$$Y(n) = [(Y(0))^c - g(S).n]^{1/c} \qquad (26.3)$$

which can be expressed in terms of n as

$$n = [(Y(0))^c - S^c]/g(S), \tag{26.4}$$

where, initially $n = 0$, and $Y(0)$ is the initial strength, also called the ultimate strength of a composite material.

For deriving a probability model for the r.v. N, the fatigue life, the following remarks are useful.

1. If $Y(0)$ is a r.v. and n is known then $Y(n)$, the residual strength at loading cycle n, is also a r.v. and its distribution is determined by the distribution of $Y(0)$.

2. The number of cycles n at which the residual strength $Y(n)$ cannot hold the applied stress level S is a value of a random variable N (fatigue life) at which the breaking of the specimen occurs. Since the residual strength $Y(n)$ in equation (26.3) decreases monotonically with respect to n it follows that, in any particular experiment, at a predetermined level of S the value of $N = n$, if and only if $Y(n) = S$. Therefore, if $Y(0)$ is a r.v., then using (26.3), the r.v. N can be expressed as

$$N = [(Y(0))^c - S^c]/g(S). \tag{26.5}$$

Now if the distribution of $Y(0)$ is known or can be modeled then the distribution of N can be derived using the random variable transformation technique.

3. The function $g(S)$ is a characteristic of a composite material and depends on the stress level, the parameters k, b, and c and the parameters of the distribution of N.

As an illustration of the above remarks (1)–(3) we record below the results from Ratnaparkhi and Park (1986). These results are also useful in Section 26.3.

First, we note that as discussed in Yang (1977, 1978) we ignore the term $S^c/g(S)$ in the following derivations.

Suppose that $Y(0)$ has a lognormal distribution with parameters μ and σ, then

$$Z = [\ln(Y(0))/\beta]/\sigma, \tag{26.6}$$

where $\mu = \ln \beta$ $(\beta > 0)$, has the standard normal distribution.

Now using (26.4) and (26.6) it can be shown [Ratnaparkhi and Park (1986)] that the r.v.

$$Z' = [\ln(N/\beta')]/\sigma' \tag{26.7}$$

has the standard normal distribution, where $\beta' = \beta^c/g(S)$ and $\sigma' = c\sigma$ (or $c = \sigma'/\sigma$). Further, as shown in Yang (1977, 1978) the scale parameter β'

represents the characteristic life of a material given by equation (26.2) and the function $g(S)$ can be expressed as

$$g(S) = \beta^c k S^b, \tag{26.8}$$

and then using (26.7) we get $\beta' = 1/(kS^b)$ which after taking natural logarithms becomes

$$\mu' = \ln \beta' = -\ln k - b \ln S. \tag{26.9}$$

Thus, we have shown that if $Y(0)$ has a lognormal distribution, then the r.v. N has a lognormal distribution with parameters μ' and σ'. We note that the definition of μ' in (26.9) plays a key role in the next section where we consider the lognormal distribution as an error model for the analysis of fatigue data using the theory of generalized linear models.

26.3 Application of GLIM: Estimation of Parameters (k, b, c) of Graphite/Epoxy Laminates

In this section we demonstrate the use of generalized linear models for estimating the parameters k, b, and c. The lognormal and gamma distributions are considered here for representing the fatigue data. First, as an example we describe here the fatigue data reported in Kim and Park (1980). A brief description of the experiments, that are conducted for obtaining these data, is also included. We have also considered the gamma distribution as an alternative to the lognormal distribution for analyzing the fatigue data. We note that two series of data are needed for estimating the parameters. These data are from the experiments conducted for studying the fatigue behavior of graphite/epoxy laminates.

26.3.1 Fatigue data for graphite/epoxy laminates and related data analysis problems

The series of data for graphite/epoxy laminates are as follows:

1. The observations on r.v. $Y(0)$, the ultimate strengths of the specimens,

2. The observations on r.v. N for five predetermined values of the stress level (S).

For obtaining the data on $Y(0)$, each specimen is applied an increasing tension loading until it breaks. The maximum tension stress at which the breaking occurs is the observation on $Y(0)$. The data on $Y(0)$ for 29 specimens are shown in Table 26.1. The summary statistics for these data are also provided.

The data on r.v. N at stress level S are obtained by subjecting the specimens to a constant tension stress S repeatedly until it breaks. The number of cycles (n) at which the breaking occurs in an observation on r.v. N. The data on N for five stress levels and their summary statistics are recorded below in Table 26.2.

The summary statistics for the data in Table 26.1 are given below

Sample mean $=84.2583$ ksi,	Sample s.d.$=5.6757$ ksi,
Sample mean of $\ln(y(0)) = 4.4316$,	Sample s.d. for $\ln(y(0)) = 0.068$

Table 26.1: Ultimate static strength $y(0)$ in ksi of G/E[0/90/$\pm$45]s (29 specimens)

72.04	72.44	76.61	771.9	77.36	79.36	80.48	81.52	81.97	82.20
82.51	83.35	83.90	84.99	85.70	85.79	86.43	86.61	86.77	87.98
88.86	89.60	90.17	91.44	91.73	92.37	93.33	82.22	88.54	

Table 26.2: Fatigue data, number of cycles (n) at levels of S for T300/5208 G/E[0/90/$\pm$45]s and summary statistics

	Fatigue Failure cycles (n) at five stress levels				
$S(ksi)-$	70	65	60	55	50
	1150	2620	10300	715050	412000
	1850	4920	21270	108550	614960
	2436	6490	22550	168700	764680
	3768	7000	28760	169480	1333390
	6898	9020	78720	325780	1367890
Sample Mean:	3220.40	6010.00	32320.00	168712.	898584.00
Sample s.d.	2269.61	2395.25	26777.18	97247.47	431405.04
Sample c.v.	0.7047	0.3985	0.8285	0.5764	0.4801

As mentioned in Section 26.2, if we transform the r.v. N using the logarithmic transformation then the sample means, sample variances and the sample c.v.'s for the transformed data are as shown below in Table 26.3.

Table 26.3: Sample means, s.d.'s and c.v.'s for logarithms of (n) of Table 26.2

	Stress Level $S(ksi)$				
	70	65	60	55	50
Sample mean	7.88	8.62	10.15	11.91	13.61
Sample s.d.	0.68	0.47	0.73	0.57	0.51
Sample c.v.	0.086	0.055	0.072	0.048	0.03

Ratnaparkhi and Park (1986) considered the simple least square method for the sample means (as dependent variable) and the stress levels S of Table 26.3 (as independent variable) and obtained the estimates of k and b. Instead of such ad-hoc method applied to the secondary data of Table 26.3 we have considered below the applications of the theory of generalized linear models for the analysis of the original data reported in Table 26.2.

For ready reference we have described below the components of the generalized linear model. Then, using the lognormal distribution as a model for the data of Table 26.2 we have obtained the estimates of k and b. To demonstrate the availability of different generalized linear models, we have considered the gamma distribution as a model for the same data and have obtained the estimates of k and b using two link functions. The procedure for estimating c is described for the lognormal model.

For the computational task we used the GENMOD Procedure from the SAS (1993) package. A brief discussion of the results is included for completeness.

26.3.2 Generalized linear models for the analysis of fatigue data [McCullagh and Nelder (1989)]

The components of the generalized linear models are as follows:

Notation:

N_{ij} – the r.v. representing the number of fatigue failure cycles for the j-th specimen subject to the i-th stress level, $j = 1, 2, \ldots, m_i, i = 1, 2, \ldots, p$,

S_i – the i-th stress level, $i = 1, 2, \ldots, p$,

$\mu_{ij} = E[N_{ij}]$ and $\mu_i =$ the mean fatigue life at stress level S_i,

$\mathbf{N}$ – vector of r.v. N_{ij}'s,

$\mu = E[\mathbf{N}]$,

ε – vector of random errors ϵ_i.

Then the generalized linear model can be written as

$$\mathbf{N} = \mu + \varepsilon, \tag{26.10}$$

where ε_i has a distribution belonging to the exponential family. Further, an important component of the generalized linear model is the link function

$$\eta = u(\mu_i), \tag{26.11}$$

a monotone, differentiable function of μ_i.

For various aspects of the data analysis using the generalized linear model theory we refer the reader to McCullagh and Nelder (1989) or Fahrmeir and Lutz (1994).

26.3.3 Results

The analysis of data of Table 26.2, under different generalized linear models can be performed as discussed below. For illustration purpose we have considered here only three models for the data of Table 26.2.

Estimation of the parameters k and b

The following Tables 26.4(a) and 26.4(b) show the summary of the analyses of data of Table 26.2. The estimates of k and b and values of the scaled deviance and scaled Pearson X^2 are shown in Table 26.4(a). The confidence intervals for k and b are presented in Table 26.4(b).

Table 26.4(a): Model description, estimates of k and b, and the deviance

Variable	Distribution	Link function	Estimate of k	Estimate of b	Scaled Pearson X^2	Scaled Deviance (DF=23)
ln N	Normal	Identity	5.4928e-38	18.4559	25.0000	25.0000
N	Gamma	$\eta = \ln \mu$	2.3797e-36	17.4788	25.8101	26.7237
ln N	Gamma	Identity	1.2790e-37	18.2498	23.5940	25.0310

Table 26.4(b): 95% confidence intervals for k and b

Variable	Distribution	95% C.I. for k	95% C.I. for b
ln N	Normal	(3.9867e-42, 7.5670e-34)	(16.1251, 20.7867)
N	Gamma	(5.6227e-40, 1.0072e-32)	(15.4367, 19.5209)
ln N	Gamma	(7.0201e-43, 2.2585e-32)	(15.3225, 21.1770)

Estimation of parameter c

The estimation procedure for the parameter c for the first model (normal distribution case) described in Table 26.4(a) is as follows:

The estimate of the scale parameter σ' using the GENMOD procedure is 0.7072. Now we notice that the estimate of σ based on the data on $Y(0)$ of Table 26.1 is 0.068. Therefore, using the relation $\sigma' = c\sigma$ of Section 26.2 the suggested estimate of c is $0.7072/0.068 = 10.4$. The estimates of c for the other models described in Table 26.4(a) can be carried out using the results for the gamma distribution.

26.3.4 Discussion

1. In the above analysis we have considered three models. These models demonstrate the flexibility and the availability of a wider class of models for analyzing the fatigue data. The objective of the paper is to demonstrate the use of GLIM for fatigue data and not the comparison of different models for such data. However, in our example, if the comparisons of the results for this data are desired then the remark (2) given below could be useful. The merits and demerits of each model should be studied using the procedures for model checking, for example, the procedures discussed in McCullagh and Nelder (1989).

2. Since the models are different, the estimates of k for the three models given in Table 26.4(a) cannot be compared directly. However, the 95% confidence intervals for k from Table 26.4(b) give some idea regarding the closeness of these estimates. Particularly, the confidence intervals for k overlap indicating the closeness of the values of the estimates. Similar observation can be made for the estimates of b. We note that, the choice of 95% confidence interval is arbitrary and is solely for demonstration. The choice of the confidence interval for the above purpose should be made keeping in view the engineering applications of the composite material that is of interest.

3. The results of Ratnaparkhi and Park (1986) for the above data using the lognormal distribution are given below:

$$k = 1.77e - 36, b = 17.86, c = 9.26.$$

These results are not exactly the same as the corresponding results shown in Table 26.4(a). However, the above analysis using the generalized linear models is based on more systematic theory than the ad-hoc method used in Ratnaparkhi and Park (1986). Therefore, the results of Table 26.4(a) could be considered as more reliable. We note that the above estimates of k and b fall in the respective confidence intervals shown in Table 26.4(b).

4. Yang and Lui (1977) had considered the Weibull distribution for the analysis of data for the composite material G/E[0/90±45] s. The series of data they analyzed had the values of S from 62 to 85 which are not the same as in our data given in Table 26.2. They used the method of moments for estimating k, b, and c. Their results are as follow:

$$k = 1.8285e - 36, b = 17.78, c = 10.818.$$

The above estimates of k and b are comparable to the results for the gamma model given in Tables 26.4(a) and 26.4(b). However, the estimates based on the generalized linear models can be studied more systematically than the moment estimators.

26.4   Summary and Concluding Remarks

In this paper we discussed briefly the use of the generalized linear models for studying the reliability of composite materials and demonstrated the use of three such models in the analysis of fatigue data for the specimens of the composite material T300/5208 G/E [0/90/±45]s. Our results show that for a given practical situation there could be two or more generalized linear models that can be considered for the analysis of fatigue data. In particular, we may have the models with the same distribution but different link functions. The comparison of such models could be an interesting research problem in view of selecting an appropriate link function that is meaningful for the physical parameters that are of interest.

Acknowledgements. The authors would like to thank the referee and editor for their constructive comments.

References

1. Dowling, N. E. (1993). *Mechanical Behavior of Materials*, Englewood Cliffs, NJ: Prentice Hall.

2. Fahrmeir, L. and Tutz, G. (1994). *Multivariate Statistical Modelling Based on Generalized Linear Models*, New York: Springer-Verlag.

3. Kim, R.Y. and Park, Won J. (1980). Proof testing under tension - tension. *Journal of Composite Materials*, **14**, 69–79.

4. McCullagh, P. and Nelder, J. A. (1989). *Generalized Linear Models*, London: Chapman and Hall.

5. Ratnaparkhi, M. V. and Park, Won J. (1986). Lognormal Distribution - Model for fatigue life and residual strength of composite materials, *IEEE Transactions on Reliability*, **35**, 312–315.

6. SAS (1993). SAS/STAT software: *The GENMOD Procedure*, SAS Institute Inc.

7. Wu, W., Lee, L. J. and Choi, S. T. (1996). A study of fatigue damage and fatigue life of composite laminates, *Journal of Composite Materials*, **30**, 123–137.

8. Yang, J. N. (1977). Reliability prediction of composites under periodic proof tests in service, *Composite Materials: Testing and Design*, **ASTM-STP #617**, 272–294.

9. Yang, J. N. (1978). Fatigue and residual strength degradation for graphite /epoxy composite under tension-compression cyclic loadings, *Journal of Composite Materials*, **12** 19–39.

10. Yang, J. N. and Lui, M. D. (1977). Residual strength degradation model and theory of periodic proof tests for graphite/epoxy laminates, *Journal of Composite Materials*, **2**, 176–200.

27

Bivariate Failure Rates in Discrete Time

G. Asha and N. Unnikrishnan Nair

Cochin University of Science and Technology, Cochin, India

Abstract: In this paper an expository analysis of three definitions of bivariate failure rates in the discrete time domain is attempted. Conditions under which each rate determines the distribution of failure times uniquely are investigated. Further, some characterizations of bivariate distributions based on the functional forms of these rates are established.

Keywords and phrases: Bivariate failure rates, characterization, discrete life time, geometric distribution

27.1 Introduction

The concepts of failure rate and mean residual life are extensively applied in modelling equipment behaviour and in defining various criteria for aging. When the specification of the functional form of the failure rate is possible based on the physical characteristics of the process governing the failure of a system or device, the result that the failure rate uniquely determines a distribution helps the identification of the failure time model. The theoretical discussions and applications of failure rates in most studies treat time as continuous. Recently some attempts are made in literature towards the understanding of failure rates in the discrete domain in the works of Xekalaki (1983), Salvia and Bollinger (1982) and Shaked et al. (1989). The need for a detailed study in the discrete case arises from the fact that variables such as 'number of cycles to failure' [Gupta (1985)] are discrete in nature or from the inaccuracy of the measuring devices [Xekalaki (1983)] that lead to discontinuous models. Further, distributions arising from discrete data can provide good approximations to continuous models to render the study of discrete time models worthwhile.

In this paper we examine three forms of bivariate failure rate and investigate the conditions under which these rates determine the corresponding distribution uniquely. The basic properties of the rates along with the inter-relationships among them are presented. It is well known that in the univariate case the geometric law is characterized by a constant failure rate. Since, we have more than one analogue of the univariate concept of failure rate in higher dimensions, one can expect a bivariate geometric model that corresponds to each definition. We present such models along with various characterizations that lead to them.

27.2　Scalar Failure Rate

Let $\mathbf{X} = (X_1, X_2)$ be a discrete random vector representing the failure times of a two-component system in the support of $I_2 = [(x_1, x_2)|x_1, x_2 = 0, 1, 2, \cdots]$ with joint survival function $R(\mathbf{x}) = P(\mathbf{X} \geq \mathbf{x})$ and probability mass function (pmf) $f(\mathbf{x}) = P(\mathbf{X} = \mathbf{x})$, where $\mathbf{x} = (x_1, x_2)$ and $\mathbf{X} \geq \mathbf{x}$ means $X_j > x_j$, $j = 1, 2$.

The scalar failure rate (SFR) is defined at those points for which $R(\mathbf{x}) > 0$, by

$$a(\mathbf{x}) = \frac{f(\mathbf{x})}{R(\mathbf{x})} \qquad (27.1)$$

The properties of $a(\mathbf{x})$ are listed below.

(i)　In general $a(\mathbf{x})$ does not determine the distribution of $\mathbf{X}$ uniquely. This is seen from

$$R(\mathbf{x}) = (0.3)^{x_1}(0.7)^{x_2}$$

and

$$R(\mathbf{x}) = \left(\frac{1}{2}\right)[(0.3)^{x_1}(0.7)^{x_2} + (0.7)^{x_1}(0.3)^{x_2}]$$

both of which have the same SFR, $a(\mathbf{x}) = 0.21$. The question of what additional requirements will be needed to ensure the uniqueness of $R(\mathbf{x})$ is settled in the following discussion.

(ii)　One of the marginal failure rates,

$$h_j(x_j) = \frac{P(X_j = x_j)}{P(X_j \geq x_j)}, \qquad j = 1, 2$$

(or equivalently a marginal distribution) along with $a(\mathbf{x})$ determines $R(\mathbf{x})$ uniquely. This follows from one of the recurrence relations derived from (27.1),

$$R(\mathbf{x}) = R(x_1, x_2 - 1) - \sum_{t_1=x_1}^{\infty} R(t, x_2 - 1)a(t, x_2 - 1),$$

or

$$R(\mathbf{x}) = R(x_1 - 1, x_2) - \sum_{t_2=x_2}^{\infty} R(x_1 - 1, t)a(x_1 - 1, t).$$

　　　　　　　　　　　　　　　　　　　　　　　　　　　　(27.2)

For example, the first equation in (1.2) can be used iteratively starting with $x_2 = 1$ and employing

$$R(x_1, 0) = \prod_{y=0}^{x_1-1} [(1 - h_1(y)].$$

The roles of x_1 and x_2 are to be interchanged if $h_2(x_2)$ is given.

(iii) $a(x_1, x_2) = h_1(x_1)h_2(x_2)$ if and only if X_1 and X_2 are independent. This gives a meaningful relationship between the bivariate failure rate and the univariate failure rates in the case of independence, which is a reasonable requirement for a bivariate definition.

PROOF. The "if" part is obvious from (27.1). To prove the "only if" part we use (27.2) and $R(0, x_2) = \prod_{y=0}^{x_1-1} [1 - h_2(y)]$ recursively to find $R(x_1, x_2) = R(x_1, 0)R(0, x_2)$ for every (x_1, x_2) in I_2^+. ∎

(iv) $a(x_1, x_2) = c$, and $h_1(x) = h_1$ where c and h_1 are constants for all x_1, x_2 such that $c < (1 - h_1)$, if and only if X_1 and X_2 are independent geometric variables with parameters h_1 and $1 - c(1 - h_1)^{-1}$ respectively. However $a(\mathbf{x}) = c$ alone does not guarantee neither independence nor geometric marginals, [see second example in (i)].

(v) The random vector $\mathbf{X}$ has bivariate geometric law with survival function (Sf)

$$R(x_1, x_2) = \begin{cases} p^{x_2}(p_1)^{x_1-x_2} & x_1 \geq x_2 \\ p^{x_1}(p_2)^{x_2-x_1} & x_2 \geq x_1 \end{cases} \qquad (27.3)$$

$1 + p \geq p_1 + p_2, 0 < p \leq p_j < 1, \quad j = 1, 2$ if and only if the SFR is of the form

$$a(\mathbf{x}) = \begin{cases} c_1 & x_1 > x_2 \\ c_2 & x_1 < x_2 \\ c_3 & x_1 = x_2 \end{cases} \qquad (27.4)$$

where c_i's are constants such that $0 < c_i < 1$, $\quad i = 1, 2, 3$ and the marginal distribution of X_j is geometric with parameter p_j, $j = 1, 2$.

PROOF. When the distribution of $\mathbf{X}$ is specified by (27.3), X_j follows geometric law with parameter p_j and further

$$a(\mathbf{x}) = \begin{cases} p_1^{-1}(1 - p_1)(p_1 - p) & x_1 > x_2 \\ p_2^{-1}(1 - p_2)(p_2 - p) & x_1 < x_2 \\ (1 + p - p_1 - p_2)(1 - p)^{-1} & x_1 = x_2 \end{cases} \qquad (27.5)$$

which is of the form (27.4). Conversely suppose that (27.4) holds. Then from (27.2) after some algebra we can deduce the form (27.3).   ∎

Notice that (27.3) mentioned in Block (1977) in a different context, is the discrete analogue of the Marshall and Olkin (1967) bivariate exponential distribution. We have now characterized it by a SFR which is piece-wise constant in the partitions of the sample space defined by $X_1 < X_2, X_1 > X_2$ and $X_1 = X_2$.

(vi) If $\mathbf{X}$ is as mentioned in (v) then the piece-wise constancy of the SFR is equivalent to the bivariate no-aging property

$$P[X_1 \geq x_1 + t, X_2 \geq x_2 + t | X_1 \geq t, X_2 \geq t] = P[\mathbf{X} \geq \mathbf{x}] \qquad (27.6)$$

for all $\mathbf{x}$ and $t = 0, 1, 2, \cdots$ as both characterize the bivariate geometric law (27.3). To prove the last assertion we note that (27.6) is equivalent to

$$R(x_1 + t, x_2 + t) = R(x_1, x_2)R(t, t) \qquad (27.7)$$

for all $\mathbf{x}$ and $t = 0, 1, 2, \cdots$.

Setting $x_1 = x_2 = x$ and $R(x, x) = G(x)$, (27.7) becomes the Cauchy functional equation

$$G(x + y) = G(x)G(y)$$

whose only solution that satisfies the conditions of a Sf is

$$G(x) = p^x, \qquad \text{for some} \quad 0 < p < 1.$$

Now, taking $x_2 = 0$ in (27.7),

$$R(x_1 + y, y) = R(x_1, 0)p^y = p^y p_1^{x_1}, \qquad 0 < p_1 < 1.$$

Thus

$$R(x_1, x_2) = p^{x_2} p_1^{x_1 - x_2} \qquad \text{for} \ x_1 \geq x_2$$

and a similar argument for $x_1 \geq x_2$ results in (27.3).

(vii) It is natural to explore the relationship between the SFR and the bivariate mean residual life function (BMRL) defined as

$$r(\mathbf{x}) = (r_1(\mathbf{x}), r_2(\mathbf{x}))$$

where

$$r_j(\mathbf{x}) = E[(X_j - x_j)|\mathbf{X} > \mathbf{x}], j = 1, 2.$$

Then

$$R(x_1 + 1, x_2 + 1)r_1(x_1, x_2) = \sum_{t=x_1+1}^{\infty} R(t, x_2 + 1) \qquad (27.8)$$

and
$$\frac{r_1(x_1, x_2) - 1}{r_1(x_1 + 1, x_2)} = \frac{R(x_1 + 2, x_2 + 1)}{R(x_1 + 1, x_2 + 1)}. \tag{27.9}$$

Similarly
$$\frac{[r_2(x_1, x_2) - 1]}{r_2(x_1, x_2 + 1)} = \frac{R(x_1 + 1, x_2 + 2)}{R(x_1 + 1, x_2 + 1)}. \tag{27.10}$$

Also
$$1 - a(x_1 + 1, x_2 + 1) = \frac{[R(x_1 + 2, x_2 + 1) + R(x_1 + 1, x_2 + 2) - R(x_1 + 2, x_2 + 2)]}{[R(x_1 + 1, x_2 + 1)]} \tag{27.11}$$

and
$$\left[\frac{R(x_1 + 2, x_2 + 2)}{R(x_1 + 1, x_2 + 1)}\right] = \frac{r_2(x_1 + 1, x_2)[r_1(x_1, x_2) - 1]}{r_2(x_1 + 1, x_2 + 1)r_1(x_1 + 1, x_2)}. \tag{27.12}$$

Thus, the required identity connecting SFR and BMRL is
$$1 - a(x_1 + 1, x_2 + 1) = \frac{r_1(x_1, x_2) - 1}{r_1(x_1 + 1, x_2)} + \frac{r_2(x_1, x_2) - 1}{r_2(x_1, x_2 + 1)}$$
$$+ \frac{r_1(x_1 + 1, x_2)[r_1(x_1 + 1, x_2) - 1]}{r_2(x_1 + 1, x_2 + 1)r_1(x_1 + 1, x_2)}.$$

27.3 Vector Failure Rate

In a variant approach, the failure rate of $\mathbf{X}$ can be defined, [Nair and Nair (1990)] as the vector
$$b(\mathbf{x}) = (b_1(\mathbf{x}), b_2(\mathbf{x}))$$

where
$$b_j(\mathbf{x}) = P(X_j = x_j | \mathbf{X} \geq \mathbf{x}), \qquad j = 1, 2. \tag{27.13}$$

Unlike (27.1), where the joint variation in the failure times is considered in the present case, the local behaviour of each of the components given the survival of both are separately assessed. The vector failure rate (VFR) in (27.13) satisfy the following properties.

(i)
$$1 - b_1(\mathbf{x}) = \frac{R(x_1 + 1, x_2)}{R(x_1, x_2)}. \tag{27.14}$$

$$1 - b_2(\mathbf{x}) = \frac{R(x_1, x_2 + 1)}{R(x_1, x_2)}. \tag{27.15}$$

(ii)   The VFR determines the distribution of **X** uniquely through the formula

$$R(x_1, x_2) = \prod_{r=1}^{x_1} [1 - b_1(x_1 - r, x_2)] \prod_{r=1}^{x_2} [1 - b_2(0, x_2 - r)] \qquad (27.16)$$

(iii)   The marginal failure rates $h_j(x_j)$ can be deduced from

$$b(\mathbf{x}) \quad \text{as} \quad h_1(x_1) = b_1(x_1, 0) \quad \text{and} \quad h_2(x_2) = b_2(0, x_2).$$

(iv) X_1 and X_2 are independent geometric variables if and only if either

$$b(\mathbf{x}) = (b_1(x_1), b_2(x_2)) = (b_1, b_2)$$

where b's are constants or

$$R(\mathbf{x} + t) = R(\mathbf{x})R(\mathbf{t})$$

where $t = (t_1, t_2) \in I_2^+$.

The proof is straightfoward. For the bivariate discrete models characterized by the local constancy and piece-wise constancy of $b(\mathbf{x})$ we refer to Nair and Nair (1990) and Nair and Asha (1993).

(v)   The BMRL defined as in (27.8) is related to VFR as

$$1 - b_1(x_1 + 1, x_2 + 1) = \frac{[r_1(\mathbf{x}) - 1]}{r_1(x_1 + 1, x_2)}$$

and

$$1 - b_2(x_1 + 1, x_2 + 1) = \frac{[r_2(\mathbf{x}) - 1]}{r_2(x_1, x_2 + 1)}. \qquad (27.17)$$

(vi)   One cannot choose the components $b_1(\mathbf{x})$ and $b_2(\mathbf{x})$ arbitrarily as they should satisfy the consistency condition

$$[1 - b_1(x_1, x_2 + 1)][1 - b_2(x_1, x_2)] = [1 - b_2(x_1 + 1, x_2)][1 - b_1(x_1, x_2)].$$

This is verified in Section 27.5.

27.4   Conditional Failure Rate

In this section we consider a third definition proposed by Kotz and Johnson (1991) as the vector

$$c(\mathbf{x}) = [c_1(\mathbf{x}), c_2(\mathbf{x})]$$

where

$$c_j(\mathbf{x}) = P(X_j = x_j | X_k = x_k, X_j > x_k), \quad j, k = 1, 2. \text{ and } j \neq k. \quad (27.18)$$

The properties of the conditional failure rate (CFR) are summarised as follows.

(i)
$$1 - c_1(\mathbf{x}) = \frac{[R(x_1 + 1, x_2) - R(x_1 + 1, x_2 + 1)]}{[R(x_1, x_2) - R(x_1, x_2 + 1)]}$$

$$1 - c_2(\mathbf{x}) = \frac{[R(x_1, x_2 + 1) - R(x_1 + 1, x_2 + 1)]}{[R(x_1, x_2) - R(x_1 + 1, x_2)]} \quad (27.19)$$

(ii) The BMRL is related to CFR by

$$1 - c_1(x_1 + 1, x_2 + 1)$$
$$= \frac{r_1(x_1, x_2) - 1}{r_1(x_1 + 1, x_2)} \left\{ \frac{r_2(x_1 + 1, x_2)}{r_2(x_1 + 1, x_2 + 1)} - 1 \right\} \left\{ \frac{r_2(x_1, x_2) - 1}{r_2(x_1, x_2 + 1)} - 1 \right\}^{-1} (27.20)$$

(iii) In general, the failure rate $c(\mathbf{x})$ determines the conditional distributions of $\mathbf{X}$ uniquely.

PROOF. By definition

$$1 - c_1(\mathbf{x}) = \frac{H(x_1 + 1, x_2)}{H(x_1, x_2)} \quad (27.21)$$

where

$$H(\mathbf{x}) = R(x_1, x_2) - R(x_1, x_2 + 1).$$

Solving (27.21)

$$H(x_1, x_2) = \prod_{r=0}^{x_1 - 1} [1 - c_1(r, x_2)] P(X_2 = x_2).$$

Thus

$$P(X_1 \geq x_1 | X_2 = x_2) = \prod_{r=0}^{x_1 - 1} [1 - c_1(r, x_2)].$$

$$P(X_1 = x_1 | X_2 = x_2) = \prod_{r=0}^{x_1 - 1} [1 - c_1(r, x_2)] c_1(x_1, x_2). \quad (27.22)$$

Similarly

$$P(X_2 = x_2 | X_1 = x_1) = \prod_{r=0}^{x_2 - 1} [1 - c_2(x_1, r)] c_2(x_1, x_2). \quad (27.23)$$

Thus $c(\mathbf{x})$ determines the conditional distribution of $\mathbf{X}$ given $X_k = x_k$ uniquely. See Gourieroux and Monfort (1979) for the results concerning the determination of joint distributions based on their conditional densities. To see when $c(\mathbf{x})$ provides a unique Sf we note that a necessary and sufficient condition for the conditional densities $f(x_1|x_2)$ and $f(x_2|x_1)$ in (1.22) and (1.23) to determine the joint pmf is that

$$\frac{f(x_1|x_2)}{f(x_2|x_1)} = \frac{A_1(x_1)}{A_2(x_2)},$$

where $A_j(x_j)$ are non-negative with finite sums over the set of non-negative integers satisfying

$$\sum_{x_1=0}^{\infty} A_1(x_1) = \sum_{x_2=0}^{\infty} A_2(x_2).$$

The proof of this result follows from an approach similar to that in Abrahams and Thomas (1984).   ■

(iii) X_1 and X_2 are independent geometric variables if and only if

$$c(\mathbf{x}) = [(h_1(x_1), h_2(x_2)] = (c_1, c_2)$$

where c's are constants. The proof is straightfoward.

(iv)   For all $\mathbf{x}$ and t_j, $s_j = 0, 1, 2, \cdots$,the following statements are equivalent.

1. $c(\mathbf{x}) = [c_1(x_2), c_2(x_1)]$.

2.

$$P[X_j \geq t_j + s_j | X_j \geq s_j, X_k = x_k] = P[X_j > t_j | X_k = x_k]. \qquad (27.24)$$

(conditional lack of memory property)

3. (X_1, X_2) has a bivariate distribution speciefied by pmf

$$f(x_1, x_2) = \alpha p_1^{x_1} p_2^{x_2} \theta^{x_1 x_2}, 0 < p_j < 1, \qquad 0 \leq \theta \leq 1, \qquad (27.25)$$

where

$$\alpha^{-1} = \sum_{r=0}^{\infty} \frac{p_1^r}{(1 - p_1\theta^r)} = \sum_{s=0}^{\infty} \frac{p_2^s}{(1 - p_2\theta^s)}.$$

PROOF.   The equivalence of (2) and (3) are established in Nair and Nair (1991). We can easily verify that (3) implies (1). Lastly if

$$c(\mathbf{x}) = [c_1(x_2), c_2(x_1)],$$

then from (27.19),

$$P(X_1 = x_1 | X_2 = x_2) = [1 - c_1(x_2)]^{x_1} c_1(x_2) \qquad (27.26)$$

and similarly

$$P(X_2 = x_2 | X_1 = x_1) = [1 - c_2(x_1)]^{x_2} c_2(x_1). \qquad (27.27)$$

The equations (27.26) and (27.27) together lead to the conditional lack of memory property and therefore, from Nair and Nair (1991), the distribution of **X** should be of the form (27.25). ∎

(**v**)   **X** has the bivariate gemetric distribution (27.3) if and only if

$$c(\mathbf{x}) = \begin{cases} (c_{11}, c_{12}) & x_1 > x_2 \\ (c_{21}, c_{22}) & x_1 < x_2 \\ (c_{31}, c_{32}) & x_1 = x_2 \end{cases} \qquad (27.28)$$

where c_{ij}'s are constants in $(0,1)$. An outline of the proof is as follows.

If $c(\mathbf{x})$ is of the form (27.28), then from (27.19)

$$\begin{aligned} P[X_1 &\geq x_1 + 1, X_2 = x_2] \\ &= (1 - c_{11})^{x_1 - x_2}(1 - c_{31})P[X_1 \geq x_2, X_2 = x_2], x_1 \geq x_2. \end{aligned} \qquad (27.29)$$

Also

$$P[X = x_1, X \geq x_2 + 1] = (1 - c_{11})^{x_2} P[X_1 = x_1, X_2 \geq x_2], \quad x_1 < x_2.$$

or

$$R(x_1, x_2) = (1 - c_{12})^{x_2} R(x_1, 0). \qquad (27.30)$$

Setting $x_2 = 0$ in (27.29) and using (27.30),

$$R(x_1, 0) = c_{12}^{-1}(1 - c_{11})^{x_1 - 1}(1 - c_{13})P(X_2 = 0). \qquad (27.31)$$

Since $R(0,0) = 1$,

$$1 - c_{11} = c_{12}^{-1}(1 - c_{13})P(X_2 = 0).$$

Thus

$$R(x_1, 0) = (1 - c_{11})^{x_1}.$$

Substituting in (27.30)

$$R(x_1, x_2) = (1 - c_{12})^{x_2}(1 - c_{11})x^{-1}, \qquad x_1 \geq x_2. \qquad (27.32)$$

Proceeding in the same manner with $c_2(\mathbf{x})$ in the region $x_1 < x_2$,

$$R(x_1, x_2) = (1 - c_{22})^{x-2}(1 - c_{21})^{x_1}, \qquad x_1 < x_2. \qquad (27.33)$$

From (27.32) and (27.33)

$$(1 - c_{22})(1 - c_{21}) = (1 - c_{12})(1 - c_{11}) = R(1,1)$$

and

$$R(x, x) = [R(1,1)]^x.$$

Writing

$$p_1 = (1 - c_{11}), \qquad p_2 = (1 - c_{22})$$

and

$$p = (1 - c_{12})(1 - c_{11}) = (1 - c_{22})(1 - c_{21})$$

we obtain the form (27.3). By direct calculations we find that for (27.3)

$$c(\mathbf{x}) = \begin{cases} \left[(1 - p_1), \ (1 - pp_1^{-1})\right] & x_1 > x_2 \\ \left[(1 - pp_2^{-1}), \ (1 - p_2)\right] & x_1 < x_2 \\ \left[\frac{1+p-p_1-p_2}{1-p_2}, \ \frac{1+p-p_1-p_2}{1-p_1}\right] & x_1 = x_2 \end{cases}$$

This completes the proof.   ∎

27.5 Inter-Relationships and Their Implications

1. Direct calculations using the definitions, yield

$$c_1(\mathbf{x})b_2(\mathbf{x}) = c_2(\mathbf{x})b_1(\mathbf{x}) = a(\mathbf{x}), \qquad (27.34)$$

$$a(\mathbf{x}) = b_1(x_1, x_2) - b_1(x_1, x_2 + 1) + b_1(x_1, x_2 + 1)b_2(x_1, x_2), \qquad (27.35)$$

$$a(\mathbf{x}) = b_2(x_1, x_2) - b_2(x_1 + 1, x_2) + b_1(x_1, x_2)b_2(x_1 + 1, x_2)$$

From the last two identities it becomes apparent that

$$[1 - b_1(x_1, x_2 + 1)][1 - b(x_1, x_2)] = [1 - b_1(x_1, x_2)][1 - b_2(x_1 + 1, x_2)] \quad (27.36)$$

must be a consistency condition for $b(x)$ to be a failure rate.

2. For all $\mathbf{x} \in I_2^+$, the relationship

$$a(\mathbf{x}) = b_1(\mathbf{x})b_2(\mathbf{x}) = c_1(\mathbf{x})c_2(\mathbf{x})$$

holds if and only if X_1 and X_2 are independent.

PROOF. The "if" part is straightfoward. Conversely, the given condition reduces to the functional equation

$$G(x_1, x_2 + 1) = G(x_1, x_2)$$

where

$$G(x_1, x_2) = \frac{R(x_1 + 1, x_2)}{R(x_1, x_2)}$$

which is satisfied if and only if $R(x_1, x_2) = R(x_1, 0)R(0, x_2)$ for all $x \in I_2^+$ and accordingly X_1 and X_2 are independent. ∎

27.6 Summary and Concluding Remarks

In this paper, we discussed three forms of bivariate failure rates and investigated the conditions under which these forms determine the corresponding distribution uniquely. Some inter-relationships among the rates were also considered.

References

1. Abrahams, J. and Thomas, J. B. (1984). A note on the characterization of bivariate densities by conditional densities, *Communications in Statistics, Theory and Methods*, **13**, 395–400.

2. Block, H. W. (1977). A family of bivariate life distributions, In *Theory and Applications of Reliability* (Eds., Tsokos, C.P. and Shimi, I.N.), **1**, 349–372.

3. Gourieroux, C. and Monfort, A. (1979). On the characterization of a joint probability distribution by conditional distributions, *Journal of Econometrics*, **10**, 115–118.

4. Gupta, P. L. (1985). Some characterizations of distributions by truncated moments, *Statistics*, **16**, 465–473.

5. Kotz, S. and Johnson, N. L. (1991). A note on renewal (partial sums) distributions for discrete variables, *Statistics and Probability Letters*, **12**, 229–231.

6. Marshall, A. W. and Olkin, I. (1967). A multivariate exponential distribution, *Journal of the American Statistical Association*, **62**, 808–812.

7. Nair, N. U. and Nair, K. R. M. (1990). Characterizations of a bivariate geometric distribution, *Statistica*, **50**, 247–253.

8. Nair, N. U. and Asha, G. (1993). Characteristic properties of a bivariate geometric law, *Journal of the Indian Statistical Association*, **32**, 111–116.

9. Nair V. K. R. and Nair, N. U. (1991). On conditional lack of memory property, *Proceedings of the Symposium on Distribution Theory*, Cochin, India, 25–30.

10. Salvia, A. A. and Bollinger, R. C. (1982). On discrete hazard functions, *IEEE Transactions on Reliability*, **R-31**, 458–459.

11. Shaked, M., Shanthikumar, J. G. and Valdez-Torres, J. B. (1989). Discrete hazard rate functions, Personal communication.

12. Xekalaki, E. (1983). Hazard functions and life distributions in discrete time, *Communications in Statistics, Theory and Methods*, **12**, 2503–2509.

28

A General Approach of Studying Random Environmental Models

Pushpa L. Gupta and Ramesh C. Gupta

University of Maine, Orono, ME

Abstract: The consequences of introducing random effects have been a major issue in models for survival analysis. Survival models that incorporate random effects provide a useful framework for determining the effectiveness of interventions. In this paper we shall present a general approach of incorporating the random environmental effect in both univariate and multivariate models. The reliability measures, namely, the failure rate, the survival function and the mean residual life function are compared with or without the environmental effect. Examples are presented to illustrate the results.

Keywords and phrases: Gamma and inverse Gaussian distributions, random effects, reliability measures, series system, univariate and bivariate models

28.1 Introduction

In assessing the reliability of a system of components, it is rarely possible to test the entire system under the actual operational environment. Therefore, the component reliabilities are often determined by life tests conducted under the controlled environment. The operating environmental conditions are generally harsher or gentler than the test environment. Life tests conducted under the controlled environment ignore the influence of the operating environment which may lead to inappropriate results.

More specifically, consider an aircraft or space vehicle whose lifetime is to be modeled. Before launching into the air, the component is tested under the laboratory or controlled environment. The operating environmental conditions are not known and cannot be determined from the data. The operating environment comprises of several stresses (covariates) which could be instantaneous, and whose presence and intensities change over time in an unpredictable manner.

Cox (1972) suggested to introduce a parameter which will exhibit the changing nature of the operating environment and by mixing over the parameter, one obtains a dependent life distribution which describes the system in a more appropriate way. Lindley and Singpurwalla (1986) considered a two component system where components have independent exponential lifetimes with failure rates $Z\lambda_1$ and $Z\lambda_2$. A gamma environmental effect on the common paramter Z yields a dependent bivariate Pareto model which can be reduced to the bivariate Pareto model considered by Mardia (1962). $Z > 1$ shows that the operating environment is harsher while $Z < 1$ exhibits the gentler environment. Nayak(1987) extended the work of Lindley and Singpurwalla (1986) to the case of multicomponent systems and derived a $k-$variate Lomax distribution. He studied various properties of this distribution and indicated their usefulness in reliability. Nayak's work was further generalized by Roy and Mukherjee (1988). Concurrently Gupta and Gupta (1990) and Bandyopadhyay and Basu (1990) considered Marshall and Olkin's (1967) bivariate exponential model with parameters $Z\lambda_1, Z\lambda_2$ and $Z\lambda_{12}$, where Z is considered random with a gamma distribution and studied the structural properties of the resulting model. Gupta and Gupta (1990) also studied the relative errors incurred in the reliability measures (the survival function, the failure rate and the mean residual life function) under the model assumption of Lindley and Singpurwalla (1986) when in fact Marshall and Olkin's model incorporating the environmental effect would have been used.

In this paper we present a general approach to consider the effect of environment in the univariate as well as in the multivariate case. In Section 28.2, we obtain the relationships between the conditional and unconditional failure rates and show that these failure rates cross at most at one point. Section 28.3 contains the study of general bivariate models along with some examples. The bivariate models under series system are studied in Section 28.4 and the crossings of the failure rates are examined for various bivariate models under the gamma and the inverse Gaussian environmental effects. Finally, in Section 28.5, some general results are presented for the crossings of the survival curves and the mean residual life functions in relation to the crossing of the failure rates.

28.2   The General Approach – Univariate Case

Let us consider a continuous nonnegative random variable T with baseline hazard function $\lambda(t)$. To introduce the heterogeneity, let the hazard rate for an equipment become $Z\lambda(t)$, where Z is a nonnegative random variable representing the environmental effect. $Z\lambda(t)$ can be considered as the conditional failure rate of a random variable T given Z. Thus, the conditional survival function of T given $Z = z$ is given by $S(t|z) = \exp(-z\Lambda(t))$, where $\Lambda(t)$ is the integrated hazard corresponding to $\lambda(t)$. Denoting the probability density function (pdf) of Z by $g(z)$, the unconditional survival function of T is given by

$$S^*(t) = \int_0^\infty \exp(-z\Lambda(t))g(z)dz. \tag{28.1}$$

The unconditional failure rate is given by

$$\lambda^*(t) = -\frac{d}{dt}\ln S^*(t). \tag{28.2}$$

The relationship between the baseline hazard and the unconditional hazard is given by

$$\frac{\lambda^*(t)}{\lambda(t)} = E(Z|T > t). \tag{28.3}$$

For details, see Gupta and Gupta (1996). The following theorem (Gupta and Gupta (1997)) shows that $\lambda(t)$ and $\lambda^*(t)$ cross atmost at one point.

Theorem 28.2.1 $\lambda(t)$ and $\lambda^*(t)$ cross atmost at one point and the crossing point is a solution of the equation

$$M_Z'(-\Lambda(t)) = M_Z(-\Lambda(t)), \tag{28.4}$$

where $M_Z(t)$ is the moment generating function of Z.

Corollary 28.2.1 If $E(Z) \leq 1$, then $\lambda^*(t) \leq \lambda(t)$.

Remark. The above results show that under gentler environments ($E(Z) \leq 1$), the failure rate under the operating environment does not exceed the baseline failure rate. In the case of harsher environment ($E(Z) > 1$), the failure rates under the operating environment exceeds the baseline failure up to a certain time. In otherwords, the effect of the harsher environment washes out after certain finite amount of time t^*. We now present two examples, one with a gamma environmental effect and the other with an inverse Gaussian environmental effect.

Example 28.2.1 Suppose Z has a gamma distribution given by

$$g(z) = \frac{1}{\beta^\alpha\Gamma(\alpha)}z^{\alpha-1}\exp(-z/\beta), z, \alpha, \beta > 0.$$

It can be verified that

$$\frac{\lambda^*(t)}{\lambda(t)} = E(Z)(1 + \beta\Lambda(t))^{-1}.$$

Thus the two hazards cross at the point $t^* = \Lambda^{-1}((\alpha\beta-1)/\beta)$ if $E(Z) = \alpha\beta > 1$.

In case T has a Weibull distribution with baseline hazard $\lambda(t) = \gamma t^{\gamma - 1}$, the crossing point becomes $t^* = ((\alpha\beta - 1)/\beta)^{1/\gamma}$.

Example 28.2.2 Suppose Z has an inverse Gaussian distribution given by

$$g(z) = \left(\frac{1}{2\pi a z^3}\right)^{1/2} \exp\left\{-(bz - 1)^2/2az\right\}, z, a, b > 0.$$

It can be verified that

$$\frac{\lambda^*(t)}{\lambda(t)} = (b^2 + 2a\Lambda(t))^{-1/2}.$$

Thus the two hazards cross at the point $t^* = \Lambda^{-1}((1 - b^2)/2a)$. If the baseline distribution is Weibull, the crossing point becomes $t^* = ((1 - b^2)/2a)^{1/\gamma}$.

28.3   The General Approach – Bivariate Models

Suppose (T_1, T_2) is a nonnegative continuous random variable with baseline survival function $S(t_1, t_2) = \exp(-\psi(t_1, t_2))$ and the hazard components

$$\lambda_1(t_1, t_2) = -\frac{\partial}{\partial t_1} \ln S(t_1, t_2)$$

$$\lambda_2(t_1, t_2) = -\frac{\partial}{\partial t_2} \ln S(t_1, t_2),$$

$\lambda_1(t_1, t_2)$ and $\lambda_2(t_1, t_2)$ are called the components of the hazard gradient. For definition, see Johnson and Kotz (1975). $\lambda_1(t_1, t_2)$ is the hazard rate of the conditional distribution of T_1 given $T_2 > t_2$; similarly for $\lambda_2(t_1, t_2)$. Assume that both the components are subjected to the same environmental conditions, so that the unconditional survival function of (T_1, T_2) given $Z = z$ is given by

$$S^*(t_1, t_2) = \int \exp(-z\psi(t_1, t_2))g(z)dz = M_z(-\psi(t_1, t_2)).$$

Let $\lambda_1^*(t_1, t_2)$ and $\lambda_2^*(t_1, t_2)$ be the components of the hazard gradient incorporating the environmental effect. Then

$$\lambda_1^*(t_1, t_2) = -\frac{\partial}{\partial t_1} \ln S^*(t_1, t_2)$$

$$= \lambda_1(t_1, t_2)E(Z|T_1 > t_1, T_2 > t_2).$$

This gives

$$\frac{\lambda_1^*(t_1, t_2)}{\lambda_1(t_1, t_2)} = E(Z|T_1 > t_1, T_2 > t_2).$$

Consider now

$$\frac{\partial}{\partial t_2} E(Z|T_1 > t_1, T_2 > t_2) = \frac{\partial}{\partial t_2} \left(\frac{\int_0^\infty z \exp(-z\psi(t_1, t_2)) g(z) dz}{\int_0^\infty z \exp(-z\psi(t_1, t_2)) g(z) dz} \right)$$

$$= -\frac{\partial}{\partial t_2} \psi(t_1, t_2) \mathrm{Var}(Z|T_1 > t_1, T_2 > t_2) < 0.$$

Hence $\lambda_1^*(t_1, t_2)/\lambda_1(t_1, t_2)$ is a decreasing function of t_2. In a similar manner we see that $\lambda_2^*(t_1, t_2)/\lambda_2(t_1, t_2)$ is a decreasing function of t_1. We now present two examples.

Example 28.3.1 Suppose the baseline random variables (T_1, T_2) are independent with conditional survival function

$$S(t_1, t_2|z) = \exp(-z(H_1(t_1) + H_2(t_2))).$$

The unconditional survival function is, therefore,

$$S^*(t_1, t_2) = M_Z(-(H_1(t_1) + H_2(t_2))).$$

We now assume that Z has a gamma distribution, as before. In this case

$$S^*(t_1, t_2) = (1 + \beta H_1(t_1) + \beta H_2(t_2))^{-\alpha}.$$

The first component of the hazard gradient, taking into account the environmental effect, is given by

$$\lambda_1^*(t_1, t_2) = -\frac{\partial}{\partial t_1} \ln S^*(t_1, t_2) = \frac{\alpha\beta h_1(t_1)}{1 + \beta H_1(t_1) + \beta H_2(t_2)},$$

where $h_1(t_1) = \frac{d}{dt_1} H_1(t_1) = \lambda_1(t_1, t_2)$. This gives

$$\frac{\lambda_1^*(t_1, t_2)}{\lambda_1(t_1, t_2)} = \frac{\alpha\beta}{1 + \beta H_1(t_1) + \beta H_2(t_2)} = E(Z|T_1 > t_1, T_2 > t_2),$$

and

$$\mathrm{Var}(Z|T_1 > t_1, T_2 > t_2) = \frac{\alpha\beta^2}{(1 + \beta H_1(t_1) + \beta H_2(t_2))^2}.$$

The above two expressions yield the square of the coefficient of variation, (c.v.) ^{2}, of Z given $T_1 > t_1$ and $T_2 > t_2$ as $1/\alpha$ (constant). It can be verified that a constant value of the c.v. occurs only in the case of a gamma environmental effect.

Example 28.3.2 Here we assume that Z has an inverse Gaussian distribution as in Section 28.2. In this case

$$S^*(t_1, t_2) = \exp\left\{ \frac{b}{a} \left(1 - \left(1 + \frac{2a}{b^2} (H_1(t_1) + H_2(t_2)) \right)^{1/2} \right) \right\}.$$

The first component of the hazard gradient, taking into account the environmental effect, is given by

$$\lambda_1^*(t_1, t_2) \;=\; -\frac{\partial}{\partial t_1} \ln S^*(t_1, t_2) = \frac{h_1(t_1)}{(b^2 + 2a(H_1(t_1) + H_2(t_2)))^{1/2}}.$$

This gives

$$\frac{\lambda_1^*(t_1, t_2)}{\lambda_1(t_1, t_2)} \;=\; \frac{1}{(b^2 + 2a(H_1(t_1) + H_2(t_2)))^{1/2}} = E(Z|T_1 > t_1, T_2 > t_2)$$

and

$$\mathrm{Var}(Z|T_1 > t_1, T_2 > t_2) \;=\; \frac{a}{(b^2 + 2a(H_1(t_1) + H_2(t_2)))^{3/2}}.$$

The above two expressions yield $(c.v.)^2$ of Z given $T_1 > t_1, T_2 > t_2$ as

$$\frac{a}{(b^2 + 2a(H_1(t_1) + H_2(t_2)))^{1/2}},$$

which is a decreasing function of both t_1 and t_2.

28.4   Bivariate Models Under Series System

Suppose the two components whose lifetimes (T_1, T_2), in a bivariate model, are arranged in a series system so that only $T = \min(T_1, T_2)$ is observed. Let $\Lambda(t) = \psi(t, t)$ be the integrated hazard of T. The conditional survival function of T given $Z = z$ is given by

$$S(t|z) = \exp(-z\psi(t, t)).$$

The unconditional survival function of T is

$$S^*(t) = M_Z(-\psi(t, t)).$$

Proceeding as in the univariate case, we obtain

$$\frac{\lambda^*(t)}{\lambda(t)} \;=\; \frac{M_Z'(-\psi(t, t))}{M_Z(-\psi(t, t))} = E(Z|T > t).$$

It can be verified that $\frac{\lambda^*(t)}{\lambda(t)}$ is a decreasing function of t and hence

$$\frac{\lambda^*(t)}{\lambda(t)} \;=\; E(Z|T > t) \le E(Z|T > 0) = E(Z), \quad \text{for all } t.$$

We now discuss the following two cases:

1. If $E(Z) \leq 1$ (gentler environment), then $\lambda^*(t) \leq \lambda(t)$.

2. If $E(Z) > 1$ (harsher environment), then $\lambda^*(t)$ and $\lambda(t)$ cross at one point and the crossing point is a solution of

$$M_Z'(\psi(t,t)) = M_Z(-\psi(t,t)).$$

Example 28.4.1 (i) Suppose Z has a gamma distribution. Proceeding as before, the crossing point is given by $t^* = \Lambda^{-1}\left(\frac{\alpha\beta-1}{\beta}\right)$ if $E(Z) = \alpha\beta > 1$. (ii) If Z has an inverse Gaussian distribution, then $t^* = \Lambda^{-1}((1-b^2)/2ab)$.

Now we would like to see the crosssing point of $\lambda^*(t)$ and $\lambda(t)$ under different bivariate models incorporating gamma or inverse Gaussian environmental effect. We provide below the details for Marshall and Olkin model only. In this case

$$S(t_1, t_2) = \exp(-\lambda_1 t_1 - \lambda_2 t_2 - \lambda_{12} \max(t_1, t_2)), \quad \lambda_1, \lambda_2, \lambda_{12} > 0.$$

The survival function of T is then

$$S(t) = \exp(-(\lambda_1 + \lambda_2 + \lambda_{12})t).$$

This gives $\Lambda(t) = (\lambda_1 + \lambda_2 + \lambda_{12})t = \psi(t,t)$.

1. Under gamma environmental effect, proceeding as before,

$$t^* = (\alpha\beta - 1)/\beta(\lambda_1 + \lambda_2 + \lambda_{12}), \text{if } E(Z) = \alpha\beta > 1.$$

2. Under inverse Gaussian environmental effect

$$t^* = (1 - b^2)/2ab(\lambda_1 + \lambda_2 + \lambda_{12}).$$

The results for independent bivariate exponential model can be obtained by putting $\lambda_{12} = 0$. In addition to the above two cases, we present the crossing points of $\lambda^*(t)$ and $\lambda(t)$ for various other bivariate models in Table 28.1. Notice that the various models in Table 28.1 have been given for $T = \min(T_1, T_2)$, i.e. under the series system.

Table 28.1: Crossing point of $\lambda^*(t)$ and $\lambda(t)$ for various distributions under Gamma and Inverse Gaussian Environmental Effects

	$Z \sim \Gamma(\alpha,\beta) = \frac{z^{\alpha-1}e^{-z/\beta}}{\Gamma(\alpha)\beta^\alpha}$ $E(Z) = \alpha\beta > 1$	$Z \sim \mathrm{IG}(a,b) = \frac{\exp(-\frac{(bz-1)^2}{2az})}{(2\pi az^3)^{1/2}}$ $E(Z) = 1/b > 1$
1. Independent Model: $S_1(t) = e^{-(\lambda_1+\lambda_2)t}$	$t^* = \frac{\alpha\beta-1}{\beta(\lambda_1+\lambda_2)}$	$t^* = \frac{1-b^2}{2ab(\lambda_1+\lambda_2)}$
2. Marshall Olkin (1967): $S_2(t) = e^{-(\lambda_1+\lambda_2+\lambda_{12})t}$	$t^* = \frac{\alpha\beta-1}{\beta(\lambda_1+\lambda_2+\lambda_{12})}$	$t^* = \frac{1-b^2}{2ab(\lambda_1+\lambda_2+\lambda_{12})}$
3. Marshall Olkin (1995): $S_3(t) = e^{-(2+\sqrt{2}\lambda(1+\theta)^{1/2})t}$	$t^* = \frac{\alpha\beta-1}{\beta(2+\lambda(2+2\theta)^{1/2})}$	$t^* = \frac{1-b^2}{2ab(2+\lambda(2+2\theta)^{1/2})}$
4. Block & Basu (1974): $S_4(t) = e^{-(\lambda_1+\lambda_2+\lambda_{12})t}$	$t^* = \frac{\alpha\beta-1}{\beta(\lambda_1+\lambda_2+\lambda_{12})}$	$t^* = \frac{1-b^2}{2ab(\lambda_1+\lambda_2+\lambda_{12})}$
5. Gumbel I (1960): $S_5(t) = e^{-(\lambda_1+\lambda_2+\lambda_{12}t)t}$	$t^* = \frac{1}{2\lambda_{12}}\{(\lambda_1+\lambda_2)+\sqrt{(\lambda_1+\lambda_2)^2+4\frac{(\alpha\beta-1)}{\beta}\lambda_{12}}\}$	$t^* = \frac{1}{2\lambda_{12}}\{(\lambda_1+\lambda_2)+\sqrt{(\lambda_1+\lambda_2)^2+4\frac{(1-b^2)}{2ab}\lambda_{12}}\}$
6. Absolutely cont. biv. Rayleigh dist. (Roy 1993): $S_6(t) = e^{-(a+b+ct^2)t^2}$	$t^{*^2} = \frac{1}{2c}\{-(a+b)+\sqrt{(a+b)^2+4\frac{(\alpha\beta-1)}{\beta}c}\}$	$t^{*^2} = \frac{1}{2c}\{-(a+b)+\sqrt{(a+b)^2+4\frac{(1-b^2)}{2ab}c}\}$
7. Bivariate Lomax (Roy 1989): $S_7(t) = \frac{e^{p\ln\delta}}{(\delta+(\lambda_1+\lambda_2)t)}$	$t^* = \frac{\delta}{\lambda_1+\lambda_2}(e^{(\alpha\beta-1)/p\beta}-1)$	$t^* = (e^{(1-b^2)/2abp}-1)\frac{\delta}{\lambda_1+\lambda_2}$
8. Bivariate Pareto $S_8(t) = e^{-(\lambda_1+\lambda_2+\lambda_{12})\ln\frac{t}{\beta}}$	$t^* = \beta e^{(\alpha\beta-1)/(\lambda_1+\lambda_2+\lambda_{12})\beta}$	$t^* = \beta e^{(1-b^2)/2ab(\lambda_1+\lambda_2+\lambda_{12})}$

28.5 Crossings of Survival Curves and Mean Residual Life Functions

As we have seen earlier, there are many practical situations where the two hazard rates cross at a point and even at more than one point. This phenomenon of crossing hazards arises in several medical studies and in the study of frailty models, see Gupta and Gupta(1997) for some references and discussion.

Generally, for analyzing survival (or reliability) data, three measures namely the failure rate (hazard rate), the survival function (reliability function) and the mean residual life function (life expectancy) are used. Theoretically these measures are equivalent, in the sense that the knowledge of one of them determines the other two, see Gupta (1981). In this section we shall consider two hazard rates which cross at one or two points and study the crossing of the corresponding survival function and the mean residual life function (MRLF).

Let X be a non negative continuous random variable with survival function $S(t)$. Then the MRLF of X is defined as

$$\mu(t) = E(X - t | X > t) = \int_t^\infty S(x) dx / S(t).$$

The failure rate $\lambda(t)$ and the MRLF $\mu(t)$ are connected by the relation $\lambda(t) = (1 + \mu'(t)) / \mu(t)$. Suppose we have two groups with failure rate, survival function and MRLF as $\lambda(t), S(t)$ and $\mu(t)$ for one group and $\lambda^*(t), S^*(t)$ and $\mu^*(t)$ for the other group. Then

$$\frac{S^*(t)}{S(t)} = \exp(-\int_0^t (\lambda^*(u) - \lambda(u)) du).$$

Assume $\lambda^*(t) > \lambda(t)$ for $0 < t < t^*$, and $\lambda^*(t) < \lambda(t)$ for $t > t^*$. i.e. $\lambda(t)$ and $\lambda^*(t)$ cross at the point t^*. Then

1. $S(t)$ and $S^*(t)$ cross at almost one point and this point (if it exists) occurs on the right of t^*.

2. $\mu(t)$ and $\mu^*(t)$ cross at almost one point and this point (if it exists) occurs on the left of t^*. For details and proofs, see Gupta and Gupta (1997).

Remark. For hazard functions and survival functions, such a phenomenon has been noticed by Pocock et. al. (1982) in comparing two groups, one of menopausal women and the other of post menopausal women. The hazard rates cross after about 4 years since initial treatment while the survival curves cross at about 6 years. In analyzing the same data set from a different point of view Gore et. al. (1984) have observed that in comparing two groups, one with tumor size ≤ 2 cms and the other with tumor size 3–4 cms, the hazard rates cross while the survival curves do not cross at all.

In case the failure rates cross at two points, we have the following results. Assume

$$
\begin{aligned}
\lambda^*(t) &> \lambda(t) \text{ for } t < t_1^* \\
\lambda^*(t) &< \lambda(t) \text{ for } t_1 < t < t_2^* \\
\lambda^*(t) &> \lambda(t) \text{ for } t > t_2^*
\end{aligned}
$$

i.e. the two failure rates cross at t_1^* and t_2^*. Then

1. $S(t)$ and $S^*(t)$ cross at atmost two points and atmost one crossing point occurs in each of the intervals $t_1 < t < t_2^*$ and $t \geq t_2^*$.

2. $\mu(t)$ and $\mu^*(t)$ cross at atmost two points and atmost one crossing point occurs in each of the intervals $t < t_1^*$ and $t_1 < t < t_2^*$.
   For details and proofs, see Gupta and Gupta (1997).

28.6   Summary and Conclusions

In this paper we have presented a general approach of studying random environmental models. Incorporating the random environmental effect provides a useful framework for determining the effectiveness of interventions. It is shown that under certain conditions, the conditional and the unconditional failure rates cross at a point. The crossings of the survival functions and the mean residual life functions are studied when the failure rates cross at one or two points.

References

1. Bandyopadhyay, D. and Basu, A. P. (1990). On a generalization of a model by Lindley and Singpurwalla, *Advances in Applied Probability*, **22**, 498–500.

2. Block, H. W. and Basu, A. P. (1974). A continuous bivariate exponential extension, *Journal of the American Statistical Association*, **69**, 1031–1037.

3. Cox, D. R. (1972). Regression models and life tables (with discussion), *Journal of the Royal Statistical Society (B)*, **34**, 187–220.

4. Gore, S. M., Pocock, S. J. and Kerr, G. R. (1984). Regression models and non proportional hazards in the analysis of breast cancer survival, *Applied Statistics*, **33(2)**, 176–195.

5. Gumbel, E. J. (1960). Bivariate exponential distribution, *Journal of the American Statistical Association*, **55**, 698–707.

6. Gupta, P. L. and Gupta, R. D. (1990). A bivariate random environmental stress model, *Advances in Applied Probability*, **22**, 501–503.

7. Gupta, P. L. and Gupta R. C. (1996). Ageing characterestics of Weibull mixtures, *Probability in the Engineering and Informational Sciences*, **10**, 591–600.

8. Gupta, R. C. (1981). On the mean residual life function in survival studies, *Statistical Distribution in Scientific Work 5 (eds. C. Taillie, G. P. Patil and B. A. Baldessari)*, 327–334, Reidal Publishing Co., Boston.

9. Gupta, R. C. and Gupta P. L. (1997). On the crossings of reliability measures, Unpublished manuscript, Department of Mathematics and Statistics, University of Maine, ME.

10. Johnson, N. L. and Kotz, S. (1975). A vector multivariate hazard rate, *Journal of Multivariate Analysis*, **5**, 53–66.

11. Lindley, D. V. and Singpurwalla, N. D. (1986). Multivariate distributions for the life lengths of components of a system sharing a common environment, *Journal of Applied Probability*, **23**, 418–431.

12. Mardia, K. V. (1962). Multivariate Pareto distributions, *Annals of Mathematical Statistics*, **33**, 1008–1015.

13. Marshall, A. W. and Olkin, I. (1967). A multivariate exponential distribution, *Journal of the American Statistical Association*, **62**, 30–40.

14. Marshall, A. W. and Olkin, I. (1995). Multivariate exponential and Geometric distributions with limited memory, *Journal of Multivariate Analysis*, **53**, 110–125.

15. Nayak, T. K. (1987). Multivariate Lomax distribution: properties and usefulness in reliability theory, *Journal of Applied Probability*, **24**, 170–177.

16. Pocock, S. J., Gore, S. M. and Kerr, G. R. (1982). Long term survival analysis: The curability of breast cancer, *Statistics in Medicine*, **1**, 93–104.

17. Roy, D. and Mukherjee, S. P. (1988). Generalized mixtures of exponential distributions, *Journal of Applied Probability*, **25**, 510–518.

18. Roy, D. (1989). A characterization of Gumbel's bivariate exponential and Lindley and Singpurwalla's bivariate Lomax distributions, *Journal of Applied Probability*, **27**, 886–891.

19. Roy, D. (1993). An absolutely continuous bivariate Rayleigh distribution, *Journal of Indian Statistical Association*, **31**, 125–127.

29

Testing for Change Points Expressed in Terms of the Mean Residual Life Function

Emad-Eldin A. A. Aly

Kuwait University, Safat, Kuwait

Abstract: Change point problems have received considerable attention in the last two decades. They are often encountered in the fields of quality control, reliability and survival analysis. A typical situation occurs in quality control when one observes the output of a production line and is interested in detecting any change in the quality of a product. In this article we consider the problem of testing against change points when the change is expressed in terms of the mean residual life function as described next. Let $X_1, X_2, ..., X_n$ be independent observations. We consider the problem of testing the null hypothesis of no change, H_o : The $X's$ have an unknown common distribution function F against the at most one-change point alternative, H_1 : $X_i, 1 \leq i \leq [n\lambda]$ have common distribution function F_1 and $X_i, [n\lambda] < i \leq n$ have common distribution function F_2, where $\lambda \in (0,1)$ and $F_1 \neq F_2$ are unknown and $F_1 \underset{MR}{\succ} F_2$ (i.e., $M_1(t) \geq M_2(t)$ for all t, where $M_i(t) = \int_t^\infty \overline{F_i}(x)dx/\overline{F_i}(t)$). We propose some nonparametric tests for this problem. We obtain the limiting distributions of the proposed tests.

Keywords and phrases: Brownian bridge, empirical distribution function, failure rate function, Kiefer process, Wiener process

29.1 Introduction

Change point problems have received considerable attention in the last two decades. They are often encountered in the fields of quality control, reliability and survival analysis. A typical situation occurs in quality control when one observes the output of a production line and is interested in detecting any change in the quality of a product and in estimating the change point. The quality of a product can be measured in terms of a specific parameter(s) (e.g., a location parameter, a scale parameter or a shape parameter). In reliability and survival analysis the quality of a device is often measured in terms of an ageing property. Aly and Kochar (1997) considered the problem of testing against a change point when the change is expressed in terms of the failure rate function. In this article we consider the problem of testing against a change point when the change is expressed in terms of the mean residual life function. For additional results and references on change point analysis we refer to Zacks (1983), Bhattacharyya (1984), Csörgő and Horváth (1988), Sen (1988), Lombard(1989) and Hušková and Sen (1989).

Let $\mathcal{L}$ be the class of all continuous life distribution functions (DF) with finite mean. Let X be a random variable with DF $F \in \mathcal{L}$. The mean residual life (MRL) function of F is defined as

$$M_F(t) = E(X - t \mid X > t) \tag{29.1}$$
$$= \begin{cases} \frac{1}{\overline{F}(t)} \int_t^\infty \overline{F}(x)dx, & 0 \le t < T_F \\ 0 & t \ge T_F, \end{cases}$$

where $\overline{F}(t) = 1 - F(t)$ is the survival function of $F(\cdot)$ and

$$T_F = \inf\{t : \overline{F}(t) = 0\}. \tag{29.2}$$

Note that $M_F(.)$ uniquely defines $F(.)$. This follows from the well known result

$$\overline{F}(x) = exp\left\{-\int_0^x \left(M_F'(t) + 1\right)/M_F(t)dt\right\}.$$

Let $F(.)$ be the DF of the life time of a device or a patient receiving a certain treatment. The proportion of similar devices/patients who survive up to time t_o is $\overline{F}(t_o)$. Given that the device/patient has survived up to time t_o, its expected remaining life is $M_F(t_o)$. The functions $\overline{F}(\cdot)$, and $M_F(\cdot)$ are of interest in reliability and survival analysis and they complement each other.

Let $F_i(\cdot)$, $i = 1,2$ be two DF's in $\mathcal{L}$. For $i = 1,2$, let $M_i(\cdot) = M_{F_i}(\cdot)$, where $M_F(\cdot)$ is as in (29.1). The DF F_1 is said to be larger than F_2 in the mean residual life order ($F_1 \underset{MR}{\succ} F_2$) if $M_1(t) \ge M_2(t)$ for $t \ge 0$ (see Alzaid (1988) and Shaked and Shanthikumar (1994)).

Let $X_1, X_2, \ldots, X_n$ be independent random variables. We consider the problem of testing the null hypothesis of no change

$$H_o : \text{The } X_i\text{'s have an unknown common DF, } F \tag{29.3}$$

against the change point alternative,

$H_1 :$ $X_i, 1 \le i \le [n\lambda]$ have a common DF, $F_1(\cdot)$, $X_i, [n\lambda] < i \le n$ have a common DF $F_2(\cdot)$, where $\lambda \in (0,1)$ and $F_1 \ne F_2$ are unknown and $F_1 \underset{MR}{\succ} F_2$ (i.e., $M_1(t) \ge M_2(t)$ for all t with strict inequality for some t).

$$\tag{29.4}$$

In the rest of this article we will use the following notation. A Brownian bridge, $B(.)$, is a mean zero Gaussian process with covariance function $EB(s)B(t) = t \wedge s - ts$ when $0 \leq t, s \leq 1$, where $t \wedge s = \min(t, s)$. A Kiefer process, $K(., .)$, is a mean zero Gaussian process with covariance function $EK(s, x)K(t, y) = x \wedge y(t \wedge s - ts)$ when $0 \leq t, s \leq 1$ and $0 \leq x, y < \infty$.

29.2   Testing for Change Points

The tests developed in this article are based on the mean residual life comparison-plot (MRLC-plot) function of Aly (1997) which is defined below.

Let $F_1(\cdot)$ and $F_2(\cdot)$ be two DF in $\mathcal{L}$. The MRLC-plot function of $F_1(\cdot)$ against $F_2(\cdot)$ is defined as

$$\gamma_{1,2}(t) = \sqrt{\overline{F}_1(t) \cdot \overline{F}_2(t)} \left\{ M_1(t) - M_2(t) \right\}, 0 \leq t < T, \tag{29.5}$$

where $T = T_1 \wedge T_2$, $T_i = T_{F_i}$, $i = 1, 2$, and T_F is as in (29.2). The deviation from $F_1 = F_2$ in favour of $F_1 \underset{MR}{\succ} F_2$ can be measured by appropriate functionals of $\gamma_{1,2}(.)$ like, for example,

$$\sup_{0 \leq t < T} \gamma_{1,2}(t)$$

and

$$\int_0^T \gamma_{1,2}(t)dt.$$

Let $F_{1,k}(\cdot)$ be the empirical distribution function based on $X_1, \ldots, X_k$ and $F_{2,k}(\cdot)$ be the empirical distribution function based on $X_{k+1}, \ldots, X_n$. Define the MRLC-change point process $\{T_n(\cdot, \cdot)\}$ as $T_n(s, t) = 0$ if $\overline{F}_{1,[ns]}(t) \cdot \overline{F}_{2,[ns]}(t) = 0$ and

$$T_n(s, t) = \frac{[ns](n - [ns])}{n^{3/2}} \sqrt{\overline{F}_{1,[ns]}(t).\overline{F}_{2,[ns]}(t)} \left(M_{1,[ns]}(t) - M_{2,[ns]}(t) \right), \tag{29.6}$$

where $M_{1,k}(t) + t$ is the average of those $X_1, \ldots, X_k$ that are greater than t and $M_{2,k}(t) + t$ is the average of those $X_{k+1}, \ldots, X_n$ that are greater than t.

Assume that the change point λ of (29.4) is known and $[n\lambda] = k$. In this case the problem at hand becomes a two-sample problem based on the samples $X_1, \ldots, X_k$ and $X_{k+1}, \ldots, X_n$. For this two-sample problem we may use any of the two test statistics of Aly (1997) for testing against the MRL ordering. These statistics are based on the MRLC-plot function $T_n(\frac{k}{n}, t)$ of (29.6). In the change point set-up the value of k is unknown, for this reason, tests for H_o of (29.3) against H_1 of (29.4) can be based on the MRLC-change point process $\{T_n(\cdot, \cdot)\}$ of (29.6).

Let $\hat{\sigma}^2(t)$ be the sample variance of those $X_1, \ldots, X_n$ that are greater than t and define

$$\hat{\sigma}_c^2 = \sum_{j=1}^{n-1} (1 - \frac{j}{n}) \hat{\sigma}^2(X_{j:n})(X_{j:n}^2 - X_{j-1:n}^2), \tag{29.7}$$

where $X_{0:n} = 0$ and $X_{j:n}$ is the jth order statistic of $X_1, \ldots, X_n$. Note that, under $H_\circ$ of (29.3) $\hat{\sigma}^2(t)$ is a consistent estimator of $\sigma_F^2(t) = Var(X|X > t)$ and $\hat{\sigma}_c^2$ is a consistent estimator of

$$\sigma_c^2 = 2 \int_0^\infty t \overline{F}(t) \sigma_F^2(t) dt. \tag{29.8}$$

To test $H_\circ$ of (29.3) against H_1 of (29.4) we propose the following test statistics

$$q_{1,n} = \hat{\sigma}^{-1}(0) \max_{1<k<n-1} \max_{1 \le j \le n} T_n(\frac{k}{n}, X_{j:n}), \tag{29.9}$$

$$q_{2,n} = \hat{\sigma}_c^{-1} \max_{1<k<n-1} \xi_n(\frac{k}{n}), \tag{29.10}$$

and

$$q_{3,n} = \frac{\sqrt{12}}{n \hat{\sigma}_c} \sum_{k=2}^{n-2} \xi_n(\frac{k}{n}), \tag{29.11}$$

where

$$\begin{aligned}
\xi_n(\frac{k}{n}) &= \int_0^\infty T_n(s, t) dt \\
&= \sum_{i=0}^{n-1} T_n(\frac{k}{n}, X_{i+1:n})(X_{i+1:n} - X_{i:n}). \tag{29.12}
\end{aligned}$$

Lemma 29.2.1 *Assume that $EX_1^2 < \infty$ and $H_\circ$ of (29.3) holds. Then, as $n \to \infty$*

$$q_{1,n} \xrightarrow{\mathcal{D}} \sup_{0 \le s,t \le 1} K(s,t) = q_1, \tag{29.13}$$

$$q_{2,n} \xrightarrow{\mathcal{D}} \sup_{0 \le s \le 1} B(s) = q_2, \tag{29.14}$$

and

$$q_{3,n} \xrightarrow{\mathcal{D}} N(0,1), \tag{29.15}$$

where "$\xrightarrow{\mathcal{D}}$" denotes convergence in distribution.

The proof of Lemma 29.2.1 is given in Section 29.3. It is well known that

$$P\{q_2 \ge u\} = exp\{-2u^2\}, u \ge 0.$$

Consequently, we reject H_o in favor of H_1 of (29.4) at approximate level α if

$$q_{2,n} > \sqrt{-\frac{1}{2}\ln\alpha}$$

Alternatively, we reject H_o at approximate level α if $q_{3,n} > z_{1-\alpha}$. The DF of q_1 of (29.13) is not available in a closed form. We conducted a Monte Carlo simulation to approximate the critical values of q_1. We have used 800 realizations of $K(\cdot,\cdot)$ generated using Cholesky's method (IMSLSTAT Library) on a grid of 400 points on $[0,1] \times [0,1]$. Estimates of the upper critical values of q_1 are 1.945, 1.623 and 1.233 for $\alpha = 0.01$, 0.05 and 0.1, respectively.

29.3   Proof of Lemma 29.2.1

Define

$$v_{1,n}(s,t) = \sqrt{[ns]}\,\overline{F}_{1,[ns]}(t)\left(M_{1,[ns]}(t) - M(t)\right)$$

and

$$v_{2,n}(s,t) = \sqrt{n-[ns]}\,\overline{F}_{2,[ns]}(t)\left(M_{2,[ns]}(t) - M(t)\right).$$

Let $K(\cdot,\cdot)$ be a Kiefer process and define the two-parameter Gaussian process $\varphi(\cdot,\cdot)$ by

$$\varphi(s,t) = \int_t^\infty K\left(F(x),s\right)dx - M(t)K\left(F(t),s\right).$$

By the discussion following (4.2) of Csörgő et al. (1986),

$$\{\varphi(s,t);\ t\geq 0,\ 0\leq s\leq 1\} \overset{\mathcal{D}}{=} \{\sigma(0)W(s,R(t));\ t\geq 0,\ 0\leq s\leq 1\}, \quad (29.16)$$

where $R(t) = \overline{F}(t)\sigma^2(t)/\sigma^2(0)$, $\sigma^2(t) = Var(X|X>t)$ and $\{W(x,y); x,y\geq 0\}$ is a standard two-parameter Wiener process. Note that $\overline{R}(t)$ is a distribution function.

Adopting the proof of (iv) of Theorem 4.1 of Csörgő et al. (1986) we can prove that, under H_o and as $n\to\infty$

$$\sup_{0\leq t<\infty}\sup_{2/n\leq s\leq 1-2/n} |v_{1,n}(s,t) - s^{-\frac{1}{2}}\varphi(s,t)| \overset{P}{\longrightarrow} 0 \qquad (29.17)$$

and

$$\sup_{0\leq t<\infty}\sup_{2/n\leq s\leq 1-2/n} \left|v_{2,n}(s,t) - (1-s)^{-\frac{1}{2}}\left(\varphi(1,t) - \varphi(s,t)\right)\right| \overset{P}{\longrightarrow} 0. \qquad (29.18)$$

Note that

$$T_n(s,t) = \sqrt{[ns](n-[ns])/n^2}\,\Big\{\sqrt{(n-[ns])\overline{F}_{2,[ns]}(t)\big/n\overline{F}_{1,[ns]}(t)}\ v_{1,n}(s,t)$$

$$-\sqrt{[ns]\overline{F}_{1,[ns]}(t)\big/n\overline{F}_{2,[ns]}(t)}\ v_{2,n}(s,t)\Big\}.$$

$$(29.19)$$

By (29.17)–(29.19) and the proof of Theorem A.1 of Aly (1997) we can show that, under H_o and as $n \to \infty$

$$\sup_{0 \le t < \infty} \sup_{2/n \le s \le 1-2/n} |T_n(s,t) - \varphi(s,t) + s\varphi(1,t)| \xrightarrow{P} 0. \tag{29.20}$$

By (29.16) we have on $t \ge 0$, $0 \le s \le 1$,

$$\varphi(s,t) - s\varphi(1,t) \overset{D}{=} \sigma(0) \{W(s, R(t)) - sW(1, R(t))\}. \tag{29.21}$$

It is well known that

$$\{W(x,y) - xW(1,y); \, y \ge 0, \, 0 \le x \le 1\} \overset{D}{=} \{K(x,y); \, y \ge 0, \, 0 \le x \le 1\}. \tag{29.22}$$

By (29.20)- (29.22) we obtain

$$\sup_{0 \le t < \infty} \sup_{2/n \le s \le 1-2/n} |T_n(s,t) - \sigma(0)K(s, R(t))| \xrightarrow{P} 0. \tag{29.23}$$

By (29.23), the definition of $q_{1,n}$ and Slutsky's Theorem we obtain (29.13) of Lemma 29.2.1. By (29.23) and the definition of $\xi_n(.)$ we obtain

$$\xi_n(s) \xrightarrow{D} \sigma(0) \int_0^\infty K(s, R(t)) \, dt. \tag{29.24}$$

It is easy to show that

$$\sigma(0) \int_0^\infty K(s, R(t)) \, dt \overset{D}{=} \sigma_c B(s), \tag{29.25}$$

where σ_c is as in (29.7). By (29.24) and (29.25) we obtain

$$\xi_n(s) \xrightarrow{D} \sigma_c B(s). \tag{29.26}$$

By (29.26) and Slutsky's Theorem we obtain (29.14) of Lemma 29.2.1. It is well known that

$$\int_0^1 B(s) \, ds \overset{D}{=} \frac{1}{\sqrt{12}} \, N(0,1). \tag{29.27}$$

By (29.26), (29.27) and Slutsky's Theorem we obtain (29.15) of Lemma 29.2.1.

29.4   Summary and Conclusions

We considered the problem of testing against one change point when the change is expressed in terms of the mean residual life function. We proposed three test statistics and obtained their limiting distributions.

Acknowledgements. This research was supported by an NSERC Canada Research Grant at the University of Alberta, Research Grant SS 024 of Kuwait University and by a travel Grant from Kuwait University.

References

1. Aly, E.-E. (1997). Nonparametric tests for comparing two mean residual life functions, *Life Time Data Analysis* (to appear).

2. Aly, E.-E. and Kochar, S. C. (1997). Change point tests based on U-statistics with applications in reliability, *Metrika*, **45**, 259–269.

3. Alzaid, A. A. (1988). Mean residual life ordering. *Statistical Papers*, **29**, 35–43.

4. Bhattacharyya, G. K. (1984). Tests for randomness against trend or serial correlations, In *Handbook of Statistics,*, **V4**, *Nonparametric Methods* (Eds., P. R. Krishnaiah and P. K. Sen) North-Holland, Amsterdam.

5. Csörgő, M., Csörgő, S. and Horváth, L. (1986). Asymptotic theory for empirical reliability and concentration processes, *Lecture Notes in Statistics*, **33**, Springer-Verlag, New York.

6. Csörgő, M. and Horváth, L. (1988). Nonparametric methods for change-point problems, In *Handbook of Statistics,*, **V7**, *Quality Control and Reliability* (Eds., P. R. Krishnaiah and C. R. Rao), North-Holland, Amsterdam.

7. Hušková, M. and Sen, P. K. (1989). Nonparametric tests for shift and change in regression at an unknown time point, In *Statistical Analysis and Forecasting of Economic Structural Change* (Ed., P. Hackl), Springer Verlag, New York.

8. Lombard, F. (1989). Some recent developments in the analysis of change-point data, *South African Statistical Journal*, **23**, 1–21.

9. Sen, P. K. (1988). Robust tests for change-point models, In *Encyclopedia of Statistical Sciences*, **8** (Eds., S. Kotz and N. L. Johnson), John Wiley & Sons, New York.

10. Shaked, M. and Shanthikumar, J. G. (1994). *Stochastic Orders and Their Applications*, Academic Press, San Diego.

11. Zacks, S. (1983). Survey of classical and Bayesian approaches to the change-point problem: fixed sample and sequential procedures of testing and estimation, In *Recent Advances in Statistics* (Eds., M. H. Rizvi, J. S. Rustagi and D. Siegmund), Academic Press, New York.

30

Empirical Bayes Procedures For Testing The Quality and Reliability With Respect To Mean Life

Radhey S. Singh

University of Guelph, Guelph, Ontario, Canada

Abstract: This paper considers a problem where lots are coming independently and sequentially for testing their quality, and at each stage a lot is accepted as reliable if and only if the mean life θ of items is at least θ_0, a specified standard required by the practitioner, and the penality for making an incorrect decision is taken as proportional to the distance the true θ is away from θ_0. When the prior distribution of θ remains unknown at each stage (hence the minimum risk Bayes optimal test procedure is not available for use at any stage), this paper provides asymptotically optimal empirical Bayes procedures for situations where life time distribution has negative exponential or gamma density and investigates the rates of convergence to optimality. We also consider the case where the life time distribution has density belonging to general exponential family, and the case where one has to deal with k varieties of lots simultaneously at each stage.

Keywords and phrases: Asymptotically optimal, empirical Bayes procedure, exponential, gamma, rates of convergence, reliability, test

30.1 Introduction and Development

30.1.1 Preliminaries

Consider a situation where the quality of a lot is judged by the mean life of its items, and the lot is accepted as reliable if and only if the mean life of its items is at least θ_0 (> 0), a specified standard required by the customers. Let Y be the random variable which denotes the life times of items of the lot to be tested for its quality and reliability. Let $E(Y) = \theta$ be the (unknown) mean life of the items of the lot, where $\theta \in \Theta = (0, \infty)$. The lot is accepted as reliable if $\theta \geq \theta_0$, otherwise it is rejected as a bad lot.

In this paper, we assume that the life time distribution of Y given θ is modelled by the negative exponential density $h(y|\theta) = (1/\theta) \exp(-y/\theta) \, I(y > 0)$, so that if X is the total life time of γ randomly chosen items from the lot then X is a sufficient statistic for θ, and given θ, the distribution of X is the gamma density

$$f(x|\theta) = (1/\Gamma(\gamma))(x^{\gamma-1}/\theta^\gamma) \exp(-x/\theta) \, I(x > 0), \quad \theta \in \Theta = (0, \infty). \quad (30.1)$$

Note that the negative exponential and gamma distributions are among the commonly used distributions to model life times of a variety of random phenomena in practice, including in survival analysis, reliability theory, life testing, engineering and environmental sciences. More applications of the exponential and gamma distributions can be found, for example, in Johnson et al. (1994) and Balakrishnan and Basu (1995). Although we have assumed that Y has negative exponential, and hence X has gamma density, the development of this section is completely independent of the choice of the density of Y.

The general set up is as follows. Let X be any random variable whose p.d.f. conditional on θ is $f(\cdot|\theta)$ and based on an observation on X, one is to test $H_0 : \theta \geq \theta_0$ (the lot is acceptable as reliable) against the alternative $H_1 : \theta < \theta_0$ (the lot is unreliable and be rejected) with the penalty functions for making incorrect decision taken as $L_0(\theta) = b(\theta_0 - \theta)I(\theta < \theta_0)$ and $L_1(\theta) = b(\theta - \theta_0)I(\theta \geq \theta_0)$, where I(A) is the indicator function of set A, $b > 0$ is a (known or unknown) constant of proportionality, and for i = 0, 1, $L_i(\theta)$ is the penalty for accepting H_i incorrectly. That is if the lot is accepted (rejected) correctly, then there is no penalty, otherwise it is proportional to the distance the true θ is away from θ_0 in each case.

30.1.2 Bayesian approach

Taking the Bayesian approach, we assume that the parameter θ is a random variable distributed according to an arbitrary but fixed prior distribution G with support in $\Theta = (0, \infty)$ for which $\int \theta dG(\theta) < \infty$. If based on $X \sim f(\cdot|\theta)$, the randomized decision rule $\delta(x) = P[accepting H_0 | X = x]$ is used, then the minimum Bayes risk w.r.t. G due to using the decision rule δ is

$$R(\delta, G) = E\left[L_0(\theta)\delta(X) + L_1(\theta)(1 - \delta(X))\right].$$

As noted in Johns and Van Ryzin (1971), this risk can be expressed as

$$R(\delta, G) = b \int \alpha(x)\delta(x)dG(\theta)dx + \int L_1(\theta)dG(\theta),$$

where

$$\alpha(x) = \int (\theta_0 - \theta)f(x|\theta)dG(\theta) = \theta_0 f(x) - \int \theta f(x|\theta)dG(\theta) \quad (30.2)$$

and f(x) = $\int f(x|\theta)dG(\theta)$ is the marginal p.d.f. of X. Note that the restriction on G that $\int \theta dG(\theta) < \infty$ is necessary to ensure that $R(\delta, G) < \infty$ and that $R(\delta, G)$ depends on δ only through the term $\int \alpha(x)\delta(x)dG(\theta)$ for each X=x. Therefore, the Bayes optimal procedure which minimizes the Bayes risk $R(\delta, G)$ over all possible randomized decision functions $\delta(\cdot)$ on $(0, \infty)$ onto [0,1] is, from the line preceeding (30.2), given by

$$\delta_G(X) = I(\alpha(X) \leq 0) \tag{30.3}$$

This tells us that when X=x is observed the best procedure is to accept the lot as reliable w.p. 1 if $\alpha(x) \leq 0$ otherwise reject it as a bad lot.

We denote the (minimum) Bayes risk attained by δ_G by $R(G) = min_\delta R(\delta, G)$ = $R(\delta_G, G)$. However, this Bayes optimal procedure δ_G is available for use in practice if and only if the prior distribution G is completely known, which is undoubtly a very strong requirement. When G is completely unknown and un-specified, the Bayes optimal procedure δ_G is not available for use. To handle this problem we take the approach popularly known as empirical Bayes (EB) approach to statistical decisions, introduced originally by Robbins (1955, 1964). The EB approach has been extensively applied to various statistical problems over the past three and an half decades, and now a vast literature exists on the subject.

30.1.3   Empirical Bayes approach when the prior distribution is unknown

Consider the situation where lots are coming sequentially and independently, and, at each stage one has to test the reliability of the lot w.r.t. its mean life and decide whether to accept or reject it. Assuming that the present testing problem that we are dealing with is the $(n+1)$th problem in the sequence, we represent the past n such problems and the present problem by the n+1 independent pairs $(X_1, \theta_1), ..., (X_n, \theta_n)$ and $(X_{n+1} = X, \theta_{n+1} = \theta)$, where X_i is the total lifetime of γ randomly chosen items from the ith lot, and given $\theta_i, X_i|\theta_i$ has the conditional density $f(\cdot|\theta_i)$; $\theta_1, \theta_2, ...$ continue to be unobservable, having the same unknown prior distribution G; and $X_1, ..., X_n$ and X are i.i.d. according to the marginal density $f(\cdot) = \int f(\cdot|\theta)dG(\theta)$.

An EB procedure for testing the quality of the present lot based on $X_1, ..., X_n$ and X is obtained first by forming a statistic $\alpha_n(x) = \alpha_n(X_1, ..., X_n; x)$ which estimates appropriately the unknown function $\alpha(x)$ on $(0, \infty)$, and then adopt-ing the (EB) test procedure, following (1.3), $\delta_n(X) = I(\alpha_n(X) \leq 0)$ as against the (unknown) Bayes optimal procedure $\delta_G(X)$

In this note we develop EB test procedures δ_n which are asymptotically optimal in the sense of Robbins (1955, 1964), i.e. for which $\lim_{n\to\infty} R(\delta_n, G) = R(\delta_G, G) = R(G)$. Thus for large n, δ_n could be considered as good as δ_G, the best procedure, but unavailable for use. We also investigate the speed of convergence and show that, for any $\epsilon > 0$, procedures δ_n can be constructed for which $R(\delta_n, G) - R(G) = O(n^{-1/2+\epsilon})$ under certain moment conditions on G. The results are extended to the case where the life time distribution has density belonging to the exponential family, and also to the case where one has to deal with k varieties of lots simultaneously at each stage.

30.2   Development of Empirical Bayes Procedures

Note that $X_1, ..., X_n$ and X are i.i.d. with (marginal) p.d.f.

$$f(x) = \int u(x)c(\theta)\exp(-x/\theta)dG(\theta) ,$$

where, with $a(\gamma) = (\Gamma(\gamma))^{-1}$, $u(x) = a(\gamma)x^{\gamma-1} I(x > 0)$ and $c(\theta) = \theta^{-\gamma}$. Write $f(x) = u(x)p(x)$, where $p(x) = \int c(\theta)\exp(-x/\theta)dG(\theta)$. Throughout this paper, all the functions and integrals are over the interval $(0,\infty)$ unless stated otherwise, and the support of G is in $\Theta = (0,\infty)$. Define $\psi(x) = \int \theta c(\theta)\exp(-x/\theta)dG(\theta)$. Then, from (30.2)

$$\alpha(x) = u(x)(\theta_0 p(x) - \psi(x)) \tag{30.4}$$

To provide an appropriate estimator of the unknown function $\alpha(\cdot)$ based on $X_1, ..., X_n$ we first provide estimators of $p(\cdot)$ and $f(\cdot)$. As in Singh (1974,1977), for $s > 1$ an arbitrary but fixed integer, let K_s be a Borel measurable bounded real valued function vanishing off $(0,1)$ such that $\int y^t K_s(y)dy = 1$ for $t = 0$, and 0 for $t = 1, 2, ..., $s-1. For examples of function K_s, see Singh (1974). Let $h = h_n$ be a sequence of positive numbers such that $h \to 0$ and $nh \to \infty$ as $n \to \infty$. We estimate $p(x)$ by

$$p_n(x) = (nh)^{-1}\sum_{i=1}^{n}[K_s((X_i - x)/h)I(u(X_i) > 0)/u(X_i)] \tag{30.5}$$

and $\psi(x) = \int_x^\infty p(y)dy = E\{I[X > x, u(X) > 0]/u(X)\}$ unbiasedly by

$$\psi_n(x) = n^{-1}\sum_{i=1}^{n}\{I[X_i > x, u(X_i) > 0]/u(X_i)\} \tag{30.6}$$

where 0/0 is treated as 0. Hence, from (30.4), our proposed estimator of the function $\alpha(x)$ is

$$\alpha_n(x) = u(x)(\theta_0 p_n(x) - \psi_n(x)) \tag{30.7}$$

and from (30.3) our proposed EB test procedure for the present (the $(n+1)$th lot) is

$$\delta_n(X) = I(\alpha_n(X) \le 0), \tag{30.8}$$

i.e. we accept the lot as reliable w.p.1 if $\alpha_n(X) \leq 0$, otherwise reject it as a bad lot.

Note that the sequence δ_n is completely nonparametric in the sense that it does not depend on the knowledge or structure of the prior distribution G. Also, by arguments used in the first paragraph of subsection 30.1.2, the Bayes risks of the test procedures δ_n are given by

$$R(\delta_n, G) = b \int \alpha(x) E\delta_n(x) dx + \int L_1(\theta) dG(\theta) .$$

30.3 Asymptotic Optimality of the EB Procedures and Rates of Convergence

In this section, we prove two theorems which provide the main results of the paper concerning the property of our proposed EB test procedures δ_n. The first theorem establishes the asymptotic optimality of δ_n whereas the second one provides the speed of convergence. Proofs of the theorems are given in Appendix. Throughout the remainder of this paper all the convergence are w.r.t. n $\rightarrow \infty$, unless stated otherwise.

Theorem 30.3.1 *For prior distributions G with support in* $\Theta = (0, \infty)$ *for which* $\int \theta dG(\theta) < \infty$

$$R(\delta_n, G) - R(G) = o(1). \tag{30.9}$$

Theorem 30.3.2 *(Rate of convergence to optimality). For every prior distribution G on* $(0, \infty)$ *for which* $\int \theta^{2+\epsilon} dG(\theta) < \infty$ *for some* $\epsilon > 0$, *and* $\int \theta^{-s} dG(\theta) < \infty$ *for some integer* $s > 1$, *the procedures* δ_n *through the choice of* K_s *and* $h = h_n \propto n^{-1/(2s+1)}$), *satisfy*

$$R(\delta_n, G) - R(G) = O(n^{-s/(2s+1)}). \tag{30.10}$$

30.4 Extension of the Results and Concluding Remarks

In this paper we have provided asymptotically optimal empirical Bayes test procedure for accepting a lot as a good lot or rejecting it as a bad lot when the life time of items have negative exponential or gamma (with same θ) density and the distribution of the mean life θ is completely unknown.

We conjecture that the procedure can easily be extended to the case, where the life time distribution has density belonging to the exponential family $f(y|\theta) = u(y)c(\theta)\exp(-y/\theta)$ where $u(\cdot)$ is any arbitrary real valued function vanishing off m$(0,\infty)$, and the parameter space is $\Theta = \{\theta > 0|c^{-1}(\theta) = \int u(y)\exp(-x/\theta)dx < \infty\}$. The test procedure δ_n given in (30.8) (with u there replaced by the one taken here) will continue to be asymptotically optimal as long as $\int \theta dG(\theta) < \infty$. Further, with notations $w(x) = \int_x^\infty (p(y)/u(y))dy$, $v_\epsilon(x) = p(x)sup_{0<z<\epsilon}(u(x+z))^{-1}$ and $l(x,\beta) = \int c(\theta)\theta^{-\beta}\exp(-x/\theta)dG(\theta)$ we believe that:

If $\int_0^\infty u(x)w^{1/2}(x)dx < \infty$, $\int_0^\infty u(x)v_\epsilon^{1/2}(x)dx < \infty$, for some $\epsilon > 0$, and for some integer $s > 1$, $\int_0^\infty u(x)l(x,s)dx < \infty$ then taking this choice of s in the definition of δ_n (through the choice of K_s) and taking $h = h_n \propto n^{-1/(2s+1)}$, we will get

$$R(\delta_n, G) - R(G) = O(n^{-s/(2s+1)})$$

for this general exponential case.

The work here can also be extended to the situation where one has to deal with k different (or same) varieties of lots simultaneously and one has to accept only good ones and reject the bad ones at each stage. Let $\pi_1, ..., \pi_k$ denote the k lots to be tested for their reliability and let the life time of an item from π_j have density $h(y_j|\theta_j) = 1/\theta_j\exp(-y_j/\theta_j)I(y_j > 0)$, where θ_j is the mean life of the items from π_j. The lot π_j is acceptable as reliable if $\theta_j \geq \theta_{oj}$, a specified standard set by the practioner for the jth variety of the lot. The random vector $\boldsymbol{\theta} = (\theta_1, ..., \theta_k)$ have unknown and unspecified prior distribution on $\Theta^k, \Theta = (0,\infty)$. The total risk accross the k testing problems is $\sum_{j=1}^k E[L_0(\theta_j)\delta_j(X_j) + L_1(\theta_j)(1 - \delta_j(X_j))]$, where for j = 1, ..., k and i = 0,1, $L_i(\theta_j) = (i - 1)(\theta_{oj} - \theta_j)I(\theta_j < \theta_{oj}) + i(\theta_j - \theta_{oj})I(\theta_j \geq \theta_{oj})$ and $L_0(\theta_j)$ (and $L_1(\theta_j)$) is the penalty due to incorrectly accepting (rejecting) the lot π_j, and X_j is the total life time of γ_j randomly selected items from π_j. The problem can be handled identically along the lines discussed in this paper by treating each population π_j separately and by developing EB test procedure δ_{jn} by the way discussed in the paper.

References

1. Balakrishnan, N., and Basu, A. P. (eds). (1995). *The Exponential Distribution; Theory, Methods and Applications*, Gordon and Breach, Pennsylvania.

2. Johnson, N., Kotz, S., and Balakrishnan, N. (1994). *Continuous Univariate Distributions*, Vol. 1, second ed., John Wiley & Sons, New York.

3. Johns, M. V. Jr., and Van Ryzin, J. (1971). Convergence rates in empirical Bayes two action problems: I Discrete case, *Annals of Statistics*, **42**, 1521–1539.

4. Robbins, H. (1955). An empirical Bayes approach to statistics, *Proceedings of the 3rd Berkeley Symposium on Mathematical Statistics*, Vol I, 157–164.

5. Robbins, H. (1964). The empirical Bayes approach to statistical decision problems, *Annals of Statistics*, **35**, 1–10.

6. Singh, R. S. (1974). Estimation of derivatives of average of μ - densities and sequence compound estimation in exponential families RM-319, Department of Statistics and Probability, Michigan State University, East Lansing, MI, USA.

7. Singh, R. S. (1976). Empirical Bayes estimation with convergence rates in noncontinuous Lebesgue exponential families, *Annals of Statistics*, **4**, 431–439.

8. Singh, R. S. (1977). Improvement on some known nonparametric uniformly constant estimates of derivatives of a density, *Annals of Statistics*, **7**, 890–902.

9. Singh, R. S. (1995). Empirical Bayes linear loss hypothesis testing in a nonregular exponential family, *Journal of Statistical Planning and Inference*, **43**, 107–120.

Appendix

We now give proofs of Theorems 30.3.1 and 30.3.2 which provide main results of the paper. The symbols $C_0, C_1, \ldots$ appearing below are simply constants.

PROOF OF THEOREM 30.3.1.   From the expressions obtained for R(G) in the first paragraph of subsection 30.1.2 and for $R(\delta_n, G)$ in the last paragraph of Section 30.2, we have $0 \leq R(\delta_n, G) - R(G) = b \int \alpha(x)[E(\delta_n(x)) - \delta_G(x)]dx$. Therefore, from Johns and Van Ryzin (1971), it follows that

$$R(\delta_n, G) - R(G) \leq b \int B_n(x)dx \tag{30.11}$$

where

$$B_n(x) = |\alpha(x)|P[|\alpha_n(x) - \alpha(x)| \geq |\alpha(x)|] \tag{30.12}$$

which is bounded by $|\alpha(x)|^{1-\lambda}E|\alpha_n(x) - \alpha(x)|^{\lambda}$ for $0 \leq \lambda \leq 2$ by Markov inequality. Hence from (30.4) and (30.8),

$$B_n(x) \leq C_0\alpha(x)|^{1-\lambda}u^{\lambda}(x)[E|\psi_n(x) - \psi(x)|^{\lambda} + \theta_0^{\lambda}E|p_n(x) - p(x)|^{\lambda}]. \quad (30.13)$$

Now we show that each term of the r.h.s. of (30.13) converges to zero as n $\rightarrow \infty$ for each $x > 0$. Since ψ_n is an unbiased estimator of ψ, for $0 < \lambda \leq 2$ by Holder inequality, $(E|\psi_n(x) - \psi(x)|^{\lambda})^{2/\lambda} \leq Var(\psi_n(x))$, which is more than $n^{-1}E[I(X_1 \geq x, u(X_1) > 0)/u(X_1)]^2$. But this is bounded above by $n^{-1}\int (p(y)/u(y))dy \leq n^{-1}(\psi(x)/u(x))$.Hence, for $0 < \lambda \leq 2$ we conclude that

$$E|\psi_n(x) - \psi(x)|^{\lambda} \leq (n^{-1}\psi(x)/u(x))^{\lambda/2}. \quad (30.14)$$

Now we examine $E|p_n(x) - p(x)|^{\lambda}$. Since $p(x)$ is continously differentiable on $(0,\infty)$ and support of K_s is in $(0,1)$, it follows from the arguments given in Singh (1977) that

$$|Ep_n(x) - p(x)| \leq C_1 h^s \sup_{z>0}p^{(s)}(x + z) = C_1 h^s p(x, \gamma + s), \quad (30.15)$$

where $p^{(s)}(\cdot)$ is the sth derivative of $p(\cdot)$, and $p(x,\beta) = \int \theta^{-\beta}\exp(-x/\theta)dG(\theta)$; and that

$$Var(p_n(x)) \leq (nh)^{-1}\int K_s^2(t)(p(x + ht)/u(x + ht))dt$$

$$\leq (nh)^{-1}\{(p(x)/u(x))\}\int K_s^2(t)dt. \quad (30.16)$$

Hence, for $0 < \lambda \leq 2$, Holder inequality implies

$$E|p_n(x) - p(x)|^{\lambda} \leq C_2\left(h^{\lambda s}p^{\lambda}(x, \gamma + s) + (nh)^{-\lambda/2}(p(x)/u(x))^{\lambda/2}\right) \quad (30.17)$$

Hence, since h $\rightarrow \infty$ and nh $\rightarrow \infty$ as n $\rightarrow \infty$, from (30.11), (30.14) and (30.17), $B_n(x) \rightarrow 0$ for $x > 0$.

Thus our proof will be complete if we can show that $\int B_n(x)dxm \rightarrow 0$, but since $|B_n(x)| \leq |\alpha(x)|$ by (30.12), this immediately follows from Lebesgue dominated convergence theorem if we can show that $\int |\alpha(x)|dx < \infty$. Notice that from (30.4) $\int |\alpha(x)|dx \leq \theta_0 + \int u(x)\psi(x)dx$. But $\int u(x)\psi(x)dx = \int u(x)\int c(\theta)\int_x^{\infty}\exp(-t/\theta)dtdG(\theta)dx = \int \theta dG(\theta) < \infty$. ∎

PROOF OF THEOREM 30.3.2. From (30.11), (30.13), (30.14) and (30.17) with $h \propto n^{-1/(2s+1)}$, $0 < \lambda \leq 1$,

$$n^{\lambda s/(2s+1)}(R(\delta_n, G) - R(G))$$

$$\leq C_3\int |\alpha(x)|^{1-\lambda}[(u(x)\psi(x))^{\lambda/2}$$

$$+ (u(x)p(x))^{\lambda/2} + (u(x)p(x, \gamma + s))^{\lambda}]dx$$

$$\leq C_3(\int |\alpha(x)|dx)^{1-\lambda}[(\int (u(x)\psi(x))^{1/2}dx)^{\lambda}$$

$$+ (\int (u(x)p(x))^{1/2}dx)^{\lambda} + (\int u(x)p(x, \gamma + s)dx)) \quad (30.18)$$

by application of Holder inequality. In the proof of Theorem 30.3.1 we have seen that $\int |\alpha(x)| dx < \infty$. Now we show that the other three integrals on the r.h.s. of (30.18) also exist.

We write the first integral as, for $\epsilon > 0$,

$$\int (u(x)\psi(x))^{1/2} dx$$

$$= \int_0^1 (u(x)\psi(x))^{1/2} dx + \int_1^\infty x^{-(1+\epsilon)/2} (x^{1+\epsilon} u(x)\psi(x))^{1/2} dx$$

$$\leq \left(\int u(x)\psi(x) dx\right)^{1/2} + \left(\int_1^\infty x^{-(1+\epsilon)} dx\right)^{1/2} \left(\int_1^\infty x^{1+\epsilon}\psi(x) dx\right)^{1/2}$$

by Holder inequality. The r.h.s. of the last inequality is finite for every $\epsilon > 0$ whenever $\int \theta^{2+\epsilon} dG(\theta) < \infty$. Similarly, it can be shown that the second integral $\int (u(x)p(x))^{1/2} dx < \infty$ whenever for some $\epsilon > 0$, $\int |\theta|^{1+\epsilon} dG(\theta) < \infty$ and the third integral $\int u(x)p(x, \gamma + s) dx = \int \theta^{-s} dG(\theta) < \infty$ by our assumption on G. The proof is complete since $0 < \lambda \leq 1$ is arbitrary. ∎

31

On a Test of Independence in a Multivariate Exponential Distribution

M. Samanta and A. Thavaneswaran

The University of Manitoba, Winnipeg, Canada

Abstract: In this article the problem of testing independence in a multivariate exponential distribution with identical marginals is considered. Following Bhattacharyya and Johnson (1973) the null hypothesis of independence is transformed into a hypothesis concerning the equality of scale parameters of several exponential distributions. The conditional test for the transformed hypothesis proposed here is the likelihood ratio test. The powers of this test are estimated for selected values of the parameters, using Monte Carlo simulation and noncentral chi-square approximation. The powers of the overall test are estimated using a simple formula involving the power function of the conditional test. Application to reliability problems is also discussed in some detail.

Keywords and phrases: Conditional test of independence, likelihood ratio test, Marshall-Olkin multivariate exponential distribution

31.1   Introduction

We consider a special form of the Marshall-Olkin k-dimensional exponential distribution with identical marginals. Suppose the random variables $X_1, X_2, \ldots, X_k$ represent the component lifetimes of a k-component parallel system where the components are similar in nature. Following Proschan and Sullo (1976), we assume that $X = (X_1, X_2, \ldots, X_k)$ has the Marshall-Olkin k-dimensional exponential distribution with identical marginals given by

$$\bar{F}(x_1, x_2, \ldots, x_k) = P[X_1 > x_1, X_2 > x_2, \ldots, X_k > x_k] = \exp[-\lambda[\sum_{i=1}^{k} x_i]$$
$$- \beta \, ax((x_1, x_2, \cdots, x_k)] \tag{31.1}$$

$0 \leq x_i < \infty$, $i = 1, 2, \ldots, k$, $0 < \lambda < \infty$, $0 \leq \beta < \infty$.

The fatal shock model leading to the above distribution is a special version of that described in Marshall and Olkin (1967) and assumes the existence of $(k+1)$ mutually independent Poisson processes $\{Z_i(t); \ t \geq 0\}$, $i = 1, 2, \ldots, (k + 1)$ which govern the occurrence of events (shocks). Each of the processes $\{Z_i(t)\}$; $i = 1, 2, \ldots, k$ has the same intensity λ and the process $\{Z_{k+i}(t)\}$ has intensity β. An event (shock) in the process $Z_{k+1}(t)$ is selectively fatal to component $i, i = 1, 2, \ldots, k$, while an event (shock) in the process $Z_{k+1}(t)$ is simultaneously fatal to all k components. Hence, if $V_1, V_2, \ldots, V_{k+1}$ denote the times to the first events in the processes $Z_1(t), Z_2(t), \ldots, Z_{k+1}(t)$, respectively, then $X_i = \min(V_i, V_{k+1})$, $i = 1, 2, \ldots, k$ where $V_1, V_2, \ldots, V_{k+1}$ are independent exponential random variables with $E(V_i) = \lambda^{-1}, i = 1, 2, \ldots, k$ and $E(V_{k+1}) = \beta^{-1}$, [see Theorem 2.1 in Proschan and Sullo (1976)]. We consider the following context to motivate the above model. Suppose a jet plane that has eight identical engines can fly with one or more engines working properly. Any particular engine can fail due to the occurence of a shock that is fatal to the engine or all eight engines can fail due to a shock resulting from a shortage of fuel. If the shocks are governed by mutually independent Poisson processes as described above then we get the model given in (31.1).

Suppose $X_l = (X_{1l}, X_{2l}, \cdots, X_{kl}), l = 1, 2, \ldots, n$ are independent and identically distributed k-dimensional random variables having the probability distribution (31.1). Since $\beta = 0$ implies independence in the model (31.1) and vice versa, we consider the problem of testing the null hypothesis $H_o(\beta = 0)$ against the alternative hypothesis $H_a(\beta > 0)$. A uniformly most powerful test of H_o against H_a in the model (31.1), with $k = 2$, has been obtained by Bhattacharyya and Johnson (1973) who have also given an easily computable expression for the exact power function of their test. The same test has been derived by Bemis, Bain and Higgins (1972), using a simple heuristic reasoning. In this paper we consider the problem of testing independence in the model (31.1) with $k \geq 3$. Following Bhattacharyya and Johnson (1973) we first obtain a conditional test, independent of the nuisance parameter λ, for testing H_0 against H_a. The hypothesis H_0 is transformed into a hypothesis concerning the equality of scale parameters of k exponential distributions and the conditional test proposed in this paper is the likelihood ratio test due to Sukhatme (1936). The power function of the conditional test is appropriately combined to provide a simple expression for the power function of the overall test. The powers of the conditional test are estimated for selected values of the parameters by using Monte Carlo simulation for small samples and noncentral chi-square approximations for large samples. Finally, the powers of the overall test are estimated by combining the estimated powers of the conditional test in an appropriate manner.

31.2 Likelihood and Sufficiency

It is known [see Marshall and Olkin (1967)] that the k-dimensional exponential distribution in (31.1) is not absolutely continuous with respect to k-dimensional Lebesgue measure. However, the distribution function is absolutely continuous with respect to a suitably chosen dominating measure. Let E_k be the k-dimensional Euclidean space and $E_k^+ = \{(x_1, x_2, \ldots, x_k) \in E_k : x_i \geq 0, i = 1, 2, \ldots, k\}$. Let μ_r denote r-dimensional Lebesgue measure and B_k^+ be the Borel field in E_k^+. We define a measure μ on (E_k^+, B_k^+) by $\mu(B) = \mu_k(B) + \sum \mu_i[B \cap \{(x_1, x_2, \ldots, x_k) \in E_k^+ : x_{j_i} = x_{j_{i+1}} = \ldots = x_{j_k} = \max(x_1, x_2, \ldots, x_k)\}]$ for any set $B \in B_k^+$, where the summation is over all $(j_i, j_{i+1}, \ldots, j_k) \subseteq \{1, 2, \ldots, k\}, i = 1, 2, \ldots, k - 1$. Then, the distribution function in (31.1) is absolutely continuous with respect to μ with a probability density function (pdf) $f(x_1, x_2, \ldots, x_k)$ given by

$$f(x_1, x_2, \ldots, x_{i-1}, x_i, x_{i+1}, \ldots, x_k)$$

$$= \begin{bmatrix} \beta\lambda^{i-1} \exp[-\lambda \sum_{j=1}^{i-1} x_j - \{\beta + (k-i+1)\lambda\}x_i], \\ \qquad \text{if } x_1 < x_2 < \ldots < x_{i-1} < x_i = x_{i+1} = \ldots = x_k, i = 1, 2, \ldots, k-1, \\ (\beta + \lambda)\lambda^{k-1} \exp[-\lambda \sum_{i=1}^{k-1} x_i - (\beta + \lambda)x_k], \\ \qquad \text{if } x_1 < x_2 < \ldots < x_k. \end{bmatrix}$$

$$(31.2)$$

The above expression for the pdf can be easily verified by using the previous representation of the random variable $X = (X_1, X_2, \ldots, X_k)$ in terms of the random variables $V_1, V_2, \ldots, V_{k+1}$. Let $(j_1, j_2, \ldots, j_k)$ be a permutation of the integers $(1, 2, \ldots, k)$ and for $i = 1, 2, \ldots, k - 1$ let M_i be the number of observations X_l $l = 1, 2, \ldots, n$ such that $X_{j_1,l} < X_{j_2,l} \ldots < X_{j_{i-1},l} < X_{j_i,l} = X_{j_{i+1},l} = \ldots = X_{j_k,l}$ for all permutations $(j_1, j_2, \ldots, j_k)$ of $(1, 2, \ldots, k)$. If we define the random variables W_1 and W_2 by

$$W_1 = \sum_{l=1}^{n} \sum_{i=1}^{k} X_{il} \text{ and } W_2 = \sum_{l=1}^{n} \max (X_{1l}, X_{2l}, \ldots, X_{kl})$$

then the joint pdf of the sample is obtained as

$$f(x_l, l = 1, 2, \ldots, n) = C \ \exp \ [-\lambda w_1 - \beta w_2]$$

where

$$C = \beta^{m*}(\beta + \lambda)^{n-m*}\lambda^{(k-1)n-n*} \qquad (31.3)$$

where $m* = \sum_{i=1}^{k-1} m_i$, $n* = \sum_{i=1}^{k-1}(k-i)m_i$.

Using the factorization theorem, it follows that $(M_1, M_2, \ldots, M_{k-1}, W_1, W_2)$ is a set of minimal sufficient statistics for (λ, β) although this set is not complete.

Let $Z_{(1l)} \leq Z_{(2l)} \leq \ldots \leq Z_{(kl)}$ be the ordered observations $X_{1l}, X_{2l}, \ldots, X_{kl}$ for $l = 1, 2, \ldots, n$ and let the statistics $T_1, T_2, \ldots, T_n$ be defined as

$$T_1 = \sum_{l=1}^{n} Z_{(1l)}, \ T_j = \sum_{l=1}^{n} \{Z_{(jl)} - Z_{(j-1,l)}\}, j = 2, 3, \ldots, k$$

From the relation

$$\lambda w_1 + \beta w_2 = \sum_{j=1}^{k} \{\beta + (k - j + 1)\} t_j \tag{31.4}$$

It also follows that $(M_1, M_2, \ldots, M_{k-1}, T_1, T_2, \ldots, T_k)$ is a set of sufficient statistics although this set is not a minimal sufficient set. We prove the distributional properties of this latter set of suffcient statistics in the following lemmas. We write "Y is $G(n, \theta)$" to mean that the random variable Y has the gamma distribution with the corresponding cumulative distribution function

$$G(y; n, \theta) = \int_0^y \frac{\theta^n}{\Gamma(n)} \exp(-\theta x) x^{n-1} dx.$$

By convention we assume that $G(y; 0, \theta) = 1$.

Lemma 31.2.1 T_1 is $G(n, \beta + k\lambda)$ and is independent of $(M_1, M_2, \ldots, M_{k-1}, T_2, T_3, \ldots, T_k)$.

PROOF. The first part of the lemma follows from the definition of T_1 and (31.1). To prove the second part of the lemma we note that the joint distribution of $(M_1, M_2, \ldots, M_{k-1}, T_1, T_2, \ldots, T_k)$ can be computed from the joint distribution of the sample given in (31.3) and the relation (31.4). This joint distribution can be factored into the distribution of T_1 and the joint distribution of $(M_1, M_2, \ldots, M_{k-1}, T_2, T_3, \ldots, T_k)$. ∎

Lemma 31.2.2 $M_1, M_2, \ldots, M_{k-1}, T_2, T_3, \ldots, T_k$ jointly have the mixed distribution

$$P[M_i = m_i; T_j \leq t_j; \ i = 1, 2, \ldots, k - 1, \ j = 2, 3, \ldots, k]$$

$$= \frac{n!}{\prod_{i=1}^{k-1} m_i! (n - \sum_{i=1}^{k-1} m_i)!} [\prod_{i=1}^{k-1} p_i^{m_i}] p_k^{n-m*}$$

$$[\prod_{j=2}^{k} G(t_j : n - \sum_{s=1}^{j-1} m_s, \ \beta + (k - j + 1)\lambda)] \tag{31.5}$$

for all $m_1, m_2, \ldots, m_{k-1}, t_2, t_3, \ldots, t_k$ satisfying $0 \leq \sum_{i=1}^{k-1} m_i \leq n$, $0 < t_j < \infty, j = 2, 3, \ldots, k$ where

$$p_i = \frac{k! \beta \lambda^{i-1}}{(k - i + 1)! \prod_{j=k-i+1}^{k} (\beta + j\lambda)}, i = 1, 2, \ldots, k - 1$$

$$p_k = \frac{k!\lambda^{k-1}}{\Pi_{j=2}^k(\beta + j\lambda)}.$$

PROOF. We have

$$P[M_i = m_i; \ T_j \le t_j; \ i = 1, 2, \ldots, k-1, j = 2, 3, \ldots, k] = P[M_i = m_i;$$

$$i = 1, 2, \ldots, k-1] \ P[T_j \le t_j; \ j = 2, 3, \ldots, k | M_i = m_i; \ i = 1, 2, \ldots, k-1].$$

We first assume that $\sum_{i=1}^{k-1} m_i \le n - 1$. In this case the statistic T_j can be written where Y_{jr}'s are independent random variables and each Y_{jr} is $G(1, \beta + (k - j + 1)\lambda)$. When $\sum_{i=1}^{k-1} m_i = n$ some T_j's may assume the value zero with probability one and the remaining T_j's are as defined before. ∎

Lemma 31.2.3 $M = \sum_{i=1}^{k-1} M_i$ *has the binomial distribution*

$$P[M = m] = \binom{n}{m} p^m (1 - p)^{n-m}, m = 0, 1, \ldots, n \tag{31.6}$$

where

$$p = 1 - \frac{k!\lambda^{k-1}}{\Pi_{j=2}^k(\beta + j\lambda)}. \tag{31.7}$$

PROOF. The proof follows from Lemma 31.2.2. ∎

Lemma 31.2.4 *Conditionally, given $M = 0$, the variables $T_2, T_3, \ldots, T_k$ are independently distributed and T_j is $G(n, \beta + (k - j + 1)\lambda), j = 2, 3, \ldots, k$.*

PROOF. The proof follows from Lemma 31.2.2 and the fact that $M = 0$ if and only if $M_i = 0, i = 1, 2, \ldots, k-1$. ∎

31.3 A Test of Independence

From Lemma 31.2.3 we have $P[M > 0 | \beta = 0] = 0$. Hence, using a parallel version of Lemma 3.1 in Bhattacharyya and Johnson (1973) we formulate a test of $H_o(\beta = 0)$ against $H_a(\beta > 0)$. We reject H_o if $M > 0$. When $M = 0$ we use the following conditional test.

The joint conditional pdf of $T_1, T_2, \ldots, T_k$,, given $M = 0$, is

$$g(t_1, t_2, \ldots, t_k) = \frac{\Pi_{j=1}^k\{\beta + (k - j + 1)\lambda\}^n}{\{\Gamma(n)\}^k} \{\Pi_{j=1}^k t_j^{n-1}\}$$

$$\exp[-\sum_{j=1}^k\{\beta + (k - j + 1)\lambda\}t_j], \tag{31.8}$$

$$0 \le t_i < \infty, i = 1, 2, \ldots, k, \ 0 < \lambda < \infty, 0 \le \beta < \infty.$$

We introduce the random variables U_j and the parameters $\delta_j, j = 1, 2, \ldots, k$ defined by

$$U_j = \frac{(k - j + 1)T_j}{k!} \text{ and } \delta_j = \frac{k!\{\beta + (k - j + 1)\lambda\}}{(k - j + 1)}.$$

Then, the joint conditional pdf of $U_1, U_2, \ldots, U_k$, given $M = 0$, is

$$g(u_1, u_2, \ldots, u_k) = \frac{\Pi_{j=1}^k \delta_j^n}{\{\Gamma(n)\}^k} \cdot \{\Pi_{j=1}^k u_j^{n-1}\}[\exp \ - \sum_{j=1}^k \delta_j u_j] \qquad (31.9)$$

$$0 \le uj < \infty, j = 1, 2, \ldots, k, 0 < \delta_1 \le \delta_2 \le \ldots \le \delta_k < \infty$$

We note that $\beta = 0$ if and only if $\delta_1 = \delta_2 = \ldots = \delta_k$. Hence, we wish to test $H_o(\delta_1 = \delta_2 = \ldots = \delta_k)$ against the alternative $H_a(\delta_i \ne \delta_j, i \ne j)$. The likelihood ratio criterion V^* first derived by Sukhatme (1936) for testing H_o against H_a in the above setup is defined by

$$V^* = \frac{k^{nk}\{\Pi_{j=1}^k U_j\}^n}{\{\sum_{j=1}^k U_j\}^{nk}} = \frac{(k!k^k)^n \{\Pi_{j=1}^k T_j\}^n}{\{\sum_{j=1}^k (k - j + 1)T_j\}^{nk}} \qquad (31.10)$$

It can be easily verified that the restriction in the parameter space does not play an important role and one obtains the same test statistic without introducing the transformation. Although the exact null distribution of V^* is not known, its chi-square approximation [see Kendall and Stuart (1973, p. 246)] can be used for all practical purposes. When H_o is true $2lnV^*\{1+\frac{k+1}{6kn}\}^{-1}$ has approximately the central chi-square distribution with $(k - 1)$ degrees of freedom. When H_a is true, $-2lnV^*$ has approximately the noncentral chi-square distribution with $(k-1)$ degrees of freedom and noncentrality parameter $-2ln\theta$ where

$$\theta = (k!k^k)^n[\Pi_{j=1}^k\{\beta + (k - j + 1)\lambda\}^{-n}]$$

$$\times \left[\sum_{j=1}^k (k - j + 1)\{\beta + (k - j + 1)\lambda\}^{-1}\right]^{-nk} \qquad (31.11)$$

Thus, when $M = 0$, a conditional test of H_o against H_a can be performed using a table of the central chi-square distribution and the power of this conditional test at various alternatives can be computed using a table of the noncentral chi-square distribution. If $\pi(\lambda, \beta)$ is the power of this conditional test then the power of the overall test is

$$1 - \{\frac{k!\lambda^{k-1}}{\Pi_{j=2}^k(\beta + j\lambda)}\}^n\{1 - \pi(\lambda, \beta)\}. \qquad (31.12)$$

31.4   Power Results

An obvious test for testing the null hypothesis $H_o : (\beta = 0)$ against the alternative hypothesis $H_a : (\beta > 0)$ is given by rejecting H_o whenever $M > 0$, where M is as defined in Lemma 31.2.3. For this test procedure the probability of type I error is zero and the power is

$$P(M > 0 | \beta) = 1 - (1 - p)^n \tag{31.13}$$

where p is as given in (31.7).

We now compare the power of the test described in Section 3 with that of the test based on M. The critical value at level .05, denoted by $V_{.05}^*$, of the conditional test based on the likelihood ratio criterion is estimated by drawing 5000 random samples each of size $n = 10$ from the four-variate exponential distribution with $\lambda = 2$ and $\beta = 0$. Estimates of powers of this test are based on another 5000 random samples each of size $n = 10$ drawn from a four-variate exponential distribution with $\lambda = 2$ and $\beta = 0.05, 0.10, 0.15, 0.20$ and 0.25. This procedure is repeated with samples of sizes $n = 15$ and $n = 20$. For a sample of size n greater than or equal to 30 the power of the likelihood ratio test is approximated by the noncentral chi-square distribution. Estimates of overall powers are obtained by using formula (31.12).

Table 31.1: Estimated powers for tests of independence using simulation[1]

	$\beta = .05$		$\beta = .10$		$\beta = .15$		$\beta = .20$		$\beta = .25$	
n	$1 - p_*^n$	Pe	$1 - p_*^n$	Pe	$1 - p_*^n$	Pe	$1 - p_*^n$	Pe	$1 - p_*^n$	Pe
10	.236	.236	.415	.415	.551	.551	.654	.655	.733	.734
15	.333	.333	.553	.553	.699	.699	.797	.797	.862	.863
20	.417	.417	.658	.658	.798	.798	.881	.881	.929	.929

[1]Estimates of $V_{.05}^*$ for $n = 10$, 15 and 20 are 0.000053012, 0.000021136, 0.000023595 respectively. $p_* = 1 - p$; Pe denotes estimated power.

Table 31.2: Approximate powers of tests of independence using noncentral chi-square approximations[2]

	$\beta = .05$		$\beta = .10$		$\beta = .15$		$\beta = .20$		$\beta = .25$	
n	$1 - p_*^n$	Pa	$1 - p_*^n$	Pa	$1 - p_*^n$	Pa	$1 - p_*^n$	Pa	$1 - p_*^n$	Pa
30	.560	.576	.799	.810	.910	.914	.959	.961	.981	.982
40	.665	.677	.881	.889	.959	.962	.986	.987	.995	.995
50	.746	.753	.931	.935	.982	.983	.995	.995	.999	.999
60	.806	.811	.959	.962	.992	.992				
100	.935	.936	.995	.996	1.000	1.000				

[2]Pa denotes approximated power; $p_* = 1 - p$.

The simulation was carried out in SAS version 6 [SAS Institute (1990)] and run on AMDAHL 580/5870 dual processor CPU as its mainframe computer. RANEXP command in SAS was used to generate the exponential samples, multiplying with an appropriate constant as desired. The noncentral chi-square values were evaluated by invoking PROBCHI command in SAS. Tables 1 and 2 compare the estimated power of our test with that of the test based on M. We note from these tables that there is little improvement in the powers of our test compared to that of the simple test. A similar observation was made in Bemis et al (1972) in the case $k = 2$.

31.5 Summary

We discussed the issue of testing independence in a multivariate exponential distribution. Powers of a conditional test were calculated using Monte Carlo simulation and a non-central chi-square approximation. The powers of the overall test were estimated using a formula involving the power function of the conditional test.

Acknowledgements. The research was supported by a grant from the Natural Sciences and Engineering Research Council of Canada.

References

1. Bemis, B. M., Bain, L. J. and Higgins, J. J. (1972). Estimation and hypothesis testing for the parameters of a bivariate exponential distribution, *Journal of the American Statistical Association*, **67**, 927–929.

2. Bhattacharyya, G. K. and Johnson, R. A. (1973). On a test of independence in a bivariate exponential distribution, *Journal of the American Statistical Association*, **68**, 704–706.

3. Kendall, M. G. and Stuart, A. (1973). *The Advanced Theory of Statistics*, Volume 2, Hafner, New York, Third Edition.

4. Marshall, A. W. and Olkin, I. (1967). A multivariate exponential distribution, *Journal of the American Statistical Association*, **62**, 30–44.

5. Proschan, Frank and Sullo, Pasquale (1976). Estimating the parameters of a multivariate exponential distribution, *Journal of the American Statistical Association*, **71**, 465–472.

6. SAS Institute Inc. (1990). SAS Language: Reference, Version 6, First Edition Cary NC: Author.

7. Sukhatme, P. V. (1936). On the analysis of k samples from exponential populations with special reference to the problem of random intervals, *Statistical Research Memoirs*, **1**, 94–112.

PART V
Statistical Methods for Quality Improvement

Random Walk Approximation of Confidence Intervals

D. J. Murdoch

Queen's University, Kingston, Ontario, Canada

Abstract: Confidence intervals and confidence regions are commonly used in process improvement studies to indicate measurement uncertainty. When the model of the system is nonlinear, confidence regions based on sums of squares are often the most accurate, but their calculation is a computationally intensive task. With up to 2 or 3 parameters, an effective method is to evaluate the sums of squares over a dense grid. In higher dimensions profile-based methods [Bates and Watts (1988)] are effective when it is practical to parametrize the model in terms of the quantities of interest, but it is difficult to construct confidence intervals for general functions of the parameters. In this paper we develop an algorithm for approximation of confidence intervals which is computationally efficient in models with up to about 10 parameters. The algorithm is based on a variation on Gibbs sampling [Gelfand and Smith (1990)] of a uniform distribution on a confidence region in the full parameter space and uses extrapolated QQ plots to adjust the borders of the resulting regions.

Keywords and phrases: Gibbs sampling, nonlinear models, QQ plots, spline smoothing

32.1 Introduction

Industrial process improvement depends upon measurement of the characteristics of a system, and this often involves building nonlinear models of the underlying processes. For example, in chemical engineering, a nonlinear compartmental model may be used to model the contents of a reactor vessel. However, estimation of the parameters of the model is not enough; it is essential to know the precision of the estimates. While precision is often expressed as estimates plus or minus standard errors, in nonlinear models this is frequently inadequate: measurement errors may be substantially larger in one direction than the other, and joint measurement errors of sets of parameters may not be well approximated by the boxes formed by intersecting univariate intervals in each coordinate, even when an estimated correlation is taken into account [Bates and Watts (1988)].

A better approach with nonlinear regression models is to consider sets of parameters that give predictions not too distant from the observed data. To be specific, suppose we observe N observations $\mathbf{Y} = (Y_1, \ldots, Y_N)$ from a P-parameter nonlinear regression model

$$\mathbf{Y} = \eta(\boldsymbol{\theta}) + \boldsymbol{\epsilon}$$

where $\boldsymbol{\theta} = (\theta_1, \ldots, \theta_P)$, $\eta(\cdot)$ is some function of the parameters and other variables in the system that expresses the predictions of the nonlinear model, and $\boldsymbol{\epsilon} = (\epsilon_1, \ldots, \epsilon_N)$ is a vector of uncorrelated $N(0, \sigma^2)$ errors. We wish to construct a confidence interval for some scalar function $g(\boldsymbol{\theta})$. A common example is a confidence interval for one of the parameters, i.e. $g(\boldsymbol{\theta}) = \theta_i$; another example might be a confidence interval for a predicted response under specified conditions.

We can construct the confidence region for $g(\boldsymbol{\theta})$ by considering all possible values of $\boldsymbol{\theta}$ such that the residual sum of squares in the model does not become too large, and putting all values of $g(\boldsymbol{\theta})$ on this set together to form a "sums of squares" confidence region. We calculate

$$\mathcal{T} = \left\{ \boldsymbol{\theta} : \frac{[S(\boldsymbol{\theta}; \mathbf{Y}) - S(\hat{\boldsymbol{\theta}}; \mathbf{Y})]}{S(\hat{\boldsymbol{\theta}}; \mathbf{Y})/(N - P)} \leq F^*_{1, N-P, \alpha} \right\} \qquad (32.1)$$

where $S(\boldsymbol{\theta}) = \sum(Y_i - \eta(\boldsymbol{\theta})_i)^2$ is the residual sum of squares at $\boldsymbol{\theta}$, $\hat{\boldsymbol{\theta}}$ is the least squares estimate of $\boldsymbol{\theta}$, and $F^*_{1, N-P, \alpha}$ is the upper α point of the F distribution on 1 and $N - P$ degrees of freedom (Hamilton, 1986; Bates and Watts, 1988). The $1 - \alpha$ confidence interval for $g(\boldsymbol{\theta})$ is $g(\mathcal{T}) = \{g(\boldsymbol{\theta}) : \boldsymbol{\theta} \in \mathcal{T}\}$. If we replaced $F^*_{1, N-P, \alpha}$ by $P F^*_{P, N-P, \alpha}$ in (32.1), then $\mathcal{T}$ would give a joint $1 - \alpha$ confidence region for $\boldsymbol{\theta}$.

In the special case of a linear model, these calculations give the standard linear model confidence regions, and they have exact coverage probabilities. In general in the nonlinear regression model the coverage probabilities are approximate.

Unfortunately, computation of $\mathcal{T}$ is often difficult, especially when P is large. There is typically no algebraic simplification possible. Because of this, linearizations of $\eta(\boldsymbol{\theta})$ and quadratic approximations to $S(\boldsymbol{\theta})$ at $\hat{\boldsymbol{\theta}}$ are often used. While there are cases where these approximations give better coverage than the sums of squares regions, in most cases the latter are more accurate. They also have the advantage of parametrization invariance and would be preferable in general, were they not so difficult to compute.

One approach to calculating $\mathcal{T}$ that is effective for $P \leq 2$ is a simple grid search. A fine grid of $\boldsymbol{\theta}$ values is evaluated, and standard contouring algorithms are used to approximate a joint region [Bates and Watts (1988)]. This rapidly becomes infeasible as P increases, since the number of points in the grid needs to increase exponentially with P.

For cases where it is practical to reparametrize the model so that $g(\boldsymbol{\theta}) = \theta_i$, an efficient way to find the confidence interval for $g(\boldsymbol{\theta})$ in higher dimensions is *profiling* [Bates and Watts (1988)]. In this technique, θ_i is held fixed while $S(\boldsymbol{\theta})$ is optimized over the other parameters. This allows a one dimensional profile of the objective function to be calculated, which is sufficient for calculation of the confidence interval.

This paper deals with the case where $g(\boldsymbol{\theta})$ is a smooth function, and where reparametrization of the model for profiling is impractical. This may be because it is analytically intractable, or because confidence intervals for a large number of different functions, e.g. simultaneous predictions under a number of different conditions, are required. Section 32.2 describes a technique for efficiently obtaining a sample of points uniformly distributed in $\mathcal{T}$. Section 32.3 describes how to use these samples to approximate confidence intervals. Examples of the use of these methods are given in Section 32.4, and Section 32.5 gives some general conclusions and recommendations.

32.2 Covering the Region by Gibbs Sampling

As described in Section 32.3 below, our calculation of confidence regions depends on having a sample of points uniformly distributed within the region $\mathcal{T}$. A simple approach to generating this sample would be an acceptance-rejection algorithm based on sampling $\boldsymbol{\theta}$ values uniformly in a large P dimensional hyperrectangle, and only accepting those points which satisfy the conditions defining $\mathcal{T}$. This approach works reasonably well in low dimensions, but becomes extremely inefficient for larger P, as the proportion of the volume of the hyperrectangle that satisfies the conditions is typically very small. For example, if the true region is contained in half of the range of each coordinate, then its volume is at most 2^{-P} of the volume of the hyperrectangle.

Our approach to addressing this problem is based on the Gibbs sampling algorithm [Gelfand and Smith (1990)]. Gibbs sampling is used for generating samples from multidimensional distributions when the joint distribution is difficult to simulate, but one dimensional conditional distributions are not. Consider a three dimensional random vector (X, Y, Z). Given a starting observation (x_0, y_0, z_0), the coordinates are successively updated by sampling from the one dimensional conditional distributions. That is, x_1 is sampled from the distribution conditional on $Y = y_0, Z = z_0$, then y_1 is sampled from the distribution conditional on $X = x_1, Z = z_0$, and finally z_1 is sampled conditional on $X = x_1, Y = y_1$. This process is repeated to generate a Markov chain of samples whose distribution converges under mild regularity conditions to the desired multivariate distribution of (X, Y, Z).

In the problem at hand, we desire to sample from the uniform distribution on the set (32.1) that we are trying to approximate. We do not have an effective way to calculate the boundaries of this set, so we cannot sample directly. Even the one dimensional conditional distributions are not easily found, but it is not difficult to use the acceptance-rejection algorithm to sample from them, since they are all uniform distributions on an interval or a collection of disjoint intervals. As long as we can bound the confidence region in a hyperrectangle, this is straightforward. In fact, we needn't sample the coordinates one at a time, we may sample along a line in a randomly chosen direction, and this has certain technical advantages [Murdoch (1996)].

Thus, a simple algorithm is as follows:

1. Choose a starting point $\boldsymbol{\theta}_0$ within the interior of the set. For confidence regions, the parameter estimate is a good choice. Update from $\boldsymbol{\theta}_i$ to $\boldsymbol{\theta}_{i+1}$ using the following steps.

2. Choose a direction $\mathbf{d}$ at random. The uniform distribution on the unit sphere or some other distribution may be used.

3. Choose a point $\boldsymbol{\theta}^*$ at random on the intersection of the line $\boldsymbol{\theta}_i + r\mathbf{d}$ with the hyperrectangle.

4. Repeat 3 until $\boldsymbol{\theta}^*$ is within the set.

5. Use the final $\boldsymbol{\theta}^*$ value as $\boldsymbol{\theta}_{i+1}$, and repeat from step 2 until a sufficiently large sample is generated.

A simple modification to this algorithm gives a considerable increase in speed. If a point is chosen that is not inside the set, then future trial points need not be generated at a greater distance from $\boldsymbol{\theta}_0$ in that direction. Thus, if the intersection of the set with our search line is very small, we quickly home in on it and don't waste a lot of trials on points outside of it.

Convergence of this algorithm is said to be attained when the set of points S in the sample path of the Markov process is an adequate approximation to a random sample from a uniform distribution on $\mathcal{T}$. Judging convergence is difficult, as it is with standard Gibbs sampling. However, the methods developed in that context, e.g. Raftery and Lewis (1992), work equally well with this algorithm.

32.3   Calculation of Interval Endpoints

Using the Gibbs-like algorithm described in the previous section, we obtain a set $\mathcal{S}$ of n points which are approximately uniformly distributed within a region $\mathcal{T}$. In this section we describe how to use this set to calculate one dimensional confidence intervals. A simple approach is to take the observed minimum and maximum values of $g(\mathcal{S})$ (i.e. $g(\boldsymbol{\theta})$ for $\boldsymbol{\theta} \in \mathcal{S}$) as the limits of the confidence interval. However, this raises the question of how close these come to the true endpoints of the interval.

To address this question we note that the distribution of one dimensional linear projections of a uniform distribution on the interior of a P-dimensional ellipsoid is Beta $[(P+1)/2, (P+1)/2, L, U]$, i.e. a location-scale shifted version of a Beta distribution of the first kind, with shape parameters both equal to $(P+1)/2$ and support on $L < x < U$ [Murdoch (1996)].

This result is directly applicable to our problem in the special case where the model is linear and the function $g(\boldsymbol{\theta})$ is also linear. In this case the points $g(\mathcal{S})$ will have a Beta$[(P+1)/2, (P+1)/2, L, U]$ distribution, where L and U will be the lower and upper limits of the confidence interval.

We immediately see a limitation of the simple approach. When P is large, the Beta$[(P+1)/2, (P+1)/2, L, U]$ distribution is concentrated near $(L+U)/2$, and rarely approaches L or U (Figure 32.1). The sample size in $\mathcal{S}$ needed to use the range of $g(\mathcal{S})$ to approximate the interval $[L, U]$ increases exponentially with P, and quickly becomes impractically large. For example, Table 32.1 gives some approximate sample sizes needed to achieve 2 digit accuracy. One would not often attempt to achieve this accuracy beyond 5 or 6 parameters.

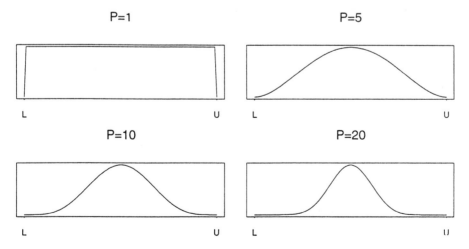

Figure 32.1: The Beta densities for $g(\mathcal{S})$ when the model and $g(\cdot)$ are both linear, for $P = 1, 5, 10$ or 20 parameters

Table 32.1: Approximate sample sizes needed from a Beta$[(P + 1)/2, (P + 1)/2, L, U]$ distribution so that the mode of the distribution of the sample minimum is less than $L + 0.01(U - L)$

Dimension P	Required Sample Size n
3	900
4	7000
5	50000
6	300000
7	2×10^6
8	1×10^7
9	5×10^7
10	3×10^8

To alleviate this problem, one might attempt to bias the sampling towards the boundary of $\mathcal{T}$. However, unless enough is known about $g(\cdot)$ to be able to bias the sampling to extreme values of g, this will have limited success. Indeed, if sampling were uniform on the surface of $\mathcal{T}$, only 2 extra dimensions would be gained, since the projection of a uniform distribution on the surface of a P-dimensional ellipsoid is Beta$[(P - 1)/2, (P - 1)/2, L, U]$ [Murdoch (1996)].

In the problems we are addressing in this paper, the regions $\mathcal{T}$ are generally not ellipsoidal, nor are the functions $g(\cdot)$ equivalent to one dimensional linear projections. However, if we assume smoothness of $\eta(\boldsymbol{\theta})$ and smoothness of $g(\cdot)$, we would expect the distribution of the simulated values $g(\boldsymbol{\theta})$ to be a smoothly distorted version of the Beta distribution. If the model and the function $g(\cdot)$ are approximately linear, there will be relatively little distortion, and the limitations described above will apply to nonlinear problems as well.

The results above suggest the following approach to improve the estimation of L and U. A plot of the quantiles of the true distribution of the simulated values of $g(\boldsymbol{\theta})$ versus the quantiles of the Beta$[(P + 1)/2, (P + 1)/2, 0, 1]$ distribution should be a fairly smooth curve, passing through the points $(0, L)$ and $(1, U)$. To estimate L and U, we extrapolate a smooth of a sample QQ plot (i.e. the ith ordered value of $g(\mathcal{S})$ plotted versus the $(i - 1/2)/n^{\text{th}}$ quantile of the Beta distribution) to the $g(\boldsymbol{\theta})$ values that would correspond to Beta quantiles of 0 and 1.

In order for the QQ plot to be valid, we need the $\mathcal{S}$ points to be a good approximation to a uniform sample. Since the random walk algorithm is often very slowly mixing, very large sample sizes may be needed to achieve this. There are various techniques for improving the mixing rate [Gilks and Roberts (1996)]; an alternative is simply culling the sample enough to reduce the clumping caused by the serial dependence, for example by using only every 100th sample. One or two hundred nearly independent observations are usually adequate for smoothing the QQ plots.

32.4   Examples

In this section we show two examples. The first has only two parameters, but is a highly nonlinear model. The second has eleven nonlinear parameters, but turns out to have relatively little nonlinearity.

32.4.1   BOD example

Bates and Watts (1988) fitted the two-parameter nonlinear model

$$y_i = \theta_1 [1 - \exp(-\theta_2 t_i)] + \epsilon_i$$

to a data set of 6 points measuring biochemical oxygen demand. Here y_i is the measured demand in mg/l, t_i is the time in days, and ϵ_i is an error term with zero mean and constant variance. This is a difficult data set, in that the 95% joint confidence region for the parameters is unbounded. However, the random walk algorithm may still be applied to find the intersection of the confidence region with a bounding rectangle.

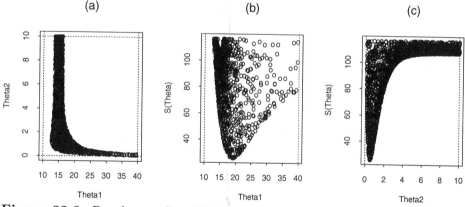

Figure 32.2: Random walk of 2000 points for BOD model. The points show the sampled steps in the walk, and the dotted lines show the bounding rectangle limiting it

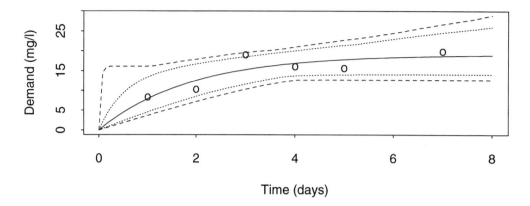

Figure 32.3: Predictions from BOD model. The points show the observations. The solid line is the predicted mean response, the inner dotted lines show pointwise 95% confidence intervals and the outer dashed lines show simultaneous 95% confidence intervals

Figure 32.1 shows the results of one such random walk. Here θ_1 was bounded by $[10, 40]$ and θ_2 was bounded by $[0, 10]$, and 2000 points were generated in the region defined by (32.1) to be the joint 95% confidence region for the parameters. Figure 32.2(a) shows the path of $\boldsymbol{\theta}$ through $\mathcal{T}$, and Figures 32.2(b) and 32.2(c) show the values of $S(\boldsymbol{\theta})$ plotted versus θ_1 and θ_2 respectively. All three plots clearly show that the bounding rectangle is too small—sampled points come right to the boundaries of it, and the value of $S(\boldsymbol{\theta})$ is well below the cut off there. They also illustrate the extreme nonlinearity of this model: $\mathcal{T}$ is clearly far from elliptical. Figure 32.2(c) gives a strong suggestion that there is no upper limit to the confidence region's extent in the θ_2 direction.

Because this model is only 2 dimensional, the sampled points approach quite closely to the boundaries of $\mathcal{T}$, and the methods of Section 32.3 are not needed; the range of values of $g(\mathcal{S})$ works well. Figure 32.3 shows estimates and pointwise 95% confidence intervals for the predicted response over the range of times. Also shown are conservative simultaneous 95% confidence intervals for the predicted responses. The latter were calculated by taking the range of predictions over all $\boldsymbol{\theta}$ in the joint region. The truncation evident in Figure 32.2 was diminished by using a larger bounding rectangle for the random walk.

32.4.2   Osborne's Gaussian mixture

Osborne (1971) presented a 65 point data set to which he fitted the 11 parameter model

$$
\begin{aligned}
y_i \;=\; & \theta_1 \exp(-\theta_5 t_i) + \theta_2 \exp[-\theta_6(t_i - \theta_9)^2] + \theta_3 \exp[-\theta_7(t_i - \theta_{10})^2] \\
& + \theta_4 \exp[-\theta_8(t_i - \theta_{11})^2] + \epsilon_i
\end{aligned}
$$

The fitted parameters in this model are quite highly correlated, and the random walk moves very slowly. In order to achieve adequate coverage of the confidence region, we used 1,000,000 steps; only every 1000th one was recorded, to save storage space. Figure 32.4 shows the results for parameters θ_4 and θ_7 (two of the more highly correlated and nonlinear parameters). Because this is an 11 parameter model, the edges of the region are much less clearly shown, and extrapolation using the extended QQ plots is necessary.

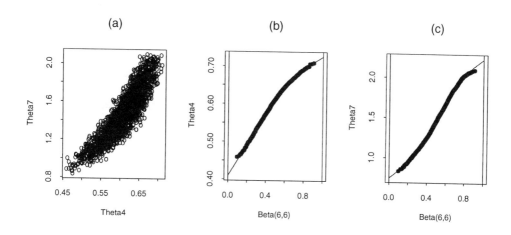

Figure 32.4: Random walk of 2000 points for Osborne model. Part (a) shows the sampled steps in the walk for parameters θ_4 and θ_7, while (b) and (c) show the QQ plot versus the Beta(6,6) distribution for the two parameters. The points are the sample values and the line is a spline smooth of the plot

Figure 32.5 shows observations and predictions for the Osborne model. For each of 50 values of t, separate QQ plot extrapolations based on the same random walk were used to calculate the confidence intervals. As it turns out, the model is quite linear, and a linearization approach would have produced similar results.

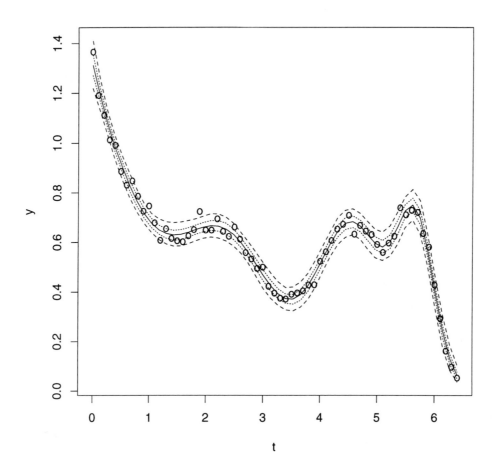

Figure 32.5: Predictions from Osborne model. The points show the observations. The solid line is the predicted mean response, the inner dotted lines show pointwise 95% confidence intervals and the outer dashed lines show simultaneous 95% confidence intervals

32.5 Conclusions

In this paper we have used the random walk algorithm to find one dimensional confidence intervals for functions of the parameters in complex nonlinear models. The method is convenient when there are a large number of functions of interest (e.g. predicted values under varied conditions), and when linear approximations are untrustworthy. Simply using the range of the observed function values appears to work well up to 3 or 4 dimensions, but the amount of simulation needed rises very rapidly with increasing dimension.

We suggested the use of extrapolated QQ plots to extend the usefulness of the method to higher dimensions. For models where the QQ plots are fairly straight, it appears quite practical to use this method up to around a dozen parameters. The QQ plots themselves may be used as diagnostics of the nonlinearity of the model.

There are many possible variations on the random walk algorithm. It is easily modified for the case where $\hat{\boldsymbol{\theta}}$ is unknown or unreliable. Here one replaces $\hat{\boldsymbol{\theta}}$ with a provisional value $\tilde{\boldsymbol{\theta}}$, which is updated when points are observed to have better sums of squares. The definition of the acceptance region shrinks down to $\mathcal{T}$ as the simulation runs. At the end, any values which would not be accepted in the final region are simply discarded.

Another class of modifications concerns the choice of **d** in step 2 of the algorithm. Considerable speed increases are possible by careful choice. For example, if a reliable estimate of the covariance matrix of $\hat{\boldsymbol{\theta}}$ is known, choosing **d** to be multivariate normal with that covariance will effectively convert the joint confidence region from an ellipsoid to a sphere.

The random walk method may also be used in other contexts besides nonlinear regression; Murdoch (1996) describes the modifications necessary for general likelihood-based models.

These variations aside, the great appeal of the random walk algorithm is its simplicity. With current desktop computer power, it is quite feasible to take several million steps within a few minutes of computer time. This allows straightforward calculation of sums of squares based confidence intervals, often in less time for the analyst than for the calculation of linearized approximations.

Acknowledgments. Financial support for this work was provided by the Natural Sciences and Engineering Research Council of Canada. Thanks are due to Colin McCulloch for his work on an early version of this paper and to Bovas Abraham for helpful comments.

References

1. Bates, D. M. and Watts, D. G. (1988). *Nonlinear Regression Analysis and its Applications*, John Wiley & Sons, New York.

2. Gelfand, A. E. and Smith, A. F. M. (1990). Sampling-based approaches to calculating marginal densities, *Journal of the American Statistical Association*, **85**, 398–409.

3. Gilks, W. R. and Roberts, G. O. (1996). Strategies for improving MCMC, In *Markov Chain Monte Carlo in Practice* (Eds., Gilks, W. R., Richardson, S., and Spiegelhalter, D. J.), pp. 89–114. Chapman and Hall.

4. Hamilton, D. (1986). Confidence regions for parameter subsets in nonlinear regression, *Biometrika*, **73**, 57–64.

5. Murdoch, D. (1996). Random walk approximation of confidence intervals, *Mathematics and Statistics Technical Report*, Queen's University.

6. Osborne, M. R. (1971). Some aspects of non-linear least squares calculations, In *Numerical Methods for Non-Linear Optimization* (Ed., Lootsma, F. A.), Academic Press.

7. Raftery, A. E. and Lewis, S. M. (1992). How many iterations in the Gibbs sampler? In *Bayesian Statistics 4, Proceedings of the Fourth Valencia International Meeting* (Eds., Bernardo, J. M., Berger, J. O., Dawid, A. P., and Smith, A. F. M.), pp. 763–773.

33

A Study of Quality Costs in a Multi Component and Low Volume Products Automated Manufacturing System

Young-Hyun Park

Kang Nam University, Korea

Abstract: In this paper, we illustrate how to design screening inspections for minimizing quality costs in a multi-stage automated manufacturing system. The total quality cost model consists of inspection costs, internal failure costs, external failure costs, and Taguchi's loss function. Although, the use of an automatic test equipment such as machine vision and CMM (coordinate measuring machine) have greatly increased the inspection speed and its accuracy, screening (100% inspection) could be considered only as a short-term method to remove nonconforming items from the population, and not as a long-term quality improvement strategy. However, screening may be used before costly operations, or after unsatisfactory operations.

Keywords and phrases: Inspection, manufacturing system, quality costs, screening procedure, Taguchi's loss function

33.1 Introduction

Quality has been measured and controlled in statistical terms. To manage the quality of products better, the quality should be converted to monetary terms since dollars are the easiest and most effective communication language. Since Feigenbaum (1961) categorized the quality cost as prevention cost, appraisal cost, internal failure cost and external failure cost, many authors have enumerated the components of the categories and investigated relationships among the categories. Recent advances in automation and computer control in manufacturing are changing the role and functions of quality control. In particular, the use of machine vision and CMM (coordinate measuring machine) have greatly increased the inspection speed and its accuracy. Consequently, screening inspection (100% inspection) is becoming an attractive practice for removing nonconforming items, Stile (1987). However, as pointed out by Deming (1986), dependence on inspection to correct quality problems is ineffective and costly, and hence screening should not be used as a long-term solution for improving the product quality or reducing the costs incurred by nonconforming items. Instead, implementing successful process control and quality improvement programs are essential for a manufacturer to survive in the competitive business world, Tang (1990). For example, proper use of statistical process control (SPC) and off-line quality control can reduce variation of proucts and improve quality of products continuously.

Two separate objectives have been commonly used to design screening procedures. One is to minimize the expected total quality cost associated with a screening procedure, and the other is to use screening to reach certain statistical goals, such as controlling the outgoing nonconforming rate of the product. In this paper, the first objective will be discussed, assuming inspection is error-free and the outgoing conforming rate is 100% after screening.

This paper is structured as follows. First, previous works related on basic models for single screening and inspection policies in a multi-stage manufacturing system is introduced. Second, screening procedure to minimize the extended quality cost is discussed.

33.2   Screening Procedures: A Review

33.2.1   Single screening procedure

In a typical single screening procedure, all the out-going items are subject to acceptance inspection. If an item fails to meet the predetermined screening specifications, the item is rejected and subjected to corrective actions. In this section two types of single screening procedures; (1) Quality costs model (2) Taguchi's loss function model, are considered.

The first model was developed by Deming (1986) based on the quality costs of Feigenbaum (1961). Suppose a lot of N items produced by a stable process is subject to attribute inspection. If an item is inspected and found to be nonconforming, it costs γ to rework or replace the item, and an unfound nonconforming item will result in an acceptance cost of β, assuming error-free inspection. And let s denote the cost of inspecting an item and f the proportion of items which are inspected. Then the expected total cost for this lot is:

$$ETC = Nsf + \gamma Npf + \beta Np(1 - f) \qquad (33.1)$$

where p is the lot proportion nonconforming. The object is to determine the value of f^* that minimizes ETC. Consequently, the value of f^* is:

$$f^* = \begin{cases} 100\% \text{ screening,} & \text{when } p > s/(\beta - \gamma) \\ 0\% & \text{otherwise} \end{cases} \tag{33.2}$$

This was proved and discussed by Papadakis (1986).

The second is tolerance design based on Taguchi's model. Consider a screening procedure where each outgoing item is subject to inspection on the performance variable. Let Y denote an N-type performace variable with the target value of m and $f(y)$ the probability density function of Y. Taguchi suggested that the loss associated with an item with $Y = y$ be determined by the following function:

$$L(y) = k(y - m)^2 \tag{33.3}$$

where k is a positive constant. For the purpose of screening, let $[m - \Delta, m + \Delta]$ be the acceptance region, and if the value y is outside this region, the item is rejected. Let r denote the cost associated with the disposition of a rejected item and s the per-item cost of inspection. Then the per-item expected total cost associated with the screening procedure is:

$$ETC = \int_{m-\Delta}^{m+\Delta} k(y - m)^2 dy + \gamma[1 - \int_{m-\Delta}^{m+\Delta} f(y)dy] + s \tag{33.4}$$

The value Δ^* that minimizes ETC is equal to $\sqrt{m/k}$ [Taguchi, Elsayed and Hsiang (1989)].

33.2.2 Screening procedures in a multi-stage system

Two types of screening procedures have been developed for a multi-stage manufacturing system. The first is how to allocate the screening efforts, and the second is how to design the process parameters to minimize the cost in a muiti-stage manufacturing system. The location of the screening procedure for minimizing costs will be discussed further.

In a multi-stage manufacturing system "where to inspect" and "how many to inspect" are important decisions for controlling costs. Two types of manufacturing systems, serial manufacturing systems and non-serial manufacturing systems, have been studied in the following literature. For example, White (1966) showed that the optimal inspection proportions should be 0 or 100%, which enables the model to be solved by dynamic programming or integer programming methods and Britney (1972) considered non-serial systems and found that "all-or-nothing" rules also apply. Chakravarty and Shtub (1987) extended White's (1966) model to multi-product situation where additional costs, setup, and inventory carrying costs are considered. Tang (1990) considered a product with multiple performance variables produced by a serial production system, where a performance variable is determined at each stage of the system. He discussed that the screening rule at any stage should be based on the quality of the product at each stage, and the total expense already on the item. The expected quality cost would be incurred in the later stages. Ballou and Pazer (1982) discussed the inspection allocation issue for a serial production system when inspection errors are present [see also Park, Peters and Tang (1991)].

Peters and Williams (1984) discussed several heurisitc rules concerning inspection location in a multi-stage manufacturing system. Some of these are:

1. Inspect after operations that are likely to produce nonconforming items.

2. Inspect before costly operations.

3. Inspect before operations where nonconforming items may damage or jam the machines.

4. Inspect before operations that cover up nonconforming items.

33.3   Model Development

33.3.1   Multi-stage manufacturing system

Inspection and testing are traditionally accomplished by using manual methods that are time consuming and costly. Consequently, manufacturing lead time and production cost are increased without adding any real value. However, with the development of autonomous inspection devices such as machine vision system and coordinate measuring machine, accurate and fast screening inspection has been possible in a computer numerical control (CNC) manufacturing system. The machining center (MC) is a machine tool capable of performing several different machining operations on a workpart in one setup under program control. The machine center is capable of milling, drilling, turning, facing, and other related operations. A manufacturing system for performing several different machining operations using maching center is considered in this paper.

Consider the case where the machining center produces parts. Let $f(y)$ be the density of the normal distribution with mean μ and standard deviation σ of the normal distribution. If the quality characteristic of products is within specification, $[\gamma - \Delta, \gamma + \Delta]$, the part will be accepted and otherwise it will be rejected. Let p be the proportion of acceptable items and $p' = 1 - p$, the proportions of defective items.

$$p = \int_{\gamma-\Delta}^{\gamma+\Delta} \frac{1}{\sqrt{2\pi}} \frac{1}{\sigma} \, exp - (\frac{1}{2})[(y - \mu)/\sigma]^2 dy \qquad (33.5)$$

Thus the general formula of the acceptance proportion is:

$$p_{(m)} = \int_{\gamma_{(m)}-\Delta_{(m)}}^{\gamma_{(m)}+\Delta_{(m)}} \frac{1}{\sqrt{2\pi}} \frac{1}{\sigma_{(m)}} \, exp - (\frac{1}{2})[(y - \mu_{(m)})/\sigma_{(m)}]^2 dy \qquad (33.6)$$

Let $N_{(0)} = N$ be the production size, $x_{(m)} = 0$ be the case of no-inspection after operation m and $x_{(m)} = 1$ the case of screening inspection after operation m. Further let $N'_{(m)}$ be the expected number of defective parts excluded from next operation and $N_{(m)}$ the expected number of passed parts when screening inspection is performed ($x_{(m)} = 1$). But when inspection is not done after operation m ($x_{(m)} = 0$), $N'_{(m)}$ is equal to 0 and $N_{(m-1)}$ is equal to $N_{(m)}$. Also let $N_{(M)}$ is the number of finished parts. Then we have:

$$N'_{(m)} = \begin{cases} N_{(m-1)} \cdot p'_{(m)}, & x_{(m)} = 1 \\ 0 & x_{(m)} = 0 \end{cases} \tag{33.7}$$

$$N_{(m)} = \begin{cases} N_{(m-1)} \cdot p_{(m)}, & x_{(m)} = 1 \\ N_{(m-1)} & x_{(m)} = 0 \end{cases} \tag{33.8}$$

$$N_{(M)} = N_{(0)} - \sum_{m=1}^{M} N'_{(m-1)} \tag{33.9}$$

33.3.2   Quality costs

Three types of quality cost are considered: inspection cost, internal failure cost and external failure cost. And the external failure cost includes the loss due to variation and defects of the products. First the inspection cost is calculated as follows. Let $C_{f(m)}$ and $C_{v(m)}$ be the fixed and variable cost components of inspection after process m respectively; then the cost of inspection after process m is:

$$E(C_{1(m)}) = \begin{cases} C_{f(m)} + C_{v(m)} \cdot N_{(m-1)}, & x_{(m)} = 1 \\ 0 & x(m) = 0 \end{cases} \tag{33.10}$$

Let $A_{(0)}$ be the material cost and $A_{(m)}$ be the value of part added due to operation m. Material cost is the cost of making all materials ready for production. Therefore, material cost includes costs of ordering, purchase, and transportation, if necessary, not only for direct materials but also for indirect materials like lubricants. Machine cost includes costs of utilites (power and fuel), maintenance of facility, repair of facility, and machine tools. Labour cost is the cost of direct labour and indirect labour involved in production activities and includes wages, salaries, and fringes.

In conventional manufacturing systems, the machine cost is an overhead item and its allocation is based on direct labour hours, because collecting data on machine hours for individual jobs is difficult and invloves additional clerical cost. In the factory automation environment, however, measuring the machine cost becomes easy as the direct linkage of computers with machines allows machine hours to be tracked quickly and reliably, as in CNC and DNC. Thus the value of part added due to operation m is:

$$A_{(m)} = Cu_{(m)} + Cm_{(m)} + Cr_{(m)} + Ct_{(m)} + Cb_{(m)} \qquad (33.11)$$

where

$Cu_{(m)}$: the utility cost processing operation m per item
$Cm_{(m)}$: the facility maintenance cost processing operation m per item
$Cr_{(m)}$: the facility repair cost processing operation m per item
$Ct_{(m)}$: the tool cost processing operation m per item
$Cb_{(m)}$: the labour cost processing operation m per item

Internal failure cost occurs when a defective part is caught within a company and then is scrapped or reworked due to actual defect. Scrap occurs when a defective part cannot be restored. Let $k_{(m)}$ be the proportion of parts restored to a good part and $R_{(m)}$ the restoring cost of parts restored to a good one. Since $p'_{(m)}$ is the proportion of defectives due to operation m, internal failure cost due to operation m is

$$E(C_{2(m)}) = \begin{cases} N_{(m-1)}p'_{(m)}\{1 - k_{(m)}\}(\{A_{(0)} + \sum_{i=1}^{m} A_{(i)} + R_{(m)})\}, & x_{(m)} = 1 \\ 0 & x_{(m)} = 0 \end{cases}$$
$$(33.12)$$

where $p'_{(m)} = 1 - p_{(m)}$. Therefore, the variance of the passed items, $v^2_{o(m)}$, is

$$v^2_{o(m)} = \frac{1}{P_{(m)}} \int_{\gamma_{(m)}-\Delta_{(m)}}^{\gamma_{(m)}+\Delta_{(m)}} \frac{1}{\sqrt{2\pi}} \frac{1}{\sigma_{(m)}} (y - \mu_{(m)})^2 exp - (\frac{1}{2})(y - \mu_{(m)})/\sigma_{(m)}]^2 dy$$
$$(33.13)$$

Thus the loss cost due to variance of the passed items for operation m is $\{N_{(m-1)} - N'_{(m)}\}\{A_{(m)}/\Delta^2_{(m)}\}v^2_{0(m)}$. The expected total cost including inspection cost, internal failure cost, and variation cost when the screening procedure is performed after operation m is the following.

$$\begin{aligned} ETC_{(m)} &= C_{f(m)} + C_{v(m)}N_{(m-1)} + N_{(m-1)}p'_{(m)}\{1 - k_{(m)}\}(\{A_{(0)} \\ &+ \sum_{i=1}^{m} A_{(i)} + R_{(m)}\} + N_{(m)}\{A_{(m)}/\Delta^2_{(m)}\}v^2_{0(m)} \qquad (33.14) \end{aligned}$$

External failure cost occurs when a defective item is not caught within a company and is then delivered to customers. Let $N_{a(M)}$ be the expected number of passed good items when screening inspection is performed after each operation and $N_{(M)}$ be the expected actual number of passed items (good and defective parts) delivered to customers. Then $N_{(M)} - N_{a(M)}$ is the expected number of defective items delivered to customers. The cost of external failure is caused by the items of imperfect quality that reach the customers. Let C_y be the direct cost per unit caused by product failure, warranty, and handling and C_z be the indirect cost per unit caused by loss in goodwill and loss in sale. Then the external cost including variation of items is

$$E(C_{3(m)}) = [N_{(M)} - N_{a(M)}][A_{(0)} + \sum_{i=1}^{M} A_{(m)} + C_z + C_y] + N_{a(M)} \sum_{i=1}^{m}$$
$$[A_{(m)}/\Delta^2_{(m)}]v_0^2(m) \tag{33.15}$$

Thus the total quality cost model for determining the allocation of screeening inspection at each stage of the system is

$$\text{Min } ETC(x_{(m)}) = \sum_{m=1}^{M} [C_{f(m)} + C_{v(m)}N_{(m-1)}] \cdot x_{(m)} \tag{33.16}$$

$$+ \sum_{m=1}^{M} N_{(m-1)}p'_{(m)}\{1 - k_{(m)}\}(\{A_{(0)} + \sum_{i=1}^{M} A_{(i)} + R_{(m)}\} \cdot x_{(m)}$$

$$+ [N_{(M)} - N_{a(M)}][A_{(0)} + \sum_{m=1}^{M} A_{(m)} + C_z + C_y]$$

$$+ N_{a(M)} \sum_{m=1}^{M} [A_{(m)}/\Delta^2_{(m)}]v_0^2(m)$$

where $N_{(M)} = N_{(0)} - \sum_{m=1}^{M} N'_{(m-1)}$, and $p'_{(m)}, A_{(m)}$ and $v_{o(m)}^2$ are as defined before.

This paper developed a model assuming no inspection error. However, it is well known that most inspection processes have inherent variability due to various factors such as variations in testing materials and inspectors. A Type I error occurs when a conforming item is classified as nonconforming, and Type II error occurs when a nonconforming item is classified as conforming. For variable inspection, inspection error is characterized in terms of bias and imprecision. Raz and Thomas (1983), Tang and Schneider (1990), and Park, Chang, and Lee (1995) studied this area. A tool wearing process, such as machining or molding is a production process that exhibits decreasing (or increasing) patterns in the process mean during the course of production. Bisgaard, et al. (1984) and Arcelus and Banerjee (1985) considered a process where the process mean changes constantly. Further study including inspection error and tool wearing process in a multi-stage manufacturing system is beyond the scope of this paper.

33.4 Concluding Remarks

In this paper, we discussed quality costs based on screening inspection policy in a multi-stage manufacturing system. The modified quality costs model includes Taguchi's loss function as well as inspection cost, internal failure cost and external failure cost. As automatic inspection equipments have been developed, screening is becoming an attractive practice for removing non-conforming items. However, as pointed out by Deming, dependence on inspection to correct quality problems is ineffective and costly, and hence screening should not be used as a long term solution for improving product quality or reducing the costs incurred by non-conforming items. Instead, implementing successful process control and quality improvement through SPC and Quality Engineering is essential for a manufacturer to survive in the competitive world.

References

1. Arcelus, F. J. and Banerjee, P. K. (1985). Selection of the most economical production plan in a tool-wear process, *Technometrics*, **27**, 433–437.

2. Ballou, D. P. and Pazer, H. L. (1982). The impact of inspector fallibility on the inspection policy in serial production system, *Management Science*, **28**, 387–399.

3. Bisgaard, S., Hunter, W., and Pallesen, L. (1984). Economic selection of quality of manufactured product, *Technometrics*, **26**, 9–18.

4. Britney, R. (1972). Optimal screening plans for nonserial production system, *Management Science*, **18**, 550–559.

5. Chakravarty, A. and Shtub, A. (1987). Strategic allocation of inspection effort in a serial, multi-product production system, *IIE Transactions*, **19**, 13–22.

6. Deming, E. W. (1986). *Out of the crisis*, MIT Press Cambridge, MA, U.S.A.

7. Feigenbaum, A. V. (1961). *Total Quality Control*, New York: McGraw Hill.

8. Papadakis, E. P. (1986). The Deming inspection criterion for choosing zero or 100% inspection, *Journal of Quality Technology*, **17**, 121–127.

9. Park, Y. H., Chang, S. J. and Lee, S. W. (1995). Quality costs in manufacturing systems, *Journal of Korean Quality Management*, **23**, 10–27.

10. Park, J. S., Peters, M. H., and Tang, K. (1991). Optimal inspection policy in sequential screening, *Management Science*, **37**, 1058–1061.

11. Peters, M. H. and Williams, W. W. (1984). Location of quality inspection stations: an experimental assessment of five normative rules, *Decision Sciences*, **15**, 389–408.

12. Raz, T. and Thomas, M. U. (1983). A method for sequencing inspection activities subject to errors, *IEEE Transactions*, **15**, 12–18.

13. Stile, E. M. (1987). Engineering the 1990s inspection function, *Quality Progress*, **20**, 70–71.

14. Taguchi, G., Elsayed, E., and Hsiang, T. (1989). *Quality Engineering in Production Systems*, New York, McGraw Hill.

15. Tang, K. (1990). Design of multi-stage screening procedure for a serial production system, *European Journal of Operational Research*, **52**, 280–290.

16. Tang, K. and Schneider, H. (1990). Cost effectiveness of using a correlated variable in a complete inspection plan when inspection error is present, *Naval Research Logistics*, **37**, 893–904.

17. White, L. S. (1966). The analysis of a simple class of multistage inspection plans, *Management Science*, **12**, 685–693.

34

Estimating Dose Response Curves

Sat N. Gupta and Jacqueline Iannuzzi

University of Southern Maine, Portland, Maine

Abstract: There are several companies in the United States that manufacture devices for detecting the presence of a variety of antibiotics in milk. All such devices that are manufactured and marketed in the USA require approval from the Food and Drug Administration (FDA), USA. For each drug whose presence in milk is to be tested, FDA has determined a critical level, and requires that all test devices show a sensitivity of at least 90% with a confidence level of at least 95% at the critical levels. This paper discusses several statistical techniques such as one-sided confidence intervals for the binomial p, and logit/probit regression analyses that are commonly used to check regulatory compliance in such cases.

Keywords and phrases: Dose-response curve, Gompertz distribution, logit, goodness-of-fit, probit, sensitivity

34.1 Introduction

This paper evolved from a consulting project handled by the first author for a local biotechnical company which produces, among other things, a test kit for detecting the presence of beta-lactum drugs such as amoxicillin, penicillin-g, ampicillin, and cloxacillin which are used for lactating cattle. For each of these drugs, the center for veterinary medicine (CVM), a division of the FDA, has determined a tolerance level in milk for human consumption. For example, 10 PPB (parts per billion) is the tolerance level for amoxicillin. The FDA requires that if a test kit is claimed to be able to detect the presence of any of these drugs, it must be able to detect it at the tolerance level with a sensitivity of at least 90% at a confidence level of 95%. The sensitivity is defined as the probability of a positive test result at a given concentration level. One way to test compliance with the FDA requirement is to spike a known number of pure milk samples with the drug to be tested at a concentration equal to the tolerance level, subject each of these samples to the test kit and record the number of positive results. With this information, one can easily find the 95% lower confidence limit for the binomial (p), the proportion of positive results, to determine if it exceeds 90%. A method preferred by the FDA is to construct a dose-response curve by subjecting multiple samples at various concentration levels (including the tolerance level) to the proposed test kit and recording the number of positive responses at each concentration. Again the lower 95% confidence limit for the probability of a positive response at the tolerance level must exceed 90%.

There are various binary response regression models that can be used to fit dose-response curves to the test results. The logistic regression model, the probit model (with or without log transformation of dose) and the complementary log-log model (with or without log transformation of dose) are some of the more commonly used models. A discussion of these models can be found in Agresti (1990). Deciding which one of these models provides the best fit for a given data set is the focus of this paper. For this discussion we use a case where the tested drug is amoxicillin and the tolerance level is 10 PPB. Samples were tested at various concentrations and the number of positive results per total number of samples tested at each concentration are given in Table 34.1.

Table 34.1: Test results for milk samples spiked with various levels of amoxicillin

Concentration level (PPB)	0	3	4	6	8	10
# positive results	0/60	0/30	0/30	3/29	15/30	29/29

Each of the binary response models has the following basic form: $p = F(\alpha + \beta x)$, where p is the probability of a response (positive test result), F is a cumulative distribution function (cdf), α and β are parameters to be estimated, and x is the dose which in our case can either be the concentration or log (concentration). The cumulative distribution function depends on the model selected; logistic for logistic regression, standard normal for probit and Gompertz for the complementary log-log. These models are special cases of generalized linear models (GLMs) introduced by Nelder and Wedderburn (1972). GLMs are specified by three components: a random component which identifies the probability distribution of the response variable; a systematic component which specifies a linear function of the explanatory variables used to predict a response, and a link function that describes the relationship between the systematic component and the expected value of the random component. In binary response data the random component is a Bernoulli random variable which in this case corresponds to a positive or negative test result. The systematic and link components define the linear relationship. From the equation above it follows that $F^{-1}(p) = \alpha + \beta x$. The link is then defined as the inverse of the cdf. The linear relationships for each model are summarized in Table 34.2.

Table 34.2: Description of linear relationships for each model type

MODEL TYPE	LINEAR RELATIONSHIP
Logistic	$\log(p/1-p) = \alpha + \beta x$
Probit	$\Phi^{-1}(p) = \alpha + \beta x$
Complementary Log-log (Gompertz)	$\log[-\log(1-p)] = \alpha + \beta x$

In order to determine which model provides the best fit, five separate models were used to analyze the data set for amoxicillin given above; 1) the logistic regression model, 2) the probit model, 3) the probit model using the log(concentration), 4) the complementary log-log model (also referred to herein as the Gompertz model) and 5) Gompertz model using the log (concentration). In models 3 and 5, x represents log (concentration) instead of the concentration. All five models were then compared using goodness of fit statistics.

34.2 Methods

Before any of these analyses are conducted one should examine scatterplots of the sample data to determine if a linear model is reasonable. The observed probability of a response (i.e. # of responses / # of test subjects) at each concentration level can be transformed using the relationships described in Table 34.2 for each model and then plotted against x to determine if the relationship looks linear. Some problems arise with this approach, however, and must be dealt with before proceeding. Firstly, the logistic transformation, called a logit, is undefined when $p = 0$ or 1. So a less biased estimator of the logit, called an empirical logit (Agresti (1990)), was calculated. The empirical logit is defined as: $\log[(\text{# of responses} + 0.5)/(\text{# of samples - # of responses} + 0.5)]$.

Secondly, when $p = 0$ or 1, the probit transformation (called a probit) is infinite, and the Gompertz transformation is undefined. In order to calculate these transformations empirical probabilities were calculated first in a manner similar to the empirical logit by letting $p/(1-p) = [(\text{# of responses} + 0.5)/(\text{# of samples - # of responses} + 0.5)]$ and then solving for p. Table 34.3 presents a summary of the data in Table 34.1 after applying appropriate transformation.

Table 34.3: Values for the logit, probit and Gompertz transformations at each concentration

CONCENTRATION (PPB)	OBSERVED PROBABILITY	EMPIRICAL LOGIT	EMPIRICAL PROBIT	EMPIRICAL GOMPERTZ
0	0	-4.80	-2.40	-4.80
3	0	-4.11	-2.14	-4.12
4	0	-4.11	-2.14	-4.12
6	0.103448	-2.02	-1.19	-2.09
8	0.500000	0	0	-0.37
10	1	4.08	2.13	1.41

The transformed data were plotted against concentration or log (concentration). Examination of the scatterplots indicated all of the five proposed models could be considered reasonable but a few things were noted. In general, only the last four data points, corresponding to concentrations 4, 6, 8 and 10 PPB, showed a linear relationship while the first one or two points did not follow this trend. At first glance it might be concluded that a linear model is unreasonable however one should keep in mind that the values of these transformations change very quickly when the probabilities are near zero. Therefore the transformed data points at concentrations 0, 3 and 4 PPB should not be considered reliable enough to either rule out a linear model or to propose a quadratic model.

Table 34.4: Fitted probabilities for each model with and without 0 PPB data

Conc	Obs. Prob.	Logistic Model		Probit Model	
		With 0	Without 0	With 0	Without 0
0	0	0.00001		0.00000	
3	0	0.00068	0.00068	0.00003	0.00003
4	0	0.00309	0.00309	0.00073	0.00073
6	0.10345	0.06039	0.06042	0.06840	0.06840
8	0.50000	0.57154	0.57154	0.58241	0.58241
10	1.00000	0.96514	0.96512	0.97154	0.97154

Conc	Obs. Prob.	Probit with Log Transformation	Gompertz Model		Gompertz with Log Transformation
		Without 0	With 0	Without 0	Without 0
0	0		0.00009		
3	0	0.00000	0.00261	0.00263	0.00024
4	0	0.00003	0.00802	0.00809	0.00256
6	0.10345	0.06486	0.07379	0.07410	0.07040
8	0.50000	0.61020	0.51802	0.51799	0.54434
10	1.00000	0.95275	0.99904	0.99901	0.99303

Another issue to be considered is that of log transformation of concentration in the probit and Gompertz models. These models are based on the concept of a tolerance distribution (not to be confused with the tolerance level of a drug in milk). The tolerance value of an individual subject is the dosage or stimulus level at and beyond which a subject will have a response. In this example the subjects are individual test kits, which will respond to certain concentrations (or dosages) of antibioitics in milk. Finney (1971) points out that the frequency distributions of tolerance levels in a population when measured in the natural scale are very often skewed to the right by individuals with very high tolerance levels. By transforming the dosage scale to a log scale the distribution is converted to the symmetrical normal distribution on which the linear model is based. However, the scatterplots indicated that the log transformation of dose in this example was not particularly helpful.

All five models were analyzed using the PROBIT procedure of the SAS statistics software package. This procedure calculates the maximum likelihood estimates of the regression parameters α and β. The models that use concentration (rather than log (concentration)) were analyzed both with and without the 0 concentration level data so that the effects of excluding 0 from the analysis for these models can be compared. Table 34.4 presents the fitted probabilities at each concentration level for each model. Table 34.5 presents the lower 95% confidence limits on sensitivity at 10 PPB for each model. It was found that excluding 0 from the analysis had only a slight impact on the fitted values and the 95% confidence limits. Therefore in all further discussions of the models it is assumed that all data were included in the analysis except of course when log (concentration) is used.

Table 34.5: 95% One-sided lower confidence limit on sensitivity at 10 PPB

MODEL	95% LOWER CONFIDENCE LIMIT	
	With 0	Without 0
Logistic Model	0.89752	0.89741
Probit Model	0.90133	0.90133
Probit (with log)	NA	0.87286
Gompertz	0.93849	0.93751
Gompertz (with log)	NA	0.92594

34.3 Discussion

Given that any of the models might be valid based on the initial assumption of linearity, there is still the problem of determining which model provides the best fit. A subjective determination can be made by comparing observed probabilities versus those predicted by each model.

A more quantitative way to evaluate the models is to assess the goodness of fit (GOF) which is a process of examining the error components of the model. According to Hosmer, Taber and Lemeshow (1991), assessing the GOF of a regression model usually entails two stages: computing statistics that provide a summary measure of the errors, and performing "regression diagnostics" to examine the role of individual subjects. For obvious reasons it is not appropriate to examine the role of individual subjects in our models so only summary statistics will be evaluated. Two GOF statistics that are computed by the SAS

Probit procedure are the Pearson chi-square test statistic and the log likelihood ratio chi-square test statistic. These statistics are appropriate when the data are grouped either initially or for the purpose of testing GOF, and the expected frequencies are relatively large. These statistics are also discussed in Cox and Snell (1989) and in Agresti (1990). The Pearson chi-square is equal to $\sum_i \sum_j (r_{ij} - n_i p_{ij})^2 / n_i p_{ij}$, where the sum on i is over grouping, the sum on j is over levels of response, the r_{ij} is the weight of the response at the jth level for the ith grouping, n_i is the total weight at the ith grouping, and p_{ij} is the fitted probability for the jth level at the ith grouping. The log likelihood ratio chi- square statistic is equal to $2 \sum_i \sum_j r_{ij} \ln(r_{ij}/n_i p_{ij})$.

If the modeled probabilities fit the data well, these statistics should be approximately distributed as chi-square with degrees of freedom equal to $(k - 1)m - q$, where k is the number of levels of response (binomial in our case), m is the number groupings and q is the number of parameters fit in the model. SAS calculates these statistics by assuming that each concentration level is a group. This is not valid in our case because the first four concentration levels have low expected values. Therefore these statistics were recalculated by grouping the data into three categories as shown in Table 34.6. We have one degree of freedom since $k = 2$, $m = 3$ and $q = 2$. In comparing these statistics, the higher the p-value the better the fit. Therefore the model exhibiting the best fit is the Gompertz model (without log). The remaining models placed in order from best to worst fit are the Gompertz (with log), the probit, the logistic and the probit (with log). This is the same order obtained by comparing the fitted probabilities with the observed probabilities at level 10 PPB in Table 34.4.

Table 34.6: Expected frequencies of models and GOF statistics

Conc.	Observed Frequency	Expected Logistic	Expected Probit	Expected Probit(log)	Expected Gompertz	Expected Gomp(log)
0,3,4,6	3	1.87	2.01	1.88	2.46	2.13
8	15	17.15	17.47	18.31	15.54	16.33
10	29	27.99	28.17	27.63	28.97	28.80
Pearson Chi-square		2.3739	2.1860	3.6428	0.1853	0.8062
Prob > Chi-square		0.1234	0.1393	0.0563	0.6668	0.3692
Likelihood Ratio Statistic		3.2694	2.9340	4.8716	0.2055	0.9663
Prob>Chi-square		0.0706	0.0867	0.0273	0.6503	0.3256

Another case can be made for accepting the Gompertz model (without log) as the best model. Table 34.4 shows that fitted probabilities approach 1 more rapidly than they approach 0. According to Agresti (1990) logistic regression and probit models are not appropriate in cases where the response curves are asymmetric and the complementary log-log model gives a better fit for data of this type. In a study of beetle mortality where the probability of death versus exposure to carbon disulfide approached 1 more rapidly than 0, Stukel (1988) showed that logistic and probit models fit the data poorly whereas the complementary log-log model fit the data quite well.

34.4 Conclusion

Obviously the choice of model can have serious implications for the manufacturer of the test kits. If the validity of the GOF assessment and other considerations are accepted then the implication is good for accepting the test kits as complying with FDA requirements based on the 95% confidence limits shown in Table 34.5. The three best fitting models, the Gompertz (with and without log transformation) and the probit (without log transformation) all indicate that the 95% confidence limit of sensitivity is greater than 90% at 10 PPB. The logistic model gives the limit as 0.89752 (almost .90) and the probit (with log transformation) model gives a 95% confidence limit of 0.87286. It is also noted that if only the samples evaluated at the 10 PPB tolerance level are considered, the 95% lower confidence limit for the binomial (p) is 0.90185 based on the fact that 29 out of 29 samples tested positive at this level. This lower 95% confidence limit is obtained by solving for p the following equation.

$$\sum_{k=x}^{n} c_k^n p^k (1 - p)^{n-k} = .05.$$

Here n is the number of samples tested (29 in our case) and x is the number of positive results (also 29 in our case).

This method also shows compliance. The only model that really shows lack of compliance is the probit model with log transformation, but this was the model that provided the worst fit.

It is therefore important that several issues be considered when choosing a model to evaluate this type of data for compliance. First, it should be determined if the proposed linear model or models are reasonable. This is easily done by examining scatterplots of transformed data. Secondly, the proposed models should be evaluated for goodness of fit. Another issue concerns the log transformation of dose which is often applied in dose response analysis. The problem with this transformation is that it ignores the 0 dose level, in our case the 0 PPB level. This is a genuinely tested data point and should not be ignored in the analysis, unless absolutely necessary. Perhaps several competing models should be evaluated and a decision on regulatory compliance could then be based on the model exhibiting the best fit, particularly if the final decision seems to be a close call.

References

1. Agresti, A. (1990). *Categorical Data Analysis*, New York: John Wiley & Sons.

2. Cox, D. R. and Snell, E. J. (1989). *Analysis of Binary Data*, Secondnd edition, London: Chapman and Hall.

3. Finney, D. J. (1971). *Probit Analysis*, Third edition, Cambridge: Cambridge University Press.

4. Hosmer, D. W., Taber, S. and Lemeshow, S. (1991). The importance of assessing the fit of logistic regression models: a case study, *American Journal of Public Health*, **81**, 1630–1635.

5. Nelder, J. and Wedderburn, R. W. M. (1972). Generalized linear models. *Journal of the Royal Statistical Society*, **A135**, 370–384.

6. Stukel, T. A. (1988). Generalized logistic models, *Journal of the American Statistical Association*, **83**, 426–431.

35

On the Quality of Preterm Infants Formula and the Longitudinal Change in Mineral Contents in Human Milk

Brajendra C. Sutradhar, Barbara Dawson and James Friel

Memorial University of Newfoundland, St. John's, Canada

Abstract: Human milk is often fortified with appropriate nutrients including minerals to allow premature infants and their families to enjoy the benefits conveyed by the feeding of breast milk while delivering an optimal nutrient supply to the baby. There is, however, considerable controversy about when, how and with what human milk should be fortified. Currently, pre-term infants' formula are prepared according to the nutrient requirements at the first or transition stage (birth to 10 days of age), a stable growing stage (10 days to 6-8 weeks following birth), and finally a post discharge stage (6-8 weeks to 12 months following birth). But, the quality of the pre-term infants' formula will be affected if there is any weekly longitudinal change in mineral content in mothers milk mainly during the first 8 to 12 weeks of lactation period following birth. Very little is known about such longitudinal changes which (if there are any) might lead to the need for appropriate changes in the preparation of pre-term infants' formula in order to meet the nutrient requirements mainly during their stable-growing period. In this paper, we consider this important issue and analyze the longitudinal change in mineral content in milk from mothers of pre-term and full-term infants. Forty three mothers from St. John's, Newfoundland participated in the study and the data were collected from them for a period of 12 weeks.

Keywords and phrases : Lag 1 correlations, mean and median plots, symmetric and asymmetric mineral concentrations

35.1   Introduction

Low birthweight infants, defined as weighing less than 2500 grams at birth, comprise 5 to 8 percent of all liveborn infants in Canada. A sizeable proportion weigh less than 1500 grams at birth and are referred to as very low birthweight infants. The majority of these infants are born prematurely, and in addition, some have experienced intrauterine growth and are of lower than expected birth weight for gestational age.

Low birthweight infants, and especially those of very low birthweight and early gestational age, have unique physiologic characteristics and are prone to pathologic conditions associated with prematurity. These characteristics and conditions may lead to mortality and both short-term and long-term morbidity. In addition, they lead to complex nutritional requirements and susceptibility to complications due to excessive or inadequate nutrient intakes.

Human milk has been widely utilized as the sole nutritional source for feeding the low birthweight premature infants. Forbes (1989) has demonstrated while it may be desirable to feed human milk to infants born prematurely, it is not adequate as the sole nutritional source for many of them. It is clear that breast milk cannot support the full complement of nutrients for very pre-term infants to deposit tissue at the same rate and composition as would have occurred had they remained in utero. Deficiency states such as hypophosphatemic rickets have been recognized in premature infants fed breast milk exclusively [Rowe et al. (1979)]. Comparative trials performed between infants fed either human milk or adopted premature formula have revealed improved growth and biochemical status with formula feeding [Brooke et al. (1982), Anderson and Bryan (1982), Gross (1983)].

A very viable solution to the above problem is to fortify human milk with the appropriate nutrients to allow premature infants and their families to enjoy the benefits conveyed by the feeding of breast milk while delivering an optimal nutrient supply to the baby [Modanlou et al. (1986), Kashyap et al. (1990)]. It may be acceptable medical practice to feed the premature infant with his or her own mother's milk alone at high volume while monitoring and appropriately treating nutritional inadequacies identified. It is preferable, however, to routinely supplement human milk fed to very low birth weight babies to achieve a more optimal nutritional profile while avoiding the effects of dietary deficiency [Fomon et al. (1977)]. There seems to be little debate about whether human milk should be fortified if it is to be the sole source of nutrition for tiny premature infants; there is considerable controversy, however, about when, how and with what. This stems from uncertainty about the appropriate standard of growth for the infant born prematurely, controversy about the long-term consequence of nutritional "inadequacy" identified in the perinatal period, potential interactions between components contained within breast milk and fortifiers, extreme variability in the composition of expressed breast milk and marked clinical heterogeneity between infants [Atkinson et al. (1990)].

The Canadian Paediatric Society has recently assembled a statement containing recommended nutrient intakes for premature infants. When differences exist between the recommended nutrient intakes for a stable, growing premature infant and the amount of nutrient that would be received if human milk alone were fed at standard fluid intakes, fortification may be recommended. The nutrient level provided by human milk is derived from the mean value determined when expressed breast milk from mothers delivering prematurely is analyzed. When used according to manufacture's instructions, the sum of the amounts of any nutrient contained within an appropriate quantity of human milk fortifier and human milk generally fed to the premature infant should come within the range of the recommended nutrient intakes and not exceed the maximum "safe" level of intake.

As the nutrient level of the milk from mothers of premature infants may vary mainly during the early stage of lactation, the use of simple mean nutrient level (determined from the expressed breast milk of the mothers of premature infants) to assess the quantity of human milk fortifier needed over time may not be valid. Consequently, the currently prescribed quantity of human milk fortifier may affect the growth of the premature infants negatively. Further, since the importance of minerals in the nutritional management of infants is now widely recognized, it seems to be quite appropriate to study the longitudinal changes in milk trace element composition during the first three months of lactation, and to evaluate the effect of gestational length on levels of macro-trace and ultra-trace elements in human milk. The macrominerals (majors) in the human milk are: calcium (Ca), magnesium (Mg), rubidium (Rb), and strontium (Sr). These macro elements are measured in micrograms per millilitre ($\mu g/ml$ or PPM). The trace elements (or minors) in the human milk are : zinc (Zn), copper (Cu), manganese (Mn), nickle (Ni), molybdenum (Mo), and cobalt (Co); and they are measured in $\mu g/ml$. The other minerals in the human milk, namely, tin (Sn), lead (Pb), cadmium (Cd), cesium (Cs), barium (Ba), lanthanum (La), and cerium (Ce) are known as ultratrace minerals and they are measured in nanograms per millilitre (ng/ml or PPB).

35.2   Analysis of Macrominerals in Human Milk

35.2.1   Longitudinal effects

Four macrominerals, Ca, Mg, Rb, and Sr were studied from the milk of 43 lactating mothers, 19 being the mothers of full-term infants ($\geq$ 37 weeks of gestation) and the remaining 24 were the mothers of preterm infants ($<$ 37 weeks of gestation). Not all mothers provided samples for all 9 weeks. A total of 288 observations over 9 weeks (weeks 1 through 8 and week 12 after birth) were collected. There were 136 observations from the 19 mothers of full-term infants, and 152 observations from the 24 mothers of premature infants. Thus there are some missing values, which for simplicity will be ignored. That is, the unbalanced data set containing these 288 observations will be treated as complete in the statistical analysis.

Histograms for each of the four macrominerals for premature and full-term groups separately and combined showed that the distributions of the three minerals Ca, Mg , and Rb were normal. The distribution of Sr did not appear to be normal. After applying the Box and Cox (1964) power transformation, it was found that $Sr^{1/2}$ followed a normal distribution. To make a preliminary assessment about the longitudinal changes of these minerals, we generated line graphs of mean values of the minerals for each week, both for premature and full-term groups. These line graphs for Ca are displayed in Figure 35.1. Values of Ca in the milk of mothers of premature infants do not appear to change significantly with respect to time. The $Sr^{1/2}$ in both preterm and full-term groups also showed similar behaviour. Generally speaking, as time passes, Ca and $Sr^{1/2}$ appear to have downs and ups in their patterns. But the Ca content in the milk of mothers of full-term infants appear to have a different pattern, with Ca level always increasing slowly with respect to time. With regard to the quantity of minerals, the mean levels of Ca appear to be higher always in the milk of mothers of full-term infants than that of premature infants. The overall mean levels of $Sr^{1/2}$ were found to be the same for both of the preterm and full-term groups.

In order to gain more insight about the pattern of mineral change with respect to time, we display an exploratory lag 1 correlation structure in Figure 35.2, for mean values of Ca and $Sr^{1/2}$. More specifically, the mean level of a mineral for week t is plotted against the mean level for week t-1. Only the data for the first eight consecutive weeks were considered. If the values in successive paired observations, for example weeks 1 and 2, followed by 2 and 3, and so on, increase, then a line with positive slope is generated with points lying above a line with 45° angle passing through the origin. If successive paired observations decrease, there again a line with positive slope is generated with points now lying below the line with 45° angle passing through the origin. In either case, there is a time effect since there is either a continuous increase or a continuous decrease in the mean concentration over time. If the amounts of mineral oscillate over consecutive time points, and the oscillations damp out as time progresses, then a line with negative slope is generated with points lying either above or below a diagonal line with angle, $\pi - 45°$, leading to negative lag 1 correlation. More specifically, if the large mineral values of the oscillating series damp out slowly, then the points representing the paired observations will fall above the diagonal line. Otherwise, that is, if the large mineral values damp out quickly, then the points will fall below the negative diagonal line. A lack of a pattern (for example, all paired observations falling on a point) or a

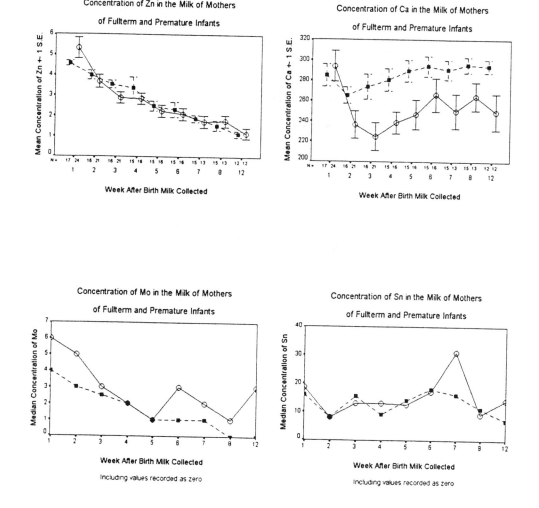

Figure 35.1: Concentrations of Zn, Ca, Mo and Sn in the milk

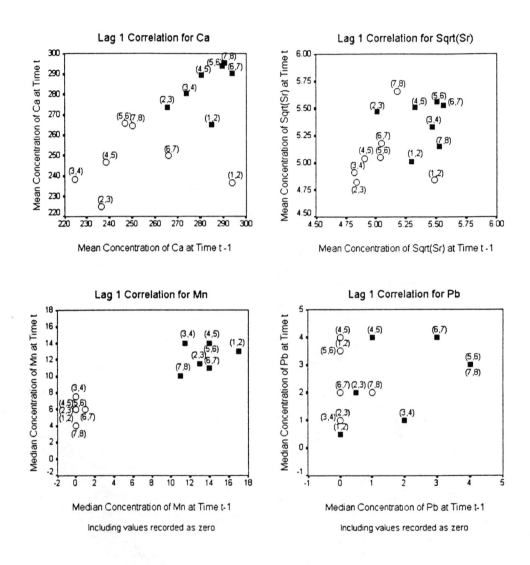

Figure 35.2: Lag 1 correlations

constant pattern (for example, all paired observations falling on a horizontal or vertical band) among the values of the minerals over the weeks will indicate the absence of time effects, leading to zero or small lag 1 autocorrelation. The lag 1 correlations for all four variables were calculated and they were found to be 0.562, 0.654, 0.971 and 0.008 for the full-term group and 0.024, 0.517, 0.890 and 0.171 for the pre-term group, for Ca, Mg, Rb and $Sr^{1/2}$, respectively. Note that the correlation values are small for Ca for the pre-term group as well as for $Sr^{1/2}$ for both pre-term and full-term groups, agreeing with the patterns discussed based on the line graphs. The mineral contents of Ca appear to be highly correlated over time for the full-term group. Similarly, the mineral contents of Mg and Rb appear to be highly correlated over time, both for pre-term and full-term groups.

Recognizing the possiblity of a time effect, we consider a multiple linear regression analysis with macromineral concentration as a dependent variable, and gestation group (pre-term or full-term) and weeks as independent indicator variables. More specifically, consider the linear model

$$y_{it} = \beta_0 + \beta_1 x_{1it} + \beta_2 x_{2it} + \ldots + \beta_8 x_{8it} + \epsilon_{it}, \tag{35.1}$$

where y_{it} is the mineral content in the milk of the i-th ($i = 1, \ldots, 43$) mother at t-th ($t = 1, \ldots, 8$) week, and

$$x_{1it} = \begin{cases} 1 & \text{if i-th mother belongs to full-term group for any t} \\ 0 & \text{otherwise,} \end{cases}$$

$$x_{jit} = \begin{cases} 1 & \text{if the milk of the i-th mother is collected in week t=j-1} \\ 0 & \text{otherwise,} \end{cases}$$

$$j = 2, 3, \ldots, 8.$$

Here β_0 is the mean effect of week 8 for the premature group, and $\beta_0 + \beta_2$ is the mean effect of week 1 for the same group. Alternatively, $\beta_0 + \beta_1$ is the effect of week 8 for the full-term group, and $\beta_0 + \beta_1 + \beta_2$ is the effect of week 1 for the same group. Thus, irrespective of the group, β_2 represents the mean difference effect between weeks 1 and 8. Similarly, β_3 represents the mean difference effect between weeks 2 and 8, and so on. Table 35.1 shows the regression estimates and their corresponding significance levels based on t-tests, for all four macrominerals. At 5% level of significance, it is seen for Ca that the mean effects of weeks 2 and 3 are different than the mean effect of week 8. For Mg, mean effects of weeks 3, 4, 5, and 6 are different than the mean effect of 8. For Rb, mean effects of weeks 1 through 5 appear to be different than the mean effect of week 8. For $Sr^{1/2}$, the mean effects of weeks 1 through 7 appear to be the same as the mean effect of week 8.

We also performed a test to examine whether the mean effects of weeks 1 through 7 are simultaneously same as the mean effect of week 8. This was done by using a partial F-test for testing the null hypothesis

$$H_0 : \beta_2 = \beta_3 = \ldots = \beta_8 = 0 \tag{35.2}$$

against the alternative hypothesis $H_1 : \beta_k \neq 0$, for at least one $k = 2, \ldots, 8$. The partial F-test statistic is defined as

$$F^* = 251(S_2 - S_1)/7S_1, \tag{35.3}$$

where $S_1 = Y'(I_n - X(X'X)^{-1}X')Y$ is the residual sum of squares of the full model (35.1) with $260 - 9 = 251$ degrees of freedom and $S_2 = Y'(I_n - Z(Z'Z)^{-1}Z')\,Y/\sigma^2$ is the residual sum of squares of the reduced model under H_0:

$$E(Y) = Z\alpha, \tag{35.4}$$

with $\alpha = (\beta_0, \beta_1)'$. In the normal set-up, under H_0,

$$F^* \sim F(7, 251),$$

where $F(7,\ 251)$ denotes the F-distribution with degrees of freedom 7 and $260 - 9 = 251$. For Ca, Mg, Rb, and $Sr^{1/2}$, the F^* values were found to be 3.003, 5.357, 17.195, and 0.858 respectively. When compared with table value $F_{7,251,0.05} = 2.05$, it is found that except for Sr, the mean week effects are different, leading to the conclusion that there is a significant longitudinal change in mineral content of Ca, Mg, and Rb in the human milk. This conclusion, in turn, indicates that this longitudinal change should be taken into consideration in the fortification process of human milk, i.e., in the preparation of pre-term infants' formula.

35.2.2   Gestation effects

Table 35.1: Regression estimates and their significance levels (in parenthesis) for the gestation and time as a specific factor for weeks 1 through 8

Mineral	$\hat\beta_0$	$\hat\beta_1$	$\hat\beta_2$	$\hat\beta_3$	$\hat\beta_4$	$\hat\beta_5$	$\hat\beta_6$	$\hat\beta_7$	$\hat\beta_8$
Ca	264.495	30.942	12.159	-28.957	-32.554	-20.693	-12.145	0.017	-9.367
		(0.359)	(0.030)	(0.016)	(0.136)	(0.381)	(0.999)	(0.510)	
Mg	33.077	-1.343	0.585	-2.388	-5.517	-5.136	-3.878	-3.814	-1.607
		(0.059)	(0.679)	(0.094)	(0.000)	(0.001)	(0.009)	(0.010)	(0.289)
Rb	0.603	-0.005	0.232	0.158	0.078	0.076	0.034	0.019	0.005
		(0.368)	(0.000)	(0.000)	(0.000)	(0.000)	(0.040)	(0.113)	(0.647)
$Sr^{1/2}$	5.266	0.220	0.055	-0.448	-0.308	-0.255	-0.104	-0.074	-0.018
		(0.116)	(0.843)	(0.111)	(0.276)	(0.382)	(0.721)	(0.799)	(0.953)
Zn	1.578	0.150	3.291	2.171	1.491	1.447	0.717	0.553	0.163
		(0.724)	(0.000)	(0.000)	(0.008)	(0.012)	(0.261)	(0.527)	(0.823)
$Cu^{1/4}$	0.727	0.018	0.109	0.084	0.054	0.057	0.011	0.031	0.004
		(0.109)	(0.000)	(0.000)	(0.016)	(0.013)	(0.628)	(0.184)	(0.873)

To test whether there is any difference between the mineral concentrations in the milk of the mothers of pre-term and full-term infants, we test the null hypothesis $H_0 : \beta_1 = 0$ in equation (35.1), by using the t test. As mentioned earlier, the p-values of the t test for Ca, Mg, Rb, and $Sr^{1/2}$ are shown in the third column of Table 35.1. It is clear that except for Ca, there is no gestation effect for the other 3 macrominerals, namely, Mg, Rb and Sr.

35.3 Analysis of Trace Elements in Human Milk

35.3.1 Longitudinal effects

The trace elements (or minors) in the human milk are : zinc (Zn), copper (Cu), manganese (Mn), nickle (Ni), molybdenum (Mo), and cobalt (Co). The application of Box-Cox power transformation suggests that except Zn and $Cu^{1/4}$, all remaining four variables are non-normally distributed. Almost all non-normally distributed minerals including Mn had a high peak at the left of the distribution due to high preponderence of zero values, with the remainder of the distribution being skewed. There were equal proportions of zero values between gestation groups for Mn and Ni, but the proportions of zero values were unequal between the gestation groups for Mo and Co. This was found by performing a binomial test for the equality of proportion of zero values between the two groups. Because of the asymmetry of the data, we studied the median plots for all non-normally distributed minerals in two parts. In one part, figures were considered for the complete data set including the zero values of the mineral, and in part 2, the data with non-zero values were considered. To save space, we show in Figure 35.1, the median plots for the complete data set including zero values, only for Mo. Gestation effects for all non-normal minerals are studied in Section 35.3.2 for the complete as well as the truncated (excluding zero values) data sets.

Since Zn and Cu were found to be normally distributed, the longitudinal and gestational nature of these variables may be studied in the manner similar to those of the macrominerals discussed in Section 35.2. The mean plots for Zn, for example, are shown in Figure 35.1. Further, the week effects of Zn and Cu are shown in Table 35.1 along with the week effects of the four macrominerals. The partial F-statistic values for testing the null hypothesis that there is no week effects on the mineral concentrations were found to be 5.357 and 6.718 for Zn and $Cu^{1/4}$ respectively. These F values are quite large as compared to the tabulated F value $F_{7,251;0.05} = 2.01$, leading to the conclusion that there is a longitudinal effect on the concentrations of minerals of these two variables. The exploratory lag 1 correlation of the mean values of these two minerals were also large, namely, 0.953 and 0.598 for Zn and Cu, respectively, for the pre-term group, and 0.968 and 0.727 for the full-term groups, indicating that there is a strong longitudinal effect on the mineral concentrations.

With regard to the non-normal variables, testing for the time effects when time is a specific factor is naturally complicated. The partial F-test (35.3) employed in Section 2 for testing similar effects for the normal variables is not valid when the distribution of the variable concerned is asymmetric. To gain some insight about the longitudinal effects on the mineral concentrations of these skewed variables, we conducted an exploratory analysis. We first studied the median plots for the minerals for each group at each time point. The median plots, for example, the plots for Mn and Mo in Figure 35.1 show how mineral concentrations are changing in mother's milk as lactation time progresses. To gain further insight about the time effects, similar to the lag 1 mean relationship for the normal variables shown in Figure 35.2, we display the lag 1 median correlations for Mn in the same Figure 35.2. More specifically, we plot the median for the t-th week against the median for the (t-1)-th week. Figure 2 clearly shows that the longitudinal changes in Mn concentrations are somewhat different for both of the pre-term and full-term groups. For the complete (including zero mineral values) full-term group, this lag 1 correlation for Mn was found to be 0.317, whereas for the complete pre-term group this correlation was found to be 0.047.

Since development of a formal statistical test of the hypothesis that the time effects are the same for every week appears to be quite difficult for this asymmetrical data, we pursue this matter further by considering time as a non-specific factor. In this approach it is argued that there will be a decay in the correlation with increasing time lags among observations collected from each mother, and the correlation structure will remain the same for all mothers. In the Gaussian case, this type of data are modeled by an autoregressive order one (AR(1)) process [see Box and Jenkins (1976)], where lag 1 correlation is of main interest. The lag 1 correlation may be modified as follows for the asymmetric data. Let there be I^* independent mothers, and n_i^* consecutive observations be collected from the ith mother. Also, let $\widetilde{Y_i^*}$ be the median of the mineral concentrations in the milk of the ith mother, for $i = 1, \ldots, I^*$. Further, let $\tilde{\phi}_i$ denotes the lag 1 correlation for the minerals of the ith mother, defined as

$$\tilde{\phi}_i = \frac{\sum_{t=2}^{n_i^*}(Y_{it}^* - \widetilde{Y_i^*})(Y_{i,t-1}^* - \widetilde{Y_i^*})}{\sum_{t=1}^{n_i^*}(Y_{it}^* - \widetilde{Y_i^*})^2}, \tag{35.5}$$

where Y_{it}^* is the mineral value collected at the tth consecutive week from the ith mother. Under the assumption that the correlation remains the same for all mothers, one may pool the I^* mothers [Quenouille (1958)] to obtain the overall lag 1 correlation estimate given by

$$\tilde{\phi} = \frac{\sum_{i=1}^{I^*} \sum_{t=2}^{n_i^*}(Y_{it}^* - \widetilde{Y_i^*})(Y_{i,t-1}^* - \widetilde{Y_i^*})}{\sum_{i=1}^{I^*} \sum_{t=1}^{n_i^*}(Y_{it}^* - \widetilde{Y_i^*})^2}. \tag{35.6}$$

The application (35.6) to the complete data set (including zero values) yielded the lag 1 correlations $\tilde{\phi}$ for the four non-normal minerals under two groups as in the following table.

<table>
<tr><th colspan="5">Lag 1 Correlation $\tilde{\phi}$</th></tr>
<tr><th></th><th colspan="4">Mineral</th></tr>
<tr><th>*Group*</th><th>*Mn*</th><th>*Ni*</th><th>*Mo*</th><th>*Co*</th></tr>
<tr><td>Full-term</td><td>-0.003</td><td>-0.010</td><td>0.099</td><td>0.040</td></tr>
<tr><td>Pre-term</td><td>0.052</td><td>0.414</td><td>-0.020</td><td>-0.009</td></tr>
</table>

For a given mineral, the $\tilde{\phi}$ value obtained by (35.6), represents the lag 1 correlation coefficient among the mineral contents of any mother under a group, either full-term or pre-term; whereas correlations shown in Figure 35.2, for example, for Mn represents an exploratory lag 1 correlation only among the median mineral contents of all mothers under the same group. In general, there does not appear to be any significant longitudinal changes for any of these four asymmetric trace elements.

35.3.2 Gestation effects

Once again as Zn and $Cu^{1/4}$ have normal distributions, the concentrations of these minerals may be modelled as in (35.1), and the difference between the mineral concentrations in the milk of the mothers of pre-term and full-term infants may be tested by using the partial F-test statistic given in (35.3). For Zn and $Cu^{1/4}$, the F^* values were found to be 0.023 and 0.000. These low values suggest that there is no gestational effect on the concentrations of these two minerals in the human milk.

Turning back to the asymmetric minerals, the overall testing for the equal effects due to gestations for any of the eight weeks appears to be complicated. This is because, as discussed previously, the observations collected in a week are naturally correlated with the observations from the previous and subsequent weeks. Consequently, in order to shed some light on the equality of gestation effects for two groups of asymmetric observations, we simply concentrate on testing gestation effects at a given week.

Although it was found from the median plots for Mn that there may be a significant difference between the medians of the pre-term and full-term groups for Mn, a nonparametric median test however showed that the medians were the same at every week. This discrepancy may be due to the fact that the median test does not incorporate the magnitude of the difference of each observation from the group median. Nevertheless, in the following table, we provide the significance levels of the median test for Mn, for all eight weeks for the complete and truncated data sets.

Data set	Week1	Week2	Week3	Week4	Week5	Week6	Week7	Week8
Complete	0.162	0.444	0.071	0.213	0.107	0.107	1.000	0.450
Truncated	0.060	0.503	0.715	0.115	0.414	0.238	1.000	0.198

Since the median test does not appear to be satisfactory in examining the gestation effect for non-normal minerals, we now propose an adhoc two sample test for testing the significance of the gestation effects for Mn, Ni, Mo, and Co, at a given week.

For a given week, let m_1 and m_2 be the medians of the mineral concentrations in the milk of the mothers of full-term and pre-term infants. Further, let s_1^* and s_2^* be the corresponding median absolute deviations defined as

$$s^* = Median[|Y_{it} - \tilde{Y}_t|/0.6745], i = 1, \ldots, n^*$$

for a given group at a given week t. Now suppose that $n^* = n_1^*$ for the full-term group, and $n^* = n_2^*$ for the pre-term group. We may then treat

$$z^* = (m_1 - m_2)/(s_1^{*2}/n_1^* + s_2^{*2}/n_2^*)^{1/2}$$

as a traditional normal test statistic. This test unlike the median test takes the magnitude of the minerals into account. The values of the z^* statistic for Mn, Ni, Mo, and Co are shown in Table 35.2. These values of $|z^*|$ are considerably larger than 1.96 for Mn for W1 (week 1), W2, W3, W4, and W5, whereas for Ni, the values of the test statistic are found to be large only for W1 and W6. For Mo, $|z^*|$ value is large for W6 only, and for Co, $|z^*|$ values are large for W1, W6, W7, and W8. The large value of $|z^*|$ indicates that the gestation effect is significant for that paricular week.

Table 35.2: The values of the adhoc normal z^* test statistic for testing the $H_0 : F_{Y_1} = F_{Y_2}$ for weeks 1 through 8

Mineral	Week1	Week2	Week3	Week4	Week5	Week6	Week7	Week8
Mn	3.20	2.60	2.05	2.29	2.72	1.91	1.37	1.91
Ni	-2.48	-0.64	-0.57	0.65	-0.74	2.12	-0.27	2.61
Mo	-1.61	-1.92	-0.62	*	*	-3.76	1.10	*
Co	-3.31	-1.34	-0.78	*	*	-2.70	-2.43	2.62
Pb	*	2.70	2.69	2.62	3.48	2.61	2.45	2.61
Sn	-0.43	*	0.55	-0.79	0.43	0.13	-1.46	0.44

* Indeterminate values i.e. $0 \div 0$

35.4 Analysis of Ultratrace Minerals in Human Milk

35.4.1 Longitudinal and gestational effects

The ultratrace minerals in the human milk are: tin (Sn), lead (Pb), cadmium (Cd), cesium (Cs), barium (Ba), lanthanum (La), and cerium (Ce) and they are measured in nanograms per millilitre (ng/ml or PPB). Among these 7 minerals,

Pb and Sn are considered more important than the others. Their longitudinal behaviour is, however, not adequately addressed in the literature. Similar to the trace elements, Mn, Ni, Mo, and Co, the two ultratrace elements Pb and Sn were also found to be asymmetrically distributed with two modal values. In order to understand the time dependence for these minerals, we display the weekly medians for Sn in Figure 35.1, and the lag 1 median correlations of Pb, for example, in Figure 35.2. We have also computed the lag 1 correlations for Pb and Sn by (35.6). The latter correlations for these two minerals for the two groups are in the following table :

	Lag 1 Correlation $\tilde{\phi}$	
	Mineral	
Group	*Sn*	*Pb*
Full-term	-0.204	-0.208
Pre-term	-0.010	0.128

It is clear from the above table that similar to the trace elements, the concentrations of these ultratrace elements in the milk do not appear to be time dependent.

Next to study their gestational effects, initially we applied the median test to examine the overall gestational difference between the full-term and pre-term groups. The p-values of the median test were 0.000 for Pb and 0.788 for Sn, indicating that unlike Sn, the Pb concentrations in the full-term group are quite different than that of the pre-term group. We also examined their distributions in two segments. In the first part, we considered the proportion of zero values for these minerals. It was found that for Pb there were 67.1% zeros in the pre-term group, and 41.1% zeros in the full-term group. These proportions were too large to be ignored. For Sn, there were 7.9% and 21.8% of zero values in the pre-term and full-term groups respectively. For the truncated part with non-zero values of the minerals, we applied the median test for the difference between the two groups and the p-values were found to be 0.012 for Pb and 0.514 for Sn. This yields the same conclusion as that was made based on the whole data set including zero values.

To examine the gestational differences at every week, we applied the 'adhoc' normal test. The results are given in Table 35.2. It is seen from the table that this test statistic is undefined when both median difference and median absolute deviation are zeros. These cases are shown by ∗ marks. It is, however, interesting to note that this test reveals that there is a gestational effect for Pb for almost all weeks, whereas there is no gestational difference for Sn between the full-term and pre-term groups.

35.5   Summary and Conclusion

In the present study, for the macro-minerals group, there appears to be significant longitudinal changes in the concentrations of Ca, Mg, and Rb, in mothers' milk, but the gestation effect is significant for Ca only. In the trace-elements group, there appears to be significant longitudinal changes for Zn and Cu, but the concentration of Mn is generally different in the milk of two groups of mothers, although there is no longitudinal change for this mineral. In the ultratrace-elements group, there does not appear to be any significant longitudinal changes for any minerals, but the amount of Pb concentrations are found to be, in general, different in the milk of the two groups of mothers.

The finding of the present paper, in particular, the difference in the longitudinal behavior of the mineral concentrations in the two groups, can be useful in improving the quality of the current pre-term formula.

Acknowledgements. The authors would like to thank a referee and the editor for constructive comments. The research was partially supported by a grant from the Natural Sciences and Engineering Research Council of Canada.

References

1. Anderson, G. H., and Bryan, M. H. (1982). Is the premature infant's own mother's milk best? *Journal Pediatric Gastroenterol Nutrition*, **1**, 157–159.

2. Atkinson, S. A., Brunton, J., Payes, B., Fraser, D., and Whyte, R. (1990). Calcium and phosphorus fortification of mother's milk and formula for premature infants: metabolic balance of calcium, phosphorus and zinc at two postnatal ages, *FASEB Journal*, **4**, 1393–1398.

3. Box, G. E. P., and Cox, D. R. (1964). An analysis of transformations (with discussion), *Journal of the Royal Statistical Society, Series B*, **26**, 211–246.

4. Box, G. E. P., and Jenkins, G. M. (1976). *Time Series Analysis, Forecasting and Control*, San Francisco: Holden-Day.

5. Brooke, O. G., Wood, C., and Barley, J. (1982). Energy balance, nitrogen balance and growth in preterm infants fed expressed breast milk, a premature infant formula and two low-solute adapted formulae, *Arch Dis Child*, **57**, 898–904.

6. Fomon, S. J., Ziegler, E. E., and Vazquez, H. D. (1977). Human milk and the small premature infant, *American Journal of the Dis Child*, **131**, 463–467.

7. Forbes, G. B. (1989). Nutritional adequacy of human breast milk for prematurely born infants, *Textbook of Gastroenterology and Nutrition in Infancy*, edited by E. Lebenthal, New York: Raven Press, pp. 27–34.

8. Gross, S. J. (1983). Growth and biochemical response of preterm infants fed human milk or modified infant formula, *New England Journal of Medicine*, **308**, 237–241.

9. Kashyap, S., Schulze, K., Forsyth, M. and Dell, R. B. (1990). Growth, nutrient retention, and metabolic responses of low-birth-weight infants fed supplemented and unsupplemented preterm human milk, *American Journal of Clinical Nutrition*, **52**, 254–262.

10. Modanlou, H. D., Lim, M. O., Hansen, J. W. and Sickles, V. (1986). Growth, biochemical status, and mininesal metabolism in very-low-birth-weight infants receiving fortified preterm human milk, *Journal of Pediatric Gastroenterol Nutrition*, **5**, 762–767.

11. Quenouille, M. H. (1958). The comparison of correlations in Time Series, *Journal of Royal Statistical Society, Series B*, **20**, 158–168.

12. Rowe, J. C., Wood, D. H., Rowe, D. W. and Raise, L. G. (1979). Nutritional hypophosphatemic rickets in a premature infant fed breast milk, *New England Journal of Medicine,* **300**, 293–296.

Subject Index

Admissible effects, 222
Alarm statistic, 189
Algorithmic process control, 61
Aliasing, 207
ANOVA, 132, 151, 263, 265, 275
Asymmetric tolerances, 79, 81
Asymptotic optimality, 375
Autocorrelation, 110, 195
Autoregressive integrated moving average (ARIMA) process, 109, 110
Autoregressive process, 64, 198
Autoregressive moving average process, 65, 67
Average adjustment interval, 100
Average cost rate, 101
Average response plots, 266
Average run length, 110, 111, 190

β-correction, 151, 155, 156, 159
Bayes risk, 373
Beta-lactum drugs, 415
Biochemical oxygen demand, 399
Bivariate chi-square distribution, 173, 176
   noncentral, 175
Bivariate normal distribution, 173
Brownian bridge, 363

Case study, 128, 154, 263, 264, 267
Central composite design, 254
Change points, 363
Chi-square
   approximation, 163, 284
   central, 199
   noncentral, 199
Clarke's constrained controller, 65–67, 69, 70
Coefficient of determination (R^2), 313
Coefficient of variation, 299, 304
Color-quality, 128

Composite materials, 327
Conditional test, 381
Confidence limits, 168, 195, 393, 416
Confirmation trials, 266
Connected effects, 223
Contaminated normal population, 173, 175
Continuous improvement, 32, 121
Control charts, 32, 51, 111, 187
Constrained controller, 61
Consulting, 45, 54, 55
Cross validation, 137
Crusher performance, 140
CUSUM, 48, 109, 136, 187, 190

D-efficiency, 225
D-optimality, 221, 229, 237, 240
Deduction, 6, 9
Design (control) factors, 250, 275
Design of experiments, 30, 31, 51, 249, 263
Desirability function, 289, 290, 299, 301
   weighted, 303
Diagnostic plots, 135
Differential equation, 156, 329
Discrete life time, 339
Distance function approach, 300
Distance-variance plot 249, 257
Distribution
   beta, 397
   binomial, 385
   bivariate geometric, 347
   bivariate Lomax, 358
   bivariate Pareto, 358
   exponential, 372
   F, 394, 430
   gamma, 327, 351, 353, 372
   geometric, 339
   Gompertz, 415
   inverse gamma, 212

inverse Gaussian, 351, 354
lognormal, 327
multinomial, 276
multivariate exponential, 381
normal, 327, 330
Weibull, 354
Dose response curve, 415

E-optimality, 221
E*-optimality, 226, 232
Electroplating, 263
EM-algorithm, 319
Empirical Bayes, 371, 373
Empirical distribution function, 363
Engineering process control, 61, 63, 65, 68
Estimating function, 109, 113
Evolutionary operations, 14, 19
EWMA, 48, 109–112, 115–117, 136
Expected total cost, 406

Factor
    block, 278
    concomitant, 277
    indicative, 275
    variation, 278
Factorial experiments, 263, 313
    fractional, 7, 32, 221
Fatigue failure experiments, 328
Fatigue life, 327
Failure rates, 352, 363
    bivariate, 339
    conditional, 344
    scalar, 346
    vector, 343
    marginal, 340
Feedback control, 98
Fixed effects model, 167
Forecasting controller, 64, 67, 69, 70
Frame soldering, 263

Gage repeatability and reproducibility
    (R & R), 29, 32
Gaussian mixture, 399
Generalized hypergeometric function, 174
Generalized linear models, 327, 331, 416
Generators, 224
Genmod procedure, 333
Geometric dimensioning, 30

Geometric mean, 301
Gestation effects, 430
Gibbs sampler, 207, 393, 395
Goodness-of-fit, 415, 419
Graphite/epoxy laminates, 327
Gum hardness data, 276, 281

Hazard function, 353

Induction, 6, 10
Industry liaison, 45
Information matrix, 223
Injection molding, 289
Innovation, 13, 14, 17, 19
Inspection, 24, 35, 38, 249, 405
    cost, 99, 103
Integrated moving average process, 97
Interactions, 207, 221, 253, 275, 283, 313
    generalized, 281
ISO-9000, 15, 23–30, 32, 47
Isotonic inference, 275
Iterative learning, 3

Kalman filter, 109
Keifer process, 363
Korean industries, 121

Lag 1 correlation, 423, 432
Latent variables, 135
Likelihood ratio, 187
    test, 381
Logit, 415
Log likelihood ratio
    chi-square, 420
Longitudinal effects, 425

Macrominerals, 425
Main effects, 210, 215
Manufacturing system, 405
Markov chain Monte Carlo, 219
Markov process, 195, 396
Mean absolute deviation, 434
Mean square error, 113–115, 313
Mean plots, 423
Mean residual life function, 359, 363
Median plots, 423
Median test, 433
Melt index, 63

Measurement systems analysis, 29, 31, 32
Mineral concentrations, 423
Mineral processing industry, 135
Minimal sufficient statistics, 383
Minimum mean square error control, 63, 64, 66, 70
Missing data, 313, 316
Mixture designs, 240
    models, 237
Moment generating function, 353
Monte Carlo, 211
Moving average, 71, 100, 115
Multi input/single output, 151
Multiple comparison procedure, 282
Multistage system, 407
Multiple quality characteristics, 289
Multivariate statistical process control, 135

NIPALS, 138
Noise factors, 250, 275
Non-linear models, 393
Non-linear time series, 114
Non-normality, 102, 104, 432
Normal probability plot, 313, 314
Nutrient intakes, 425

On-line control, 97
Optimal filter, 109
Optimal predictor, 114
Optimal smoother, 109, 112, 114
Ordinal categorical data, 283
Orthogonal arrays, 3, 9, 131, 254, 293
    effects, 223

Paradigm, 4, 5, 13, 17
Parameter design, 40, 290
Parity
    odd, 223
    even, 223
Partial F-test, 430
Partial least squares, 135–137, 140
Pearson chi-square, 420
Performance robustness, 67, 68
Petrochemical process, 61
Plackett and Burman designs, 9, 207, 221, 225
Poisson process, 382

Polymer, 62
Power, 386, 387
Preterm infants formula, 423
Principal components analysis, 136, 137
Printed circuit board, 267
Priors, 209
Probit, 415
Problem solving, 121
Process capability, 29, 31, 32
Process capability indices (C_{pk}, C_{pm}, P_{pk}, ...), 30, 79, 163, 173
Process control, 29, 30, 35, 39
Process monitoring, 61
Production part approval process (PPAP), 29
Projection properties, 221

QS-9000, 23–33, 47
Q-Q plots, 393, 398
Quality assurance, 24
Quality awards, 13, 15, 16
Quality costs, 405, 409
Quality evolution, 24
Quality function deployment, 30, 35, 41–43
Quality improvement, 45, 62, 124, 249, 313
Quality loss, 40
Quality planning, 28
Quality system requirements, 23, 30

Random effects, 351
    model, 165
Random environmental models, 351
Random walk, 97, 99, 113, 393, 401
Rates of convergence, 371, 375
Reliability, 351, 371
Regression, 130, 151, 157, 276, 305, 313
    binary response, 416
    multiple linear, 429
    nonlinear, 394
Response contours, 132
Response surface methodology, 7, 8, 30
Robust design, 31, 289
Robust product, 249
Robustness, 97, 99, 173, 189, 240

Sampling interval, 97, 155
Scheffé polynomial, 238

Scientific learning, 3–5
Search designs, 322
Sensitivity, 415
Series system, 351, 356
Shewhart-Deming cycle, 5
Shiryaev optimality, 187
Signal-to-noise ratios, 250, 265, 292
Simplex-centroid design, 239
    -lattice design, 246
Simulation, 195, 199, 387
Simultaneous optimization, 299
Single screening procedure, 406
Slutsky's theorem, 368
SO_2 gpl, 151, 155, 156,160
Special causes, 30
Specification limits, 79, 165, 196
Spline smoothing, 393
Stability concentration, 151, 157
Stability robustness, 67
Statistical Advisory Service, 55, 57
Statistical Process Control, 24, 42, 46,
    50, 61, 70, 72, 110, 121, 197,
    249, 412
Statistical Quality Control, 24, 46, 275
Statistical thinking, 35, 37

Stochastic search, 210
Strategic quality management, 24
Survival analysis, 351
SW-kerosene, 128
Systems thinking 35, 36

Taguchi index, 197
    methods, 35, 40, 41
    loss function, 405
Taste testing data, 277, 282
Tests of hypothesis, 173, 184, 430
Tolerance studies, 30
Total quality management, 35, 275
Trace elements, 425, 431
Training, 45, 53, 122

Ultratrace minerals, 434
Uniformly most powerful test, 382
Unstable production process, 163

Variation reduction, 35, 37, 42, 249

Wiener process, 363
Workshop, 45, 47, 49–52

Yates methods, 319